Human Evolution in China

Human Evolution in China

A Metric Description of the Fossils and a Review of the Sites

Xinzhi Wu
*Institute of Vertebrate Paleontology
and Paleoanthropology
Academia Sinica
Beijing, China*

Frank E. Poirier
*Department of Anthropology
The Ohio State University
Columbus, Ohio*

New York Oxford Oxford University Press 1995

Oxford University Press

Oxford New York
Athens Auckland Bangkok Bombay
Calcutta Cape Town Dar es Salaam Delhi
Florence Hong Kong Istanbul Karachi
Kuala Lumpur Madras Madrid Melbourne
Mexico City Nairobi Paris Singapore
Taipei Tokyo Toronto

and associated companies in
Berlin Ibadan

Published by Oxford University Press, Inc.,
198 Madison Avenue, New York, New York 10016

Library of Congress Cataloging-in-Publication Data
Wu, Xinzhi.
Human evolution in China : a metric description of the fossils and a review of the sites / Xinzhi Wu and Frank E. Poirier.
p. cm. Includes bibliographical references and index.
ISBN 0-19-507432-7
1. Fossil man—China.
2. Man, Prehistoric—China.
3. China—Antiquities.
I. Poirier, Frank E., 1940-
II. Title.
GN282.W79 1995 573.3′0951—dc
94-41509

1 3 5 7 9 8 6 4 2

Printed in the United States of America
on acid-free paper

Preface

Until the late 1970s, Western researchers had little contact with Chinese colleagues for a period of about 20 years. This project stems from the new openness between countries and peoples. With new collaborations and an increasing number of articles in journals and books published outside of China, there is no excuse to be ignorant of China's vast fossil wealth, including the wealth of fossils for primate evolution. Information provided here is current, but given the pace of new finds from China, new, exciting, and perhaps unexpected evidence will come to light.

Evidence from China is crucial to understanding hominoid evolution. Many stages of ape and human evolution were played out in China. Pongid materials abound, for example, *Lufengpithecus* = (*Ramapithecus*), *Dryopithecus,* and *Gigantopithecus*. Human evolutionary history in China may have begun 1.7 million years ago, judging from the fossil evidence from Yunnan province. Claims for an even earlier *Australopithecus* presence, for example from Hubei province, are unsubstantiated. In addition to the long-recognized finds from Zhoukoudian in northern China, *Homo erectus* specimens are widely distributed in China. Archaic *Homo sapiens* and *Homo sapiens sapiens* sites occur in various parts of China. Some of these materials have important implications in answering the hotly debated questions about the origins of modern humans. Were the origins of *Homo sapiens sapiens* solely African based, the Out-of-Africa model, or did these origins occur in different parts of the world, as envisioned by the Multi-Regional Evolution model? Fossils from China that are central to this argument are discussed herein.

Until about two decades ago the chronological framework for Chinese hominid sites was based on biostratigraphical correlation and the comparative morphology of the hominid remains. Since the mid-1970s paleomagnetic dating has been applied to the most important Chinese *Homo erectus* sites, such as Yuanmou, Lantian, and Locality 1 at Zhoukoudian. Since the late 1970s an increasing number of radiocarbon dates have been made available for many of the Late Paleolithic sites in China. In the early 1980s Zhoukoudian was dated utilizing such techniques as fisson track, T L, uranium series, and amino acid racemization. The results of the various techniques are fairly consistent with one another, and they basically concur with newer electron spin resonance (ESR) dates. In the mid-1980s most of the important early *Homo sapiens* sites in China were dated with uranium series, although there are problems with some of the dates obtained. The lack of volcanic activity during the time period when many of the Chinese sites were formed impedes the use of potassium-argon dating.

This book contains information not available before in English. Extensive recent reviews written in English of some of the materials described in this book can be found in Pope and Keates (1994) for the archaeology and Etler and Li (1994) for the fossils. Numerous tables provide researchers with considerable comparative data. The detailed morphological descriptions for many fossils are not available elsewhere in English. Much of the material currently available on the Chinese fossil sites is written in Chinese and dispersed in journals that are not easily accessible to a foreign audience. Although some articles have English abstracts, the abstracts do not contain the kinds of data researchers need for comparative purposes. The current book contains information not previously translated from the original Chinese. This information was concisely translated, paraphrased, and compiled by Xinzhi Wu. For those materials published in English, such as Franz Weidenreich's work on Zhoukoudian, the essence of the original morphological summaries is preserved with only minor editing. Our regret is that we cannot express our gratitude to F. Weidenreich personally for the information and insights his years of diligent work have provided.

The co-authors contributed differently to this project. All the descriptions in Chapters 2–6 were first written by Professor Wu, one of the few scientists with first-hand knowledge of many of the remains described herein. Professor Poirier amended the text. Collaboration between authors separated by thousands of miles—Columbus, Ohio, and Beijing, China—and using different word processing programs, had its own unique challenges.

The detailed entries in this book provide readers with ready access to information on the skeletal remains, associated artifacts, and the geological context of the most important Chinese hominid sites. The task of providing this data was made urgent by the need for information by paleontologists beyond China's borders. Hopefully information provided in this book will lead to new cooperation between more researchers in China and beyond.

We have relied heavily on first-hand accounts of excavations and original descriptions of the fossils, geology, and archaeology. Much of this material has been translated and summarized for the first time for publication in this book. We thank the many researchers whose work was consulted. We relied heavily on some earlier English-language treatments of the Chinese materials by Franz Weidenreich and Rukang Wu. Those having read their respective works will easily recognize their contributions to this book. Their work remains the best, and in some cases the only, source of information for some fossil remains.

This book provides paleoanthropologists ready access to all major Chinese fossils. The treatment concentrates on metrics and description, and evolutionary phylogeny is only briefly discussed. We are certain that the raw data provided here will be very useful in comparisons with fossil materials from non-Chinese sites.

There has been considerable revision of the English spelling of Chinese place names and the names of individuals. We attempted to provide various spellings of names where an older, more recognized spelling has recently been replaced by a less familiar spelling. Western names are listed with the given name first and the family name second.

Although we agree with efforts to remove gender-specific terms from general descriptions, we have retained one such reference by referring to *Homo erectus pekinensis* as *Peking Man*. This latter term is widely recognized and is still often cited not only in China but in other parts of the world. We hope our readers will understand our use of the term.

Archaeological and paleontological research in China has gone through a number of phases. Before 1949 most of the famous human fossils from China came from Zhoukoudian, where western scientists monopolized the research. It should be recognized, however, that some of the faunal and archaeological studies and much of the research was done by Chinese workers. Until 1949 three sites containing human fossils were identified in China: Locality 1 and the Upper Cave of Zhoukoudian, and Salawusu in the Ordos regon.

Since 1949 Chinese researchers have taken control of scientific research in their country. Large-scale construction after that date exposed many important fossil and archaeological remains, and people were encouraged to be aware of the importance of fossil remains and cultural relics. The data that were collected were used to help explain past Chinese social and political change. The government sponsored archaeological and paleontological research and, as a result, there was a rapid expansion of our knowledge about human evolution in China.

The time after 1949 can be separated into two periods, with the dividing line around 1970. During the 1970s interaction between Chinese scientists and their Western colleagues resumed, resulting in a period characterized ". . . as [one] of mutual influence and scholarly dynamics between China and the West" (Olsen, 1988: 284). This book is one of the many fruits of this spirit of cooperation.

ACKNOWLEDGMENTS

This book would have been impossible to produce without the help of the following individuals. The foresight of our editor at Oxford University Press, Ms. Joyce Berry, will hopefully be rewarded by a new wave of collaborative research. Mr. James Stewart, of The Ohio State University, provided Professor Poirier with the technical information and helped to "translate" computer languages. This was critical, given the fact that Professors Wu and Poirier were writing and editing in different programs that needed to be translated. Without Jim's help and skills, this book would have taken much longer to complete. He has our gratitude. Jim also prepared the final tables. Professor Poirier sincerely thanks Mr. Joggeshwar Das for his editiorial assistance. We also thank Mr. Donald Anderson for his diligence and help during copyediting. Mr. Daniel Wilson also helped copy edit.

We are indebted to the assistance of Mrs. Wenlu Zhuang and Mrs. Fang Zheng of the IVPP for their computer aid. Mrs. Yan Chen helped draw maps and Mrs. Baoyi Ren of the IVPP Reference Room lent us photographs in her care. Most of these photographs were taken by Mr. Zhefu Wang. We appreciate his work.

Contents

Human Evolution in China

1

Introduction

In Chapter 1 we briefly review some of the materials that we will be discussing in later chapters. China is very rich in hominoid and hominid fossil remains. The earliest appearing fossil hominoid from China seems to be *Homo erectus,* despite questionable claims for an earlier appearance of *Australopithecus.* The following chapters provide a comprehensive review of the Chinese fossil record. Many of the materials have not previously been published outside of China. Detailed metrical analyses and the latest information on the geological and archaeological context of important fossils are provided to allow other researchers to make useful comparisons between the Chinese and non-Chinese fossil records.

Primate evolution in China spans the geological epochs of the middle Paleocene to the Holocene. Perhaps as early as 60 million years ago (mya) primates appeared in China. There is a rather rich history of hominoid evolution beginning in the early or middle Miocene. The first Chinese hominids, which appear during the Pleistocene, may date to more than 1 mya and most likely belonged to the taxon of *Homo erectus.*

Since the earliest *H. erectus* finds at Zhoukoudian (then spelled Choukoutien), originally called *Peking Man* or *Sinanthropus pekinensis,* many other hominids have been unearthed. Recently there has been renewed interest in Franz Weidenreich's (1946) and Charlton Coon's (1962) proposals that modern Chinese populations can trace their ancestry back to Chinese *H. erectus* or early *H. sapiens* populations, challenging the commonly accepted theory of a solely African based origin for the evolution of modern *H. sapiens.*

HOMINID REMAINS (FIG. 1.1)

In 1903 the German researcher Max Schlosser made the first scientific report of a human tooth derived from the Chinese fossil record. He described a specimen collected in a Chinese traditional medicine shop in Beijing by another German researcher, G. Heberer. Red earth adhered to the root of this tooth, and because the tooth was associated with *Hipparion* teeth and other Pliocene fossils, Professor Schlosser predicted the possibility of discovering Tertiary hominids in China.

The first human fossils found in China on scientific excavations are two *Homo erectus pekinensis* teeth dating to the Middle Pleistocene. The specimens were recovered in 1921 and 1923 near Zhoukoudian village, 48 km from Beijing. Extensive excavations at the site continued until the time of World War II, producing considerable hominid remains. Two other sites, the Upper Cave at Zhoukoudian and Salawusu, dating to the late Paleolithic, yielded *H. sapiens* remains in China before

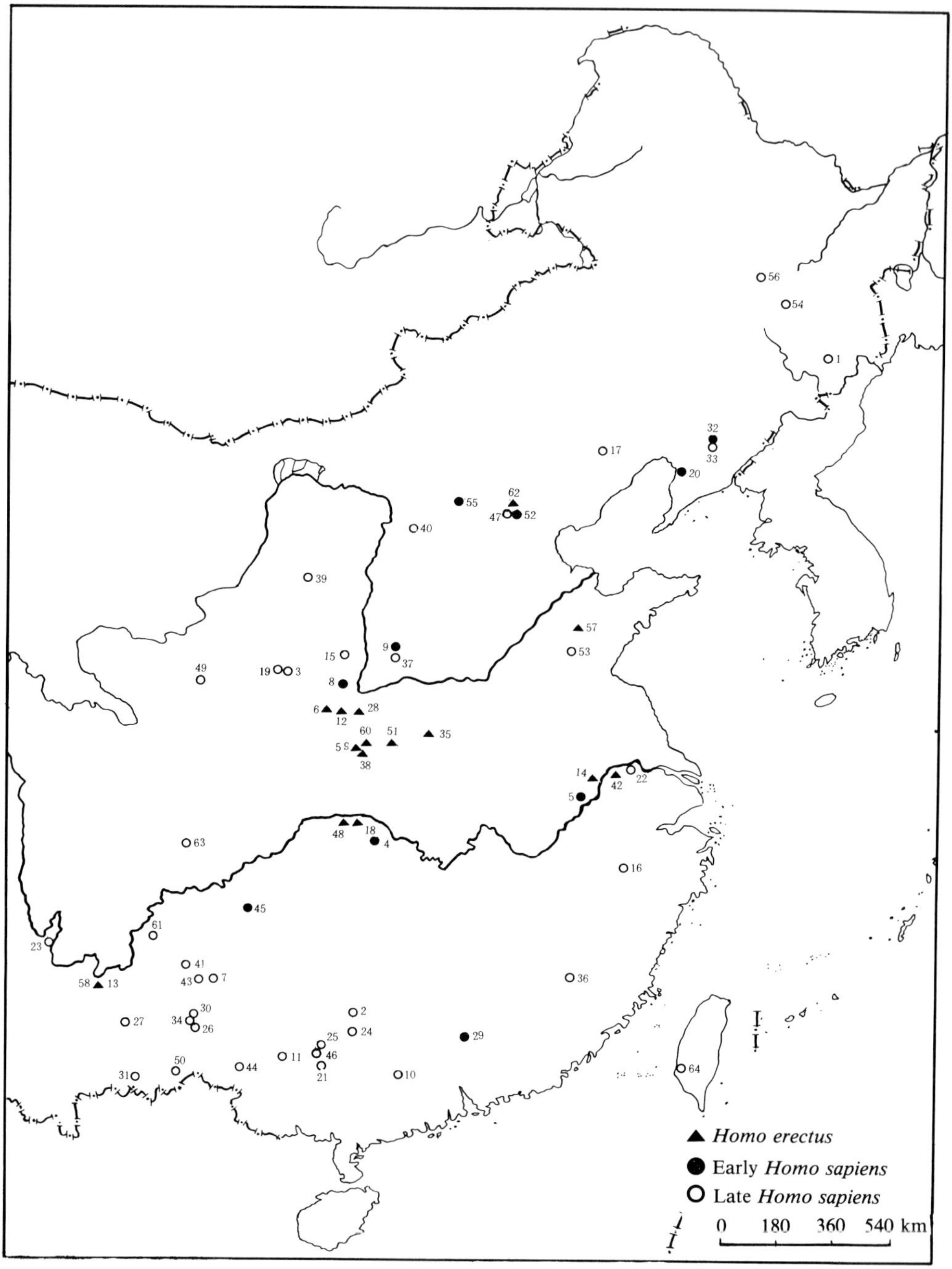

FIG. 1.1 Hominid locations in China.

FIG. 1.1 (continued)

1. Antu
2. Baojiyan
3. Changwu
4. Changyang
5. Chaoxian
6. Chenjiawo
7. Chuandong
8. Dali
9. Dingcun
10. Dongzhongyan
11. Du'an
12. Gongwangling
13. Guojiabao
14. Hexian
15. Huanglong
16. Jiande
17. Jianpin
18. Jianshi
19. Jinchuan
20. Jinniushan
21. Laibin
22. Lianhua
23. Lijiang
24. Lipu
25. Liujiang
26. Longlin
27. Longtanshan
28. Luonan
29. Maba
30. Maomaodong
31. Mengzi
32. Miaohoushan
33. Miaohoushan Dongdong
34. Nalai
35. Nanzhao
36. Qingliu
37. Quwo
38. Quyan River Mouth
39. Salawusu
40. Shiyu
41. Shuicheng
42. Taohua
43. Tiandong
44. Tongzi
45. Tubo
46. Upper Cave
47. Wushan (Damiao)
48. Yuanyang
49. Xichou
50. Xichuan
51. Xingdong
52. Xintai
53. Xuetian
54. Xujiayao
55. Yanjiagang
56. Yiyuan
57. Yuanmou
58. Yunxi
59. Yunxian
60. Zhaotong
61. Zhoukoudian
62. Ziyang
63. Zuozhen
64. Zuozhen

World War II. After 1949, excavations began anew and many more human fossils were recovered from many different regions in China.

Homo erectus

Despite persistent reports of *Australopithecus* remains from central China, none has been confirmed. The earliest hominid remains from China belong to *H. erectus*, and some have suggested these remains may date to as early as 1.7 mya. Although the paleomagnetic date of 1.7 mya for a Chinese *H. erectus* specimen has been confirmed by three laboratories in China, a paleomagnetic date of 600,000–500,000 years also appears in the literature. The specimens under discussion are two considerably worn incisors larger than those of *H. erectus pekinensis*. Their origin is the Yuanmou basin in Yunnan Province, southern China.

Until only recently most Chinese representatives of *H. erectus* came from Zhoukoudian. Locality 1, Zhoukoudian, is a limestone cave filled with deposits. The fossils were found in the early 1900s by quarry workers and called "dragon bones" (long-gu) by the local people, who showed them to J.G. Andersson during a visit to the site in 1921. The story of the finding and the eventual loss of most of the original hominid materials from Zhoukoudian has been recounted by Jia (1975) and Jia and Huang (1990), among others. Although rather late in the *H. erectus* sequence from China, dating from 500,000 to 230,000 years ago, the Zhoukoudian remains are significant because of their historical precedence and their number.

Palynology indicates that the site lay near the border zone separating the northern coniferous belt, or boreal forest, from the temperate steppe. The surrounding hills were covered in pine and spruce, and faunal remains also suggest a northern temperate zone. Evidence from different cave layers suggests climatic shifts during the long history of human occupation at the site. Jia (1978) presents a listing of the climatically sensitive fauna present at Zhoukoudian (Chapter 2).

Numerous quartz stone tools were recovered from the 40 m of sediments at Locality 1. Two major arguments concerning the *H. erectus* inhabitants are whether they made and used fire, and whether they cannibalized their own kind. Although Binford and Ho (1985) and Binford and Stone (1986) questioned many of the accepted notions on tool use, possible cannibalism, and the use of fire associated with the finds, they had little or no first-hand experience at the site. Chinese researchers such as Jia (1978), who have worked at Zhoukoudian, dismiss any claims that the cave inhabitants did not control fire. This is discussed in Chapter 2.

Two other important Chinese *H. erectus* sites located in Lantian County, Shaanxi Province, predate Zhoukoudian. The human remains from the site consist of a robust mandible, discovered in 1963 near the village of Chenjiawo, and facial bones, a tooth, and a fairly complete skullcap (all from the same individual) found in 1964 near the village of Gongwangling. Paleomagnetic dating suggests that the cranial material may date to 1,150,000 years ago and the mandible to 650,000–500,000 years ago.

According to J. Woo (1964b) the Lantian County material is closely related to the *H. erectus* fossils from Zhoukoudian, despite the fact that there are some morphological differences. Initially he referred to the material as a separate species, *Sinanthropus lantianensis,* but upon further study changed the designation to the subspecies *Homo erectus lantianensis*.

In eastern China, a heavily fossilized *H. erectus* skullcap and dental remains belonging to a young male were unearthed in 1980 from Longtandong cave (Dragon

Pool Cave), Hexian County, Anhui Province (Huang et al., 1982). The Hexian specimens document the youngest-dated *H. erectus* finds in China and are perhaps the last known occurrence of this taxon anywhere in the world. Further excavation in 1980 and 1981 exposed parts of the frontal and parietal bones, the left side of a fairly robust mandible, and 12 teeth. The materials show some resemblance to similar parts from Zhoukoudian. The uranium series date is 190,000–150,000 years ago and the thermoluminesence (TL) date on quartz is 195 ± 16,000 years ago.

Three other skulls provisionally attributed to *H. erectus* have recently been found. Two of them were found in 1989 and 1990 in fluvial deposits, at Quyuan River Mouth site, Yunxian County, Hubei Province. The third was found in 1993 in a cave near the town of Tangshan, Nanjing County, Jiangsu Province. The Quyuan River Mouth skulls are fairly large and heavily distorted. No chronometric dates have been published for any of the specimens, but faunal correlations indicate a Middle Pleistocene date.

Homo erectus fossils found in Java and in China are related; the major differences are in cranial capacity and teeth. Although both the Chinese and Javanese *H. erectus* specimens have thick skull bones and large brow ridges, there are major differences in cranial capacities. The mean cranial capacity of the Javanese specimens is estimated at 975 cc or lower; the Lantian material is estimated at 780 cc; and the adult material from Locality 1, Zhoukoudian, is 1088 cc. Compared with Javanese specimens thousands of years older, the dental arch of the Zhoukoudian sample is shorter and more rounded in front, and there is no sign of a diastema in the maxilla. The Zhoukoudian mandible is shorter and more compact.

Questions have been raised concerning the role of *H. erectus* in the evolution of *H. sapiens* and the validity of maintaining the African and East Asian *H. erectus* samples in the same species. Some of the African fossils, such as KNM-ER 3733 and 3833 from Lake Turkana in Kenya, are quite different from the classic East Asian *H. erectus*. African *H. erectus* skulls, for example, generally have thinner skull bones and, because they are older than the East Asian specimens, could represent a population ancestral to Asian *H. erectus*. In a new habitat these Asian specimens could have evolved into a new geographical subspecies.

Sohn and Wolpoff (1993) are among those who suggest a regionally based evolutionary history of Asian *H. erectus*. They compared the material from Zhoukoudian with the frontofacial fragment from Zuttiyeh, Israel. This specimen, dating between 350,000 and 250,000 years ago, usually has been regarded as either a Neandertal, an early *H. sapiens*, or a generalized ancestor to both. Sohn and Wolpoff argue that whereas the Zuttiyeh material differs from the Neandertals it shares features with *H. erectus pekinensis*, suggesting the possibility of an ancestral-descendant relationship. They write "The hypothesis of a recent unique African ancestry for all modern humans is disproved by our study, which shows Asia as a significant source area for at least some living populations" (Sohn and Wolpoff, 1993: 325).

Early *Homo sapiens*

Major archaic or early *H. sapiens* finds in northern China come from the Middle Pleistocene sites of Jinniushan and Dali. Both samples share a mosaic of archaic and more advanced traits. The Jinniushan (Gold Ox Hill) material comes from Liaoning Province, northeastern China. Although original reports (Lu, 1989) referred these materials to *H. erectus*, the *H. sapiens* designation is now accepted. The remains include a nearly complete skeleton from an approximately 30-year-old male, with bones of the hands, feet, spine, ribs, and ulna. The well-preserved skull has a relatively low cranial vault and prominent brow ridges. There are a few stone tools, and burned animal bones, burned clay, and carbon are evidence of fire. Several uranium series dates, ranging from 310,000 to 160,000 years ago, have been obtained from the animal fossils.

The Dali skull, recovered in 1978 in Shaanxi Province, is the most complete cranium dating from the Middle Pleistocene period. The face is short, with the frontal process of the zygomatic more forward and the lower margin of the zygomatic process of the maxilla more curved than those of European Neandertals. Its uranium date is 209,000 ± 23,000 years ago.

Early *H. sapiens* specimens also have been recovered at Maba, Changyang, Chaoxian, Xujiayo, and other sites. The Maba skullcap, consisting of the frontal, parietal, and nasal bones, and the lower border of the right eye socket, were recovered in a limestone cave in Guangdong Province in 1958. The specimen has more skeletal features of the so-called European classic Neandertals than any other Asian remains. Its uranium date is around 130,000 years ago.

From Changyang County, Hubei Province, part of the left maxilla and a premolar were recovered in 1956. The material is dated to about 190,000 years ago. The Chaoxian remains, a damaged occipital, a maxilla containing the P1 to M1, and three isolated maxillary teeth, were recovered in the early 1980s from Anhui Province. A uranium series gives a date of between 200,000 and 160,000 years ago. Xu et al. (1984) suggest that the material belonged to a female younger than 26 years old. These specimens should provisionally be regarded as early *H. sapiens* with some traits relating them to Asian *H. erectus*.

Xujiayao constitutes the largest single sample of Premodern Asian *H. sapiens* recovered. Twenty specimens of various ages were recovered from fluvial deposits in Shanxi Province in the 1970s. The horizontal relationship of the finds is unclear, and the dating has engendered continuing controversy. The specimens share skeletal traits with both *H. erectus* and *H. sapiens*.

Anatomically modern *Homo sapiens*

Although the Out-Of-Africa model for the origin of modern *H. sapiens* is accepted by many Western researchers, Chinese and some Western researchers reject the theory. Many cranial and dental traits link *H. erectus* and *H. sapiens* samples in China. An argument has been made that both hominid skeletal remains, and paleolithic cultures in China favor an indigenous origin of *H. sapiens* in Asia with a certain amount of gene flow between these populations and populations in other parts of the world. This evidence is discussed in Chapter 5.

Some of the most important *H. sapiens* material found in northern China is from the Upper Cave of Zhoukoudian. These specimens include three almost complete skulls and other cranial and postcranial remains representing at least eight individuals dating to around 18,000–10,000 years ago. A limestone cave in Liujiang, Guangxi Province, yielded the most important skull in southern China. This skull and the postcranials found with it date to around 67,000 years ago.

Southern China, from about 100,000 to 70,000 years ago, was probably the area of dispersal for the earliest *Homo sapiens sapiens* in Asia. The population there had such recognizable Asian traits as shovel-shaped incisor teeth. Beginning about 70,000 years ago, populations from China may have radiated east, moving across the continental shelf to Ryuku and Japan. They also radiated south, moving through insular Southeast Asia and reaching Australia perhaps by 40,000 years ago. Skeletal evidence for this radiation is discussed in Chapter 5.

HOMINOIDS (FIG. 1.2)

Although less well known, and perhaps less numerous or less diversified than their African cousins (but only time will tell), are the Miocene hominoids from China. In 1978 a partial maxilla with three teeth was discovered in eastern China in the Xiacowan Formation, northern Jiangsu Province. It dates about 19–16 million years ago and is called *Dionysopithecus* (C. Li, 1978). This specimen is almost identical to a small *Micropithecus* maxilla from Miocene deposits in Uganda, and *Dionysopithecus* is the first indication of a link between the East African early Miocene sites and locales of similar age in eastern China. The Jiangsu specimen argues for a hominoid dispersal from Africa to Asia about 20 million years ago.

The most extensive Chinese hominoid remains come from Shihuiba in the Lufeng basin of Yunnan Province in southern China. Nearly 1000 specimens dating to about 8 mya represent tens, if not hundreds, of individuals of both sexes. There are 5 reasonably complete skulls, 6 cranial fragments, 10 mandibles, 41 jaw fragments, 29 tooth rows, hundreds of isolated teeth, 1 scapula, 1 clavicle, and 2 digits (Wu et al., 1985, 1986).

The taxonomy of the Lufeng materials has been debated. They were first considered as belonging to *Sivapithecus* and *Ramapithecus,* the former in the orang-utan lineage and the latter in the hominid lineage. In 1986 Rukang Wu and his colleagues made a detailed comparsion of the materials and found that the difference between the *Sivapithecus* and *Ramapithecus* samples is comparable to differences separating male and female orang-utans. They attributed all the specimens to *Sivapithecus lufengensis.* In 1987 Rukang Wu further suggested that the Lufeng materials be placed in a new species, *Lufengpithecus lufengensis.* He is of the opinion that *Sivapithecus* is probably ancestral to orang-utans and that *Lufengpithecus* is probably the common ancestor of African apes and hominids. To date, *Lufengpithecus* has been found only in Yunnan Province in southwestern China. Other hominoid fossils closely related to *Lufengpithecus* were found at Kaiyuan and several sites at Yuanmou, Yunnan Province.

Of all primate fossils *Gigantopithecus* specimens have one of the most colorful histories. This primate, which may have weighed between 400 and 600 pounds and perhaps stood 6 feet tall, has been linked, wistfully, to many creatures, including the "Yeti" and China's Yeren or the "Hairy Wildman" (Poirier et al., 1983; Greenwell and Poirier, 1989). It seems to have ranged in time from 9 to 5 million years ago in India to as recently as 500,000 years ago in China. In China, *Gigantopithecus* remains are found in deposits yielding bones of ancestral giant pandas and of *H. erectus.* Early mention of *Gigantopithecus* was made by G. R. von Koenigswald in 1935, based on one tooth found in a Chinese herbal medicine shop in Hong Kong. The tooth, part of a "dragon bone" collection, was to be used in herbal medicines. *Gigantopithecus* is known from India and has been reported in Vietnam. The primate probably entered China from India and by most accounts was a relatively long-lived side branch of hominoid evolution.

Three mandibles and more than 1000 teeth belonging to the Chinese species of *Gigantopithecus, G. blacki,* had been recovered. No other confirmed skeletal parts are known. The remains are from four sites in Guangxi Autonomous Region (Guan-

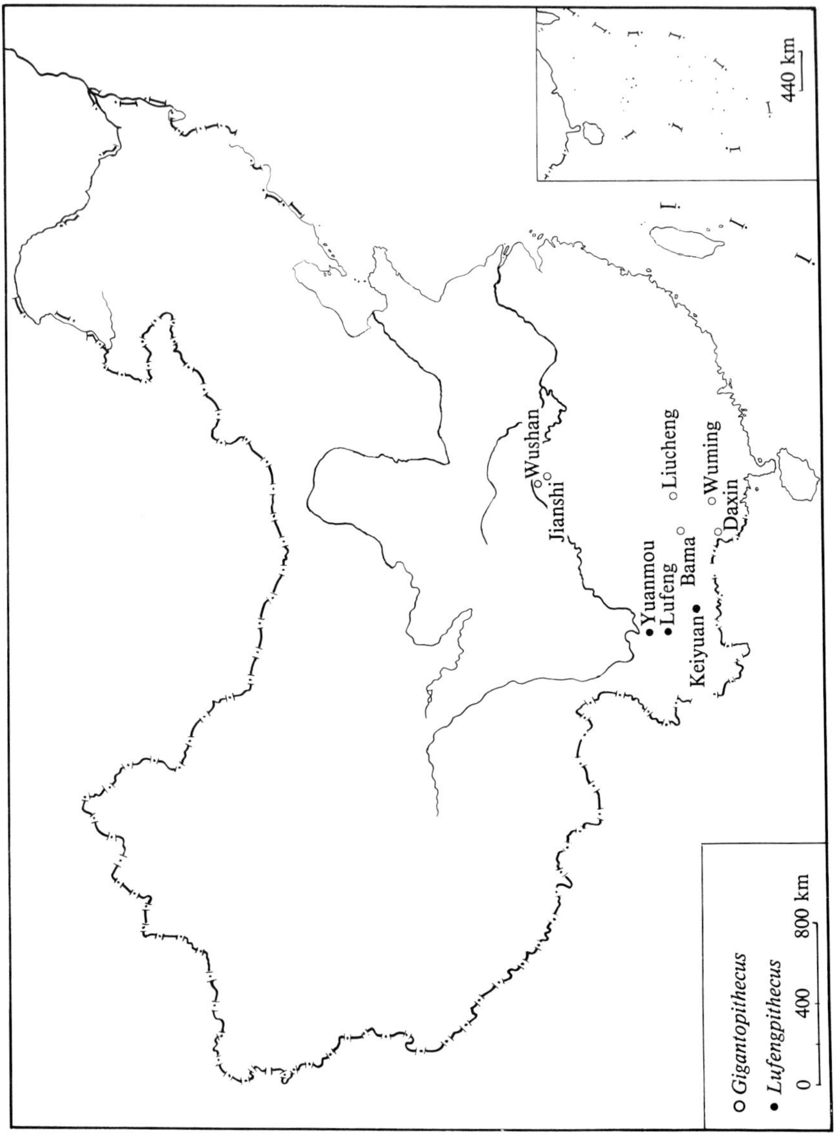

FIG. 1.2 Hominoid locations in China.

gxi A.R.) in southern China and from sites in Hubei and Sichuan provinces. In southern and central China specimens were found in warehouses for Chinese medicinal products as well in cave deposits.

The jaws are deep and very thick, and the huge molars have flat and high crowns. The first lower premolar is bicuspid, the canines are conical, and the incisors are especially thick in labio-lingual diameter. In 911 premolars and molars from Liucheng, Guangxi, the percentage of caries is 9.8%. No caries, however, are found in the panda specimens from the same cave.

Phytoliths, tiny pieces of silica or plant stones, adhering to *Gigantopithecus's* teeth, have been analyzed. Phytoliths differ in every species where they exist; therefore, their identification is an important clue to diet. An examination of the microscopic scratches on the teeth and gritty embeded plant remains suggest they ingested seeds, fruit, and bamboo (Ciochon, 1991). The high incidence of caries suggests that a major dietary component was plant foods rich in carbohydrates. Zhang (1983), citing damage on the enamel, suggested the possibility that it was due to grit eaten while chewing roots and tubers.

SUMMARY

China has many sites containing human fossils and many more with archaeological materials. The earliest Chinese hominids probably belong to the taxon of *H. erectus*. Despite efforts to link Chinese and African *H. erectus*, there is some evidence that they had rather independent evolutionary histories. Strong evidence linking the Chinese *H. erectus* directly to the evolution of *Homo sapiens* in Asia is discussed in Chapters 4 and 5.

One of the most popular scenarios in human evolution is the so-called Out-of-Africa hypothesis. However, there is evidence that the evolution of *H. sapiens* in China was mainly a regional event. It has long been recognized that populations in Southeast Asia and the ancestors of Native Americans (Paleo-Indians) trace their evolutionary roots to China. Any understanding of human evolution is woefully inadequate without a full understanding of human evolution in China. China also was involved in early hominoid evolution, including such forms as *Lufengpithecus* and the enigmatic *Gigantopithecus*.

2

Homo Erectus

The first human fossils found in China resulting from scientific excavations are two *Homo erectus pekinensis* teeth discovered from a cave deposit called Choukoutien (Zhoukoudian) dating to the Middle Pleistocene. Extensive excavations at Zhoukoudian produced considerable hominid remains prior to World War II. The oldest dated *Homo erectus* remains in China come from Yuanmou, Yunnan Province in southern China. Although the dating is disputed, the Yuanmou material dates to either 1.7 mya or 600,000 to 500,000 years ago. The array of Chinese *H. erectus* remains is discussed below.

YUANMOU (SHANGNABANG VILLAGE) (101°55′ E, 25°40′ N)

On May 1, 1965, Fang Qian, a geologist from the Institute of Geology, Chinese Academy of Geological Sciences, found two human incisors in a small hill situated between Danawu and Shangnabang villages in Yuanmou County, located about 5 km southeast of Yuanmou City. No further faunal remains were found at the site until 1973, when an excavation by a team from the Institute of Vertebrate Paleontology and Paleoanthropology (IVPP), Academia Sinica, in Beijing, found additional remains.

Human fossils

The Yuanmou specimens are left and right upper median incisors from the same adult (Fig. 2.1). They are pale gray in color, shovel-shaped, and have thin cracks filled with brown clay. The teeth have been described by Cheng-chih Hu (Chenzhi Hu) and Guoxing Zhou (Hu, 1973; Zhou and Hu, 1979).

The incisors are as large as those belonging to *H. erectus pekinensis* males, and this suggests that the Yuanmou incisors also belong to a male. The incisors are moderately worn, with a sloping plane formed by attrition on the biting edge. The crown is rather thick and swollen. The labial surface is rather flat and has fine furrows. The upper part is more convex. A very prominent basal tubercle on the upper part of the lingual surface has three fingerlike processes diverging from its lower border. The process in the middle of the lingual extends almost to the biting edge and is the longest and thickest of the three. Prominent rims appear along the lateral and medial margins of the lingual surface; the lateral rim is the more prominent. There is a small notch between the basal tubercle and the lateral rim.

The lower part of the left incisor root was preserved. In cross section it is ellipsoid and slightly flattened antero-posteriorly. The cervical line on the medial surface is more curved than that of the lateral surface.

The incisors from Yuanmou and those from Zhoukoudian share a general mor-

FIG. 2.1 Upper median incisors of *Homo erectus* from Yuanmou. (Courtesy of IVPP.)

phology in such features as the swollen basal part of the crown, the strong prominence of the basal tubercle, the parallel fingerlike processes, and the prominent marginal rim of the lingual surface (Table 2.1).

Geology

The Yuanmou Formation dates to the Lower Pleistocene. It is 673.6 m thick and has been divided into four members, including 28 layers reported by Fang Qian and others (1991). The human fossils reportedly were found in the uppermost part (fourth member), which is 122.2 m thick and includes the 24th to 28th layers. The fourth member consists of fluvial and deluvial deposits, composed of sandy gravel, silty clay, and clay. The 25th layer can be subdivided into two layers: the upper is a brownish-red silty clay 8.8 m thick with a lens of gravel. The lower layer, 10 cm thick, has a brown clay region, a yellowish-brown silty clay region with a sandy gravel lens, and a region consisting of a conglutination of pebbles 2.14 cm in diameter and sand. Most of the pebbles are sandstone; some are quartzite. The human incisors were found embedded in brown clay, and the upper part of the lower sublayer of the 25th layer in the fourth member. Molluscan and mammalian fossils and stone artifacts were also deposited in this layer.

Carbon particles recovered near the bottom of this layer are sparsely distributed within an area 3 m thick in the deposit. Some claim that these particles are evidence that ancient hominids used fire, a claim many Chinese paleoanthropologists reject. Yet to date no convincing evidence for the use of fire or evidence excluding the possibility that the carbon is the residue from a carbonized plant or the result of natural fires has been presented.

Table 2.1

Measurements of upper median incisors of *Homo erectus* from Yuanmou

		left	*right*
Crown	height	(11.2)	(11.1)
	length	11.4	11.5
	breadth	8.1	8.6
Root	height	—	(13.2)
	length	8.1	(8.1)
	breadth	7.5	(7.6)

1. Figures in parentheses are worn (crown) and/or damaged (root)
2. For this and all subsequent tables, linear measurements are given in mm, angles in degrees, and volume in cubic cm.
3. Measurements are from Zhou and Hu (1979)

Although most of the palaeomagnetic dates for the lower sublayer of the 25th layer cluster around 1.7 mya (Qian et al., 1991), Liu and Ding (1984) noted that the upper strata of the Yuanmou Formation *(sensu lato)* contain more extinct species than the lower levels. Based on the recognition of a fault disconformity in the Yuanmou Formation, they suggest that the earlier magnetostratigraphic sequence at the site may have another explanation, preferring the normal polarity shown in the segment yielding the human teeth with the Brunhes Epoch rather than the Matuyama Epoch. Therefore, they suggest that the layer yielding the human fossils is possibly only 600,000–500,000 years old.

Qian et al. (1991) published electron spin resonance (ESR) and fission track dates for the Yuanmou Formation. The ESR date is 1.3 mya, based on the enamel of deer teeth collected from the middle part of the 26th layer. Because the human teeth were found more than 20 m below this level, their ESR date is probably earlier than 1.4 mya.

Pan et al. (1991) listed the mammalian fauna from the fossil bed corresponding to that yielding the human teeth (i.e., fourth member of the Yuanmou Formation), and Han and Xu (1989) provided a faunal list associated with the hominid remains. The following list is compiled from both sources. (An asterisk identifies those animals presented in both lists, and a pound sign identifies animals listed only in Han and Xu.)

Primates:
*Homo erectus**
Lagomorpha
Ochonoides complicidens
Rodentia
Microtus sp.*
Rhizomys sp.*
Arvicola sp.#
Hystrix subcristata.*

Carnivora
Viverricula malaccensis fossilis
Hyaena sp#
Megantereon cf. *nihowanensis*
Panthera tigris
Panthera pardus
Felidae
Proboscidea
Stegodon elephantoides
Stegodon sp.*
Perssidactyla
*Equus yunnanensis**
Nestoritherium sp.*
Rhinoceros sinensis
Rhinoceros sp.*
Artiodactyla
Sus sp.*
Eostyloceros longchuanensis
Metacervulus capreolinus
Paracervulus attenuatus
*Cervocerus ultimus**
Axis cf. *rugosus*
Axis shansius
Axis sp.*
Muntiacus lacustris
*Rusa yunnanensis**
Rusa sp.
*Rusa stehlini**
Cervus sp.
*Procapreolus stenosis**
Bos sp.
Bibos sp.*
Gazella sp.*

Molluscan remains from the fourth layer include: *Viviparvus* sp., *Catheica* sp., *Gyraulus* sp., and *Corbicula* sp., among others. Flora from the member including the hominid site consists of about 70 genera or species, including *Pinus, Alnus, Castanopsis, Juglans, Ulmus,* and others (Qian et al., 1991).

Xihao Wu (1986, cited in Qian, 1991), suggests that the average annual temperature was 10–12°C when hominids inhabited Yuanmou, and Qian (1991) suggests a figure of 12–14°C. With a cooler and wetter climate than in present times, it can be assumed that the basin had ample water and rich grass growth. Forests were shrinking, limited, for the most part, to the hills surrounding the basin. Mammals seem to have inhabited the grasslands.

Archaeology

No stone artifacts were found directly associated with the human fossils. However, during an excavation 8 years after the original hominid discovery, an IVPP team unearthed six stone artifacts in situ. Another 10 pieces were found on the surface. Three of the artifacts found in situ were made of quartize and were identified by Wen (1978) as small scrapers. They include a nucleus, a red sandstone flake, and a point. Succeeding excavations found six more pieces of stone artifacts (Zhou et al., 1991).

GUOJIABAO (101°55′ E, 25°40′ N)

In December 1984, a field team led by Bo Wang of the Beijing Museum of Natural History recovered a fragment of a left tibia. It was found with other vertebrate fossils on the south slope of Guojiabao Hill, about 250 m south of the Yuanmou site. Guoxin Zhou, et al. (1991) identified the tibia as belonging to a hominid.

Human fossil

According to Zhou et al. (1991), the fragment is from the middle of the tibia. There are longitudinal cracks on both internal and external surfaces, and its reddish brown color is similar to that of other animal fossils unearthed at Yuanmou. The 227 mm long tibial fragment is rather gracile and without any robust ridges, and it probably belonged to an adolescent female. At what is assumed to be the middle point of the shaft, the circumference, transverse, and longest sagittal diameters are 78 mm, 17 mm, and 29 mm, respectively. With a cnemic index of 58.0, the tibia is considered to platycnemic.

The anterior border fragment is rounded at the upper segment and becomes thinner at the middle part, and more rounded again at the lower part. The S-shaped curve of the border is weaker than that of modern humans. There is a popliteal line on the posterior surface. The bone wall is thick. The reconstructed stature of this individual is 130.4–123.6 cm.

Geology

Zhou et al. (1991) first described the tibia as having been found in the bottom layer of slope materials. Of a deep brown color, these materials include silty clay and stones of different sizes, and overlie a layer of brownish-red silty clay. However, the tibia was later described as coming from the superficial part of the brownish-red silty clay layer instead of from the slope materials. The discrepancy concerning the point of origin of the tibia cannot be explained; therefore, the provenance of the tibia must be in doubt. Zhou et al. (1991) correlated the slope materials with the 26th layer of the fourth member of the Yuanmou Formation, and consider the tibia to be in excess of 1 million years old, dating to the later Early Pleistocene.

GONGWANGLING (LANTIAN) (109°29′ E, 34°11′ N)

Gongwangling is a small hill situated near Gongwang Village about 17 km east of the city of Lantian, Shaanxi Province, northwestern China. The hill lies north of the Qingling Range, the western boundary of the southern and northern Pleistocene faunas of China. The site is in the western part of the hill's northern face.

In May 1964 an IVPP field team, including Weiwen Huang and Maolin Wu, found an isolated left upper second molar and other fossils that were so densely embedded in the concretions that large sections of the deposits were packed in wooden boxes and transported to the IVPP in Beijing. By October pale yellow and heavily mineralized cranial fossils had been removed from the hard matrix. These fossils may represent the earliest *H. erectus* specimens from eastern Asia. The small number of isolated artifacts recovered from diverse localities, some with questionable stratigraphic provenience and conflicting dates, raise many questions. However, the artifacts appear to be more than 780,000 years old.

Perhaps of similar age to the Gongwangling (Lantian) deposits are a number of sites in the Nihewan (Nihowan) Basin located west of Beijing and astride the border of western Hebei and eastern Shanxi provinces. These are the earliest well-documented Paleolithic occurrences containing large artifact assemblages in eastern Asia (Schick and Dong, 1993).

Many scholars have studied the Gongwangling materials. The following description is from Wu et al. (1966), Chen and Qi (1978), An et al. (1990), Qi (1989), Dai (1966), Hsu (1966), and Tai and Hsu (1973).

Human fossils

The human fossils (PA 1051–6) include a complete frontal, a large part of the parietal (lacking the infero-posterior angle and the posterior part), a large part of a right temporal (without the mastoid), a large part of the left nasal, the upper part of the right nasal, a large part of the right maxilla associated with the second and third molars, the body and frontal process of the left maxilla, and an isolated left upper second molar. Judging from their location and morphology, size, color, and dental attrition, the materials probably belong to one individual.

These specimens (Figs. 2.2 and 2.3) were studied by J. Woo (1964a,b and 1965) in detail.

The skullcap was slightly deformed by the pressure of the enclosing earth. The coronal suture and the anterior part of the sagittal suture are fused but still can be detected. The obliteration of the cranial sutures in fossil populations probably occurred earlier than in modern humans. According to aging standards in modern populations, this individual was about 40 years old. Attrition on the second upper molar indicates a similiar age based on age estimates of modern northern Chinese. However, because the diet of *H. erectus* is assumed to have been grittier than the diet of modern populations, with more pronounced dental attrition, X Wu et al. (1966) estimated the individual's age at just more than 30 years. On the basis of sexual dimorphism in *H. erectus pekinensis* and the tooth size of the Gongwangling specimens, the latter probably belonged to a female. The maxilla is gracile, and the pyramid of the temporal and the middle cranial fossa are small. Cranial features and the size of the molars also support attribution to a female.

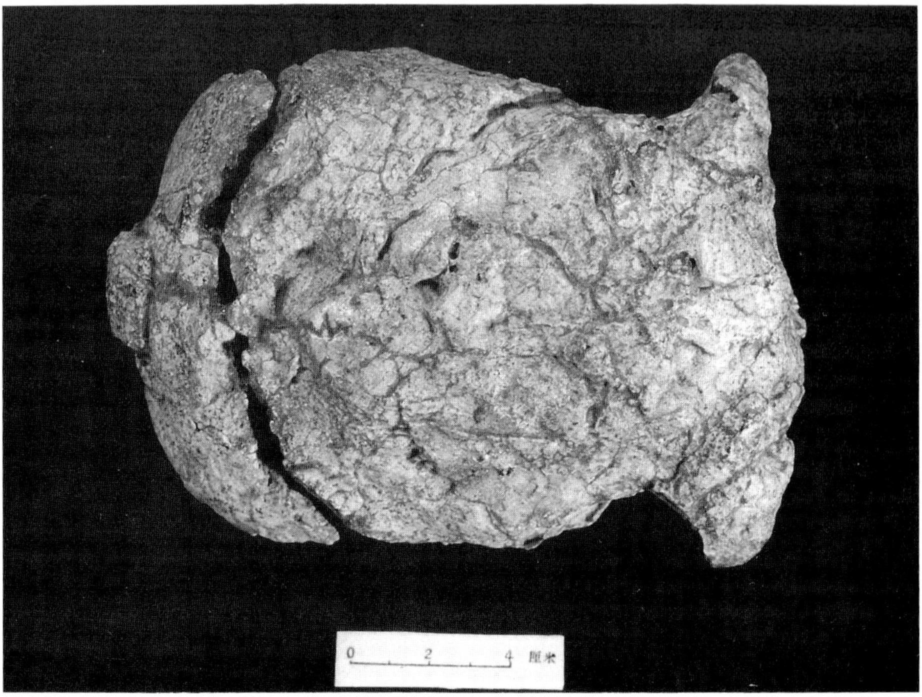

FIG. 2.2 *Homo erectus* skullcap from Gongwangling, Lantian. (Courtesy of IVPP.)

Frontal bone

The frontal bone is rather wide, approximating the upper limit of *H. erectus pekinensis*. The very robust supraorbital torus forms a transverse bony bar in top view. The glabellar region protrudes slightly forward. The lateral ends of the supraorbital torus extend laterally instead of turning slightly backward, as in *H. erectus* specimens from Zhoukoudian and from Trinil in Java. Postorbital constriction is also more obvious than these specimens. In frontal view, the medial segment of the superior margin of the torus above each orbit turns slightly downward toward the glabella region and toward the lateral end. The medial part of the torus is thicker, but it thins toward the lateral end. The morphology of the torus suggests a quadrangular orbit instead of a spherical one. The orbital roof is flat, as in *H. erectus pekinensis*, with a trace of a frontal notch on each side. There is no supraorbital foramen or lacrimal fossa.

There is no broad sulcus between the supraorbital torus and the frontal squama. In this respect, the specimen is similar to that from Trinil. The material from Zhoukoudian has a sulcus in this area.

The external surface of the bone is eroded, so that only traces of a midsagittal prominence and a cross eminence at the junction of the coronal and sagittal sutures can be detected. The middle of the anterior portion of the frontal squama is slightly bulging. The temporal line appears as a prominent ridge.

The broad and high frontal crest on the median part of the inner surface of the

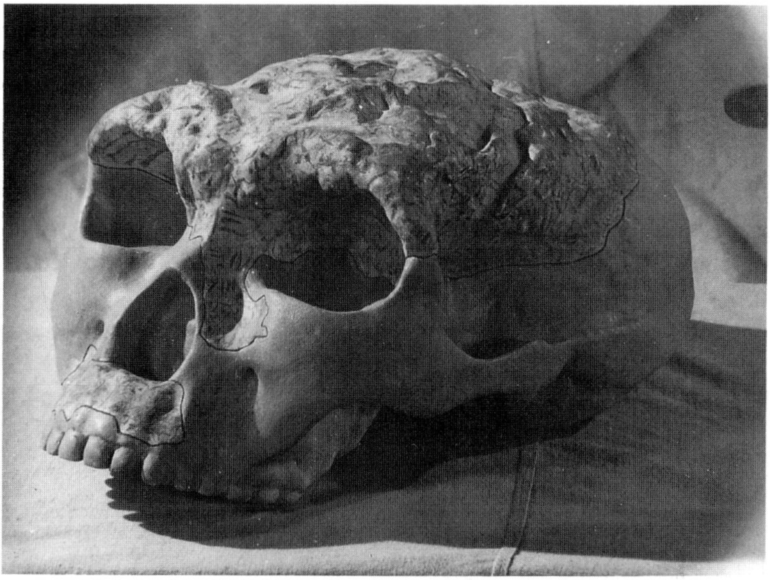

A

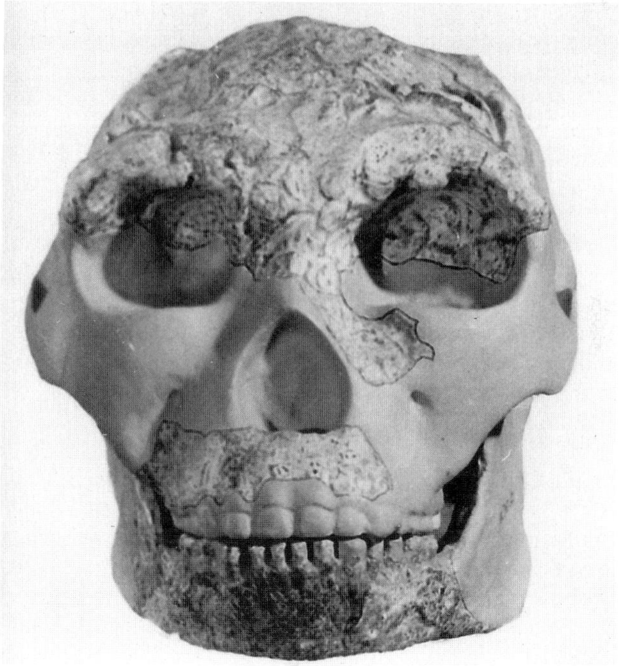

B

FIG. 2.3 Two views of the reconstructed *Homo erectus* skull and face from Gongwangling, Lantian. (A) Lateral view. (B) Anterior view. (Courtesy of IVPP.)

frontal bone disappears at the junction between the upper and middle thirds of the frontal bone without any bifurcation at its terminus. A median groove exists on the upper part of the frontal bone. No frontal sinus can be shown by X-ray.

According to the reconstruction by J. Woo (1965), bregma, located above the external auditory meatus, is much more posterior than in modern humans. The nasion is not depressed when viewed from the lateral side. The glabella region is very robust.

Parietal bone

The partietal bone is rectangular, with almost a right angle at the bregma. The sagittal border of the parietal is shorter than the median sagittal length of the frontal bone. The anterior part of the sagittal suture curves slightly. The medial part of the coronal suture is almost straight. The depression for the anterior branch of the left meningeal artery is more obvious than is its right counterpart. Depressions for the posterior branch of the left meningeal artery cannot be identified because of erosion on the internal bone surface.

Temporal bone

A large part of the pyramid of a right temporal bone is preserved. The premolar's size and shape approximate that of a modern female. Both anterior and posterior surfaces of the pyramid are steeper than that of *H. erectus pekinensis*, but the upper border of the *H. erectus pekinensis* pyramid is more obtuse (less acute). These features resemble more closely those of *H. erectus* from Trinil than the Zhoukoudian sample. The arcuate eminence is more similar to modern humans than to the Zhoukoudian *H. erectus* sample. However, a small flat facet on the posterior surface of the pyramid lateral to the opening of the internal meatus is more similar to the Zhoukoudian sample than to modern humans.

Bones of the nasal region

The upper and middle parts of the left nasal bone, the upper part of the right nasal bone, and the upper part of the frontal process of both maxilla bones are preserved. In the horizontal plane, the sutures between the frontal bone on the upper side, and the two nasal and two maxillary bones on the lower side, form a slightly curved line. The central part of this line is slightly convex upwards. The width of the upper part of the Lantian nasal bones is broader than in modern humans. The nasal saddle is slightly more protruding than that of *H. erectus pekinensis* and is flatter than in modern humans.

Maxillary bones

A large part of the right maxilla and a small part of the left maxilla are joined. These bones were so flattened anteriorally by pressure from the overlying earth that the canine alveoli are located at almost the same level as the incisors. The exposed maxillary sinus extends between the level of the canine and the anterior surface of the third molar. Its medial margin corresponds to the lateral border of the palatal process of the maxilla instead of protruding into the palatal process.

The anterior surface of the bones shows alveolar prognathism. The median sagittal contour of the surface is obviously convex, as in modern apes. This surface forms almost a right angle with the floor of the nasal cavity, and a demarcation between the two and a small but distinct anterior nasal spine clearly exists. There is a fairly marked jugum alveolare of the canine on the right maxilla, lateral to which is a sulcuslike depression. No exostosis appears on the lateral surface of the maxilla. On the right side, the second and third molars are preserved and attached to the maxilla. The roots of Pl and Ml are preserved within the alveoli. The alveoli of the two incisors, the canine, and the second premolar are distinct.

The slightly worn occlusal surface of the second upper molar is rectangular, with its width greater than its length. There is a vertical groove on the buccal surface. Although the right upper third molar is similar in morphology to the second, it has a much smaller crown length and a relatively greater width. Its occlusal surface is nearly triangular.

The exposed root segment of the second upper molar has two buccal and one lingual branch that join near the neck of the tooth. The buccal and lingual branches are 19 mm and 16.5 mm long, respectively, forming approximately a 45° angle. The cross section of the branches is elliptical. The terminal ends of the branches turn to the distal side. The root of the upper right third molar also has three branches. The bifurcation angle between the buccal and lingual branches is smaller than that of the second molar. The length between the neck line and the point of bifurcation of the root is shorter than in the second molar.

J. Woo (1965) reconstructed the skull based on the fragments (Table 2.2). Figure 2.3 is of the reconstructed face and skull from Gongwangling.

Table 2.2

Measurements of *Homo erectus* skull from Gongwangling, Lantian

N-sg arc			37
N-sg chord			33
Sg-b arc			88 ?
Sg-b chord			86
G-n arc			14.5
Min. frontal width (ft-ft)			92
Max. fontal width (co-co)			109 ?
Interorbital width			28.5
Width of two nasal bones at upper part			12.5
Reconstructed skull			
length			189
width			149
auricular height			87
Thickness of cranial bones			
frontal bone	glabella		24
	supraorbital torus	medial	17
		middle	14
	center of squama		15
	temporal suface		7
parietal bone	near bregma		16
temporal bone	center of squama		11.5

Measurements according to J. Woo (1965)

J. Woo (1965) attributed the human fossils from Gongwangling to a new subspecies, *Homo erectus lantianensis,* based on the mandible found in Chenjiawo (details follow) in 1963.

Geology

The Gongwangling hominid remains were found in a laminated layer of pale gray silty clay 2.6 m thick. The hominid and other large mammalian fossils were found in or attached to concretions varying from the size of a fist to about 1.5 m^3 in the middle and upper parts of this layer. Almost all of the larger mammalian fossils had accumulated in piles. The smaller mammalian fossils embedded in the basal part of the layer had no connection with the concretions.

There are four paleosol layers in the profile containing the Gongwangling hominid remains. Some flakes of vein quartz and quartzite appear in different sections of the profile. The sections are as follows:

12 light brownish-purple clay (paleosol), 1 m thick
11 light yellow loesslike silty clay, 1.2 m thick
10 light brownish-purple clay (paleosol), 0.5 m thick
9 light brownish-yellow sandy clay, 1.4 m thick
8 brownish-purple clay (paleosol); most artifacts and a few bone fragments derive from the lower part of this layer; 2 m thick
7 light brownish-yellow transitional layer, 0.2–.0.6 m thick
6 light yellow loesslike silty clay with many concretions in the middle and upper parts; human fossils and fossils of other large mammals are found within or attached to the concretions; fossils of small-sized mammals come from the lower part of this layer; 2.6 m thick
5 Light brownish-purple loesslike sandy clay, 0.5 m thick
4 Purplish clay (paleosol), 1 m thick
3 yellow loesslike silty clay, 1.35 m thick
2 light brownish-purple loesslike sandy clay, 1.35 m thick
1 light purplish sandy clay (paleosol), 1.8 m thick

Below this layer is basal gravel dating to the Early Pleistocene (X. Wu et al., 1966).

Several paleomagnetic dates are available for the hominid locale: 800,000–750,000 years ago (Ma et al., 1978, cited in R. Wu et al., 1989), 1 million years ago (Chen et al., 1976, cited in R. Wu et al., 1989), and 1,150,000 years ago (An et al., 1990).

The mammalian fauna found associated with the hominid fossils appears in the following list (Qi, 1989; *Rhinopithecus* citation, Gu and Jablonski, 1989). Tropical and subtropical species indicate that the habitat was warmer than today. Pollen identified as belonging to *Pinus, Celtis, Ulmus, Quercus, Carpinus, Betula,* and *Artemisia,* among others, has been found in strata corresponding to the hominid-bearing layer (Hsu, 1966).

Primates
Homo erectus
Rhinopithecus lantianensis
Insectivora
Scaptochirus moschatus
Lagomorpha
Ochotona cf. *thibetana*
Ochotonoides complicidens
Rodentia
Petaurista sp.
Bahomys hypsodonta
Allocricetus teilhardi
Cricetulus cf. *barabensis griseus*
Cricetulus sp.
Myospalax tingi
Myospalax fontanieri
Myospalax sp.
Allophaimys sp.
Microtus epiratticeps
Apodemus sp.
Hystrix cf. *subcristata*
Carnivora
Canis variabilis
Ursus cf. *etruscus*
Ailuropoda melanoleuca fovealis
Meles cf. *leucurus*
Hyaena sinensis
Megantereon lantianensis
Panthera cf. *tigris*
Acinonyx pleistocaenicus
Proboscidea
Stegodon orientalis
Perissodactyla
Equus sanmeniensis
Nestoritherium sinensis
Tapirus sinensis
Megatapirus augustus
Dicerorhinus cf. *mercki*
Dicerorhinus lantianensis
Artiodactyla
Sus lydekkeri
Elaphodus cephalophus

Megaloceros konwanlinensis
Cervus (P.) grayi
Capricornis sumatraensis qinlingensis
Leptobos brevicornis
Leptobos sp.

Archaeology

Dai (1966) and Tai and Hsu (1973) described the archaeological contents of the Gongwangling site, which yielded 20 pieces of stone artifacts: 11 nuclei, 5 flakes, and 4 scrapers. The nuclei and flakes are large. Four of the flakes were made with the hammering technique, while the other was probably made with anvil processing. The composition of the stone assemblage suggests that this site was neither a manufacturing place nor a living floor, and perhaps was only temporarily occupied.

In the vicinity of the Gongwangling site, a quartzite handaxe found on the surface looks similar to Abbevillian bifaces from the early European Paleolithic.

CHENJIAWO (109°14′ E, 34°14′ N)

This site in northwest China is situated near the town of Yehu located about 10 km west of Lantian city, or about 40 km east of Xian, the capital of Shaanxi Province. Many researchers have studied the fossils and the site, but the following description relies on the work of J. Woo (1964a, b), Zhang et al. (1978), An et al. (1990), Hsu (1966), and Qi (1989).

Human fossil

A reconstruction of *H. erectus* from Lantian is provided in Fig. 2.4. In 1963 a human mandible (PA 102) without the posterior portion of its ramus was found (Fig. 2.5). It was described by Rukang Wu (J. Woo, 1964a,b; Table 2.3). The mandibular body and most of the teeth are preserved. The crowns of the left C–M1 were missing. The right first premolar was lost before death, and there is no third molar on either side. Very heavily worn dentition indicates that the mandible belongs to an old individual and, based on the height and thickness of the mandibular body, the remains are assigned to a female. The first mandibular left molar is larger than that assigned to a female *H. erectus* from Zhoukoudian.

A weak lateral prominence that exists on both sides of the mandible divides into two branches posterior to the first molar. The superior branch (the superior lateral torus) of the right side extends over the larger mental foramen and continues with the canine jugum. The inferior branch (the marginal torus) runs along the inferior margin and terminates at an anterior marginal tubercle below the mental foramina. A shallow intertoral sulcus lies between these two branches. The superior lateral torus, marginal torus, and the intertoral sulcus are present on both sides of the mandibular body. No oblique striae are visible on the lateral surface of the mandible.

The larger of the two mental foramina on the right side of the mandible is at the level of the first premolar and is positioned slightly lower than the smaller one. The latter is situated below the septum between the first and second premolars. There

FIG. 2.4 Bust of *Homo erectus* from Lantian. (Courtesy of IVPP.)

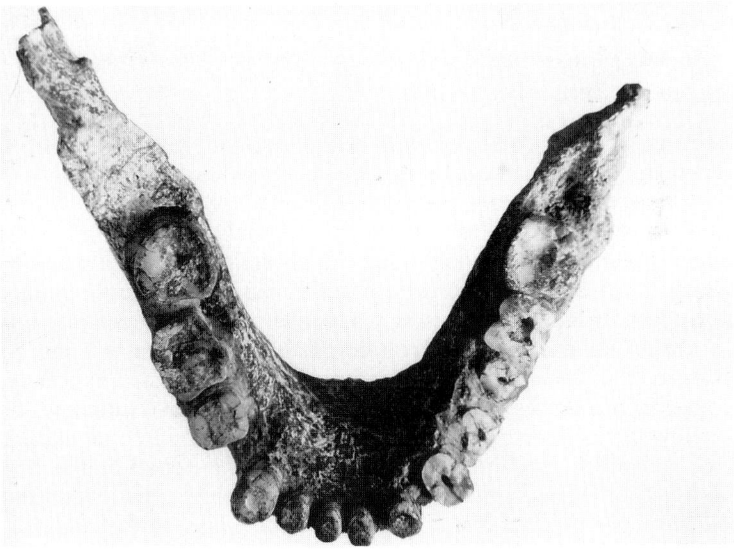

FIG. 2.5 *Homo erectus* mandible from Chenjiawo, Lantian. (Courtesy of IVPP.)

Table 2.3

Measurements of *Homo erectus* mandible from Chenjiawo

Symphyseal height	35
Height of mandibular body anterior to mental foramen	27
Height of mandibular body at level of mental foramen	26.5
Thickness of mandibular body at level of mental foramen	15.4
Thickness of mandibular body at level of M1	16
Length of alveolar arch	51.7
Width of alveolar arch	58.1
Length of anterior alveolar arch	27.3
Width of anterior alveolar arch	48
Length of basal arch	33
Width of basal arch	65.5
Length of digastric fossa	
(left)	13
(right)	15
Width of digastric fossa	
(left)	3.9
(right)	4.9
Angle of inclination of symphysis	55°
Angle of molar rows	27.5°
Index of robustness (at level of mental foramen)	58.1
Index of alveloar arch	89
Index of anterior alveolar arch	56.9
Index of basal arch	50.4

1. Measurements according to Wu (1964a, b)

are four mental foramina on the left side, and all are small and situated below the two premolars. Two of these foramina are positioned lower than the others.

There is a mental trigon on the lower part of the anterior surface of the mandibular body. Located on the posterior surface of the symphyseal portion is a weak depression situated below the alveolar border. Below this depression is a weak transverse prominence corresponding to the superior transverse torus in Great Apes. Another shallow depression appears below this prominence, and yet another prominence, corresponding to the inferior transverse torus, is found below this depression. There are three mental spines in the symphyseal region; the lower and median ones are the more obvious. On the inner surface of the posterior part of the mandibular body, the alveolar prominence is slightly developed. Below the rather distinct mylohyoid line is a rather distinct subalveolar fossa. There is no mandibular torus on either side.

The anterior margin of the digastric fossa is concordant with the lower border of the mandible. The fossa itself inclines slightly backwards and upwards. The interdigastric spine is a downward and backward protrusion of the basal trigon, which is located between the digastric fossae of both sides.

Although the posterior part of the mandibular ramus was lost, a trace of an eversion of the mandibular angle is discernable. A weak eminence on the right side of the mandibular ramus is situated at the position of the ectocondyloid crest on the lateral surface. There is also a triangular torus on the medial surface of the mandibular ramus. A triangular lingula exists in front of and above the mandibular foramen. The preangular notch is recognizable.

The alveolar arch is fundamentally similar to that of "Peking Man." This mandible shares other features with "Peking Man," such as multiple mental foramina, obvious protrusion of the symphyseal part, the location of the digastric fossa at the lower surface of the mandibular body, the robustness of the mandibular body, and the shape of the alveolar arch. Many features, such as the smaller indices of the alveolar and basal arches, are different from those seen in the Zhoukoudian *H. erectus;* these are listed in Table 2.4

Both canines, the right second premolar, first molar, both second molars, the roots of four incisors, the left premolars, and the left first molar are attached to the mandible. Their measurements are close to the average values of the male and female means of *H. erectus pekinensis*. Judging from the canine, second premolar, and second molar, the teeth belonging to the Chenjiawo and Zhoukoudian *Homo erectus* specimens differ more in width than in length. The Chenjiawo teeth have a greater breadth than those of *H. erectus* from Zhoukoudian. This is seen in Table 2.5.

There is congenital loss of the third molar in the Lantian mandible. The buccal part of the alveolum of the right first molar shows distinct periodontoclasia. Atrophy

Table 2.4

Comparison of *Homo erectus* mandibles from Lantian and Zhoukoudian

	Lantian	*Zhoukoudian*
Ecotocondyloid crest	indistinct	marked
Lateral prominence	weak	strong
Angle of inclination of the symphyseal part	more acute	less acute
Angle of molar rows	larger	smaller
Difference between symphyseal height and height at the level anterior to the mental foramen	8.0 (female)	4.8 – 4.9 (female) 5.6 (male)
Index of alveolar arch	low	higher
Index of basal arch	low	higher

1. Based on data from Wu (1964a, b)

Table 2.5

Measurements of mandibular teeth of *Homo erectus* from Chenjiawo*

	Right			*Left*		
	length	*width*	*index*	*length*	*width*	*index*
C	—	9.2	—	—	9.1	—
P2	7.2	9.6	133.3	—	—	—
M1	12.6	11.5	91.3	—	—	—
M2	12.6	13.0	103.2	12.5	13.0	104.0

*Because about half the height of the crown of the cheek teeth was worn, measurements are smaller than would be true of unworn teeth.

1. Measurements according to Wu (1964a, b).

is visible on the other alveoli. The premortem loss of the right first premolar was likely due to periodontal disease.

Geology

Y. Zhang et al. (1978) described the geology at Chenjiawo. From the surface downward the layers are as follows:

9. Light yellow loesslike sandy clay, 1 m thick
8. Light yellowish brown paleosol with scattered concretions in its upper part, 3 m thick
7. Light red paleosol with aggregated concretions, 3.4 m thick
6. Purple paleosol with sporadic concretions, 3.8 m thick
5. Three layers of reddish brown paleosol with aggregated concretions and vertebrate fossils, 7.4 m thick
4. Light red paleosol with human and other vertebrate fossils, 0.3 m thick
3. Light red paleosol, 4.4 m thick
2. Light yellow loesslike sandy clay, 3.5 m thick
1. Greyish yellow gravel, 1.5–2.0 m thick

———unconformity———

Tertiary greyish white sandstone

Magnetic studies suggest that the Chenjiawo site dates to 650,000 years ago (Ma et al., 1978, cited in R. Wu et al., 1989; An et al., 1990) or 500,000 years ago (Chen et al., 1978, cited in R. Wu et al., 1989).

A list of the fauna derived from the same layer as the hominid mandible follows (Qi, 1989):

Primates
- *Homo erectus*

Lagomorpha
- *Lepus wongi*
- *Ochotonoides complicidens*

Rodentia
- *Bahomys hypsodonta*
- *Myospalax tingi*
- *Myospalax* cf. *fontanieri*
- *Apodemys* cf. *sylvaticus*
- ? *Hystrix* sp.

Carnivora
- *Cuon alpinus*
- *Meles* cf. *leucurus*
- *Panthera tigris*

Proboscidea
Elephas sp.
Artiodactyla
Sus cf. *lydekkeri*
Megaloceros sp.
Cervus (P.) grayi

The floral assemblage includes *Biota, Betula,* Gramineae, Cyperaceae, and Chenopoidiaceae (Hsu, 1966). The flora and fauna indicate a grassland habitat and cooler climate than that existing in the time of Gongwangling hominid.

Archaeology

Only a very crudely made scraper and three quartz flakes were found during the excavation.

ZHOUKOUDIAN (LOCALITY 1) (115°55′ E, 39°41′ N)

This site was found in the summer of 1921 by J. Andersson, who visited the place known as Dragon Bone Hill (Longgushan in Chinese). A conversation with a local quarry worker led to the discovery of this large and rich fossiliferous deposit. The site became known as Locality 53 in Andersson's field notes and subsequently as Zhoukoudian Locality 1 (the "Sinanthropus" site). The Locality 1 deposit was partially exposed at the entrance of an abandoned quarry on the north side of a small outlying foothill of Ordovician limestone. During Andersson's examination of the talus of loose material that had fallen into the quarry from the face of the deposit, he was surprised to find fragments of white quartz, a mineral normally foreign to the locality. The significance of this occurrence was immediately apparent to Andersson.

In the summer of 1921, and again for a longer period in 1923, Otto Zdansky, an Austrian geologist, excavated the cave deposit. In 1923 he recovered a worn and fossilized hominid molar, which he recognized at the time of its discovery as being humanlike. Later, among material sent to Uppsala, Sweden, for preparation, a second hominid tooth, an unerupted lower permanent premolar, was recognized (D. Black et al., 1933).[1]

On 16 October, 1927, a slightly worn left lower first molar was found at Locality 1. D. Black (1927) and O. Zdansky proposed a new taxon, *Sinanthropus pekinensis,* on the basis of this tooth and formerly discovered specimens from this site. In December, 1929, a complete adolescent skullcap was discovered by Wenzhong Pei (formerly Wenchung Pei).

Before World War II, 5 skullcaps, cranial fragments belonging to 7 individuals, 6 facial fragments, 16 mandibular fragments belonging to 14 individuals, 147 teeth (among which 64 are isolated), 9 femoral fragments belonging to 7 individuals, 2

[1] Black in a 1927 article, wrote: "During the summer of 1926 Dr. O. Zdansky working in Professor Wiman's laboratory in Uppsala discovered among the rich material recovered from the Chou Kou Tien deposit two teeth, an immature left lower premolar and a worn right upper molar, which could not otherwise be named than *Homo* sp." (p. 1)

humeral fragments, 1 clavicular fragment, and a fragment of a lunate were recovered (Weidenreich, 1936a,b, 1937, 1938, 1941, 1943). During World War II all these specimens were lost (see Shapiro, 1974; Jia and Huang, 1990). Excavations resumed in 1949 and 1951, and five teeth (left upper median incisor, right upper first and second premolars, left lower first, and right lower second molars) were found. In 1951 a humeral and a tibial fragment were found in the laboratory among a collection of broken bones recovered before the war (Woo and Chia, 1954). In 1952, a right lower molar was found in the collection from the 1921 and 1923 excavations. This material was stored at Uppsala University in Sweden. A female mandibular fragment was found during the 1959 excavations (Wu and Chao, 1959). In 1966 a frontal and an occipital fragment, as well as a right lower premolar, were found (Qiu et al., 1973). All specimens recovered after World War II are stored at the Institute of Vertebrate Paleontology and Paleoanthropology, Academia Sinica, Beijing. The three teeth found in the collection stored in Sweden in 1952 are kept in the Museum of Paleontology of the University of Uppsala. A catalog of the hominid fossils from Locality 1 appears in Table 2.6.

Human fossils

The morphology of *H. erectus pekinensis* has been described in the excellent monographs and reports written by Franz Weidenreich (1936a,b, 1937, 1938, 1941, 1943), Black (1927), J. Woo and Chia (1954), J. Woo and Chao (1959), and Qiu et al. (1973). The latter three sets of researchers did major studies on the Zhoukoudian remains recovered after 1949. These publications, especially those by Weidenreich, were relied upon very heavily for the description given below. It is impossible to cite each sentence or idea taken from Weidenreich's individual works and those of the various Chinese researchers, because that would make this description very cumbersome. Therefore, the important debt to Weidenreich and others is gratefully acknowledged.

The following description of the *H. erectus pekinenses* skull (see Figs. 2.6 and 2.7) relies heavily upon Weidenreich's (1943) original summary. The maximum breadths of the skull vaults coincide with the biauricular breadth (average 145 mm). The breadth at the level of the temporal squama or at the parietal bone is distinctly smaller. The skulls are dolichocranial and are chiefly characterized by a low profile (Tables 2.7 and 2.8). The greatest breadth is at the base, at the level of the angular torus and the supramastoid crest. The breadth steadily decreases from that level to the top of the skull. Although the forehead is receding, there is a distinct bump on the frontal squama of most specimens. A sharp bend between the upper and lower scale of the occipital bone forms an angle of 103.2°. The flatness of the skull vault is also manifested in the sagittal cranial curvature. The length of the base, as represented by the nasion-basion line, amounts to about three quarters of the nasion-opisthion line. This situation is due to a more central instead of a more rearward positioning of the foramen magnum. The porion is situated on or above a horizontal plane drawn through the nasion-opisthion line (Table 2.9).

The very heavy and projecting supraorbital ridges are continuous and are connected by a robust glabellar torus. A continuous frontal torus is separated from the rising frontal squama by a well-defined supraorbital sulcus. The supraorbitals are not separated into medial and lateral portions.

A distinct sagittal prominence originates above the frontal tuberosity and disappears in the obelion region. A parasagittal depression highlights the sagittal ridge.

Table 2.6

Catalog of hominid fossils from Locality 1, Zhoukoudian

Skull No.	*Number of fragments*	*Type of specimen*	*Sex*	*Age*	*Site and year of discovery*	*Number of general catalog*	*Individual*
		A. Calvaria and Fragments					
I	2 larger and many small ones	1. Right (?) parietal bone (crushed)	M	ad.	Locus B 1928 Layer 4	PA 21	B II?
		2. Fragment of left half of frontal bone	M	ad.	Locus B 1928 Layer 4	PA 78	B II?
II	1	Calvaria (both temporals and occipitals missing)	?	ad.	Locus D (Lower fissure) 1929	PA 17	D I
III	1	Calvaria	M	juv.	Locus E (Lower fissure) Dec.2, 1929	PA 16	E I
IV	1	Fragment of a right parietal bone	M	juv. or adol.	Locus G 1931 Gezitang (Kotzetang) Quartz Horizon II (Layer 7)	PA 23	G II
V	5	1. Left temporal bone with adjacent parts	M	ad.	Locus H 1934 Layer 3	PA 74	H III
		2. Fragment of right temporal bone; tympanic region	M	ad.	Locus H 1936 Layer 3	PA 86	H III
		3.Fragment of left part of frontal squama and connecting part of greater wing of sphenoid and nasal bone	M	ad.	Locus H 1966 Layer 3	PA 109	H III

(*continued*)

Table 2.6 (continued)

Catalog of hominid fossils from Locality 1, Zhoukoudian

Skull No.	*Number of fragments*	*Type of specimen*	*Sex*	*Age*	*Site and year of discovery*	*Number of general catalog*	*Individual*
		4.Fragment of right part of frontal squama with right extremity of brow ridge and the adjoining small part of parietal bone	M	ad.	Locus H 1966 Layer 3	PA 109	H III
		5.Right half of occipital bone and connecting parts of parietal bone	M.	ad.	Locus H 1966 Layer 3	PA 109	H III
VI	4	1. Fragment of frontal bone	F	ad.	Locus I 1936 Layer 8-9	PA 92	I I
		2. Fragment of left parietal bone	F	ad.	Locus I 1936 Layer 8-9	PA 90	I I
		3. Smaller fragment of the same parietal bone	F	ad.	Locus I 1936 Layer 8-9	PA 91	I I
		4. Fragment of right temporal bone	F	ad.	Locus I 1936 Layer 8-9	PA 93	I I
VII	1	Angulus mastoideus of right parietal bone	M	adol.	Locus I 1936 Layer 8-9	PA 326	I II
VIII	1	Fragment of occipital bone	F?	juv. not older than 3 yrs.	Locus J 1936 Layer 8-9	PA 95	J I

(continued)

Table 2.6 (continued)

Catalog of hominid fossils from Locality 1, Zhoukoudian

Skull No.	*Number of fragments*	*Type of specimen*	*Sex*	*Age*	*Site and year of discovery*	*Number of general catalog*	*Individual*
IX	5	1. Fragment of frontal bone	M	juv. c. 6 yrs.	Locus J 1936 Layer 8-9	PA 315	J IV?
		2. 4 small fragments without connection	M		Locus J 1936 Layer 8-9	PA 315	J IV?
X	1	Calvaria (right temporal bone missing) (For facial bones belonging to the Calvaria see below under B I and II)	M	ad.	Locus L 1936 Layer 8-9	PA 98	L I
XI	1	Calvaria (For facial bones belonging to the Calvaria see below under B III and IV)	F	ad.	Locus L 1936 Layer 8-9	PA 99	L II
XII	1	Calvaria	M	ad.	Locus L 1936 Layer 8-9	PA 100	L III
XIII	1	Calvarium XIII represented only by left maxilla with six teeth (see below under B V)	M?	ad.	Locus O 1937 Layer 10	PA 313	O I
XIV	1	Calvarium XIV represented only by fragment of left maxilla with four teeth (see below under B VI)	M	ad.	Upper Cave 1933	PA 115	UC?

(*continued*)

Table 2.6 (continued)

Catalog of hominid fossils from Locality 1, Zhoukoudian

Skull No.	*Number of fragments*	*Type of specimen*	*Sex*	*Age*	*Site and year of discovery*	*Number of general catalog*	*Individual*
		B. Separate Facial Bones					
I	1	Frontal process of left maxilla	M	ad.	Locus L 1936 Layer 8-9	PA 98	Belongs to Skull X
II	1	Fragment of left zygomatic bone	M	ad.	Locus L 1936 Layer 8-9	PA 98	Belongs to Skull X
III	1	Fragment of left maxilla with five teeth(P1-M3)	M	ad.	Locus L 1936 Layer 8-9	PA 99	Belongs to Skull XI
IV	1	Approximately right half of palate	F	ad.	Locus L 1936 Layer 8-9	PA 98/ 99	Probably belongs to Skull XI
V	1	Fragment of left maxilla with six teeth (I2, P1-M3)	M?	ad.	Locus O 1937	PA 313	O l Skull XIII
VI	1	Fragment of left maxilla with four teeth (P1, M1-M3)	M	ad.	Upper Cave 1933	PA 115	?Skull XIV
		C. Adult Mandibles					
I	1	Right part	F	ad.	Locus A 1928 Layer 5	—	A II
II	1	Left condylar process	M?	ad.	Locus B 1928/35 Layer 4	—	B II
III	3	Fragments of left and right side	M	ad.	Locus G 1931 Layer 7	—	G I

(*continued*)

Table 2.6 (continued)

Catalog of hominid fossils from Locality 1, Zhoukoudian

Skull No.	*Number of fragments*	*Type of specimen*	*Sex*	*Age*	*Site and year of discovery*	*Number of general catalog*	*Individual*
IV	1	Anterior and right part	F	ad.	Locus H 1934 Layer 3	—	H I
V	1	Anterior and adjacent part of both sides	F?	ad.	Locus H 1934/35 Layer 3	—	H IV
VI	1	Left mandibular body	M	ad.	Locus K 1936 Layer 8-9	—	K I
VII	1	Fragment of left part	M	ad.	Locus M 1937 Layer 8-9	—	M I
VIII	1	Large part of left mandibular body	F	ad.	Locus M 1937 Layer 8-9	—	M II
IX	2	Left and right mandibular body	F	ad.	1959 Layer 10	—	PA 86
		D. Juvenile Mandibles					
I	1	Anterior and right part	F	juv.	Locus B 1928 Layer 4	—	B I
II	1	Fragment of right side	M	juv.	Locus B 1932/35 Layer 4	—	B III
III	1	Fragment of right side	F	juv.	Locus B 1928/35 Layer 4	—	B IV
IV	1	Fragment of anterior and right side	M	juv.	Locus B 1928/35 Layer 4	—	B V

(continued)

Table 2.6 (continued)
Catalog of hominid fossils from Locality 1, Zhoukoudian

Skull No.	*Number of fragments*	*Type of specimen*	*Sex*	*Age*	*Site and year of discovery*	*Number of general catalog*	*Individual*
V	1	Fragment of left ramus	F	juv.	Locus C 1929 Layer 8-9	—	C I
VI	1	Fragment of right side	M	juv.	Locus F 1930 Layer 11	—	F I
			E. Femora				
I	1	Upper part of shaft of left femur	M	ad.	Locus C 1929/38 Layer 8-9	—	C III†
II	2	Fragments of middle shaft of left femur	F	ad.	Locus J 1936/38 Layer 8-9	—	J II
III	1	Medial part of upper shaft of right femur	M	ad.	Locus J 1936/38 Layer 8-9	—	J III
IV	1	Almost complete shaft of right femur	M	ad.	Locus M 1937/38 Layer 8-9	—	M IV
V	1	Proximal part of shaft of left femur	M	ad.	Locus M 1937/38 Layer 8-9	—	M IV ?
VI	2	Fragments of middle part of left femur	M	ad.	Locus M 1937/38 Layer 8-9	—	M I
VII	1	Lateral part upper shaft of left femur	M	ad.	Locus M 1937/38 Layer 8-9	—	M III
			F. Tibia				
I	1	Lower middle part		ad.	1951*	—	

(continued)

Table 2.6 (continued)

Catalog of hominid fossils from Locality 1, Zhoukoudian

Skull No.	*Number of fragments*	*Type of specimen*	*Sex*	*Age*	*Site and year of discovery*	*Number of general catalog*	*Individual*
		G. Humerus					
I	1	Distal part of shaft of left side	M	ad.	Locus B 1928/35 Layer 4	—	B II
II	1	Shaft of left side	M	ad.	Locus J 1936/38 Layer 8-9	—	J III
III	1	Small fragment of middle part of right shaft	M	ad.	1951*	—	
		H. Clavicle					
I	1	Shaft of left side	M	ad.	Locus G 1931 Gezitang Quartz Horizon II (Layer 7)	—	G II
		I. Lunate					
I	1	Almost complete	M?	ad.	Locus B 1928 Layer 4	—	B II
		J. Teeth					
	2	Lt. lower P1, Rt. upper M3			1921 or 1923/1926		
	1	Rt. lower P2			1921 or 1923/1952**		
	5	Lt. upper I1, lower M1, M2;			Locus M ? 1949#		
		Rt. upper P1			1951		
		Rt. upper P2			1953##		
	147	52 upper permanent teeth, 82 lower permanent teeth, and 13 deciduous teeth representing 20 adults or adolescents and 12 children			1927-1937		

(continued)

Table 2.6 (continued)

Catalog of hominid fossils from Locality 1, Zhoukoudian

Skull No.	*Number of teeth*	*Type of tooth*	*Site and year of discovery*
	1	Lt. lower M1 (connected with a mandible)	1959
	1	Rt. lower P1	1966

† This femur was designated Femur CII in the explanation of Plate I-III in Weidenreich (1941).

* Discovered in laboratory in Beijing among the broken bones found before World War II.

** Recognized in 1952 in laboratory in Uppsala,Sweden, among the fossils excavated in 1921 and 1923.

Discovered from the deposit excavated formerly

In loose materials formerly excavated.

Figures according to Weidenreich (1936a, 1937, 1941, 1943) and Wu and Dong (1982).

At the bregma the sagittal prominence broadens into a crosslike eminence that extends along the bregmatic section of the coronal suture on either side. A well-developed occipital torus extends across the entire occipital bone, as well as extending laterally and forward to the mastoid region. The occipital torus is demarcated from the occipital plane by the supratoral sulcus. The mastoid angle of the parietal is occupied by its very distinct continuation—the angular torus—which continues into the supramastoid crest. In the female representative (Skull XI), the occipital plane is demarcated from the nuchal plane by a faintly developed line. In the male skull (XII) deep muscular depressions on either side below the torus are separated from each other by an external occipital crest. The central portion of the torus is continuous in its entirety and thickest in the middle.

The frontal torus, the occipital torus and its lateral continuation, the angular torus, and the sagittal ridge, form the architectual hallmarks of the skull. These components are a reinforcement system having four constituents:

1. A basal series comprising a pre- and post-otic section. From front to back this section contains the frontal torus, the frontomaxillary process, the zygomatic arch, the supramastoid crest, the angular torus, and the occipital torus.
2. A sagittal arch extending from the foramen caecum to the occipital foramen. The arch is visible partly outside and partly inside of the vault. Outside the skull the arch is the sagittal prominence that extends from the upper end of the central tuberosity to the obelion region. Inside the skull the arch is a continuation of the sagittal prominence and consists of the internal frontal crest and the vertical area of the cruciate eminence.
3. A basic center chiefly represented by the bodies of the sphenoid and occipital bones.
4. A transverse connecting system with anterior and posterior branches. The anterior branch is the Sylvian crest, which continues into the posterior edge of the lesser wing of the sphenoid. The posterior branch is the pyramid of the temporal bone.

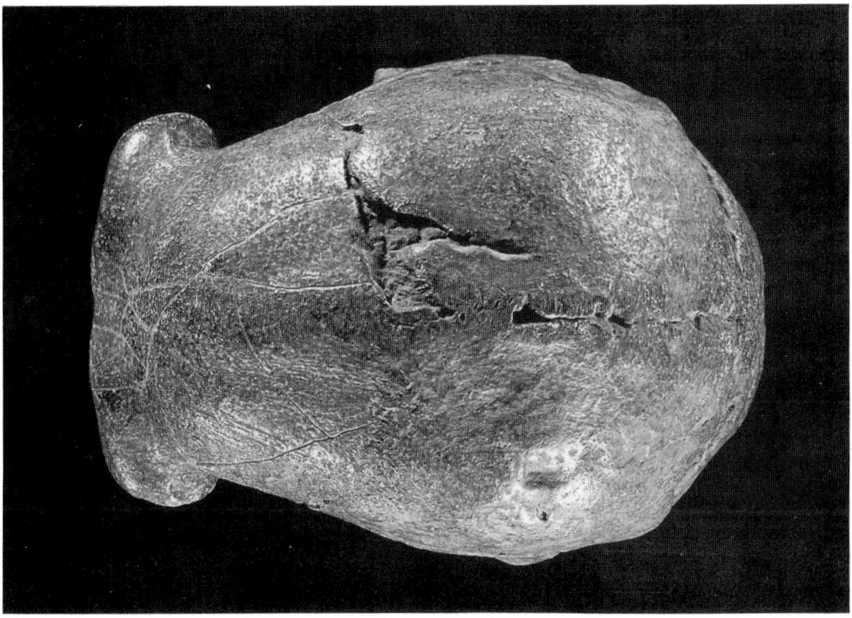

FIG. 2.6 First *Homo erectus* skull found at Zhoukoudian in 1929 at Locus E. (Courtesy of IVPP.)

FIG. 2.7 Anterior (mostly the frontal) and posterior (mostly the occipital) parts of *Homo erectus* Skull H found in 1966. (Courtesy of IVPP.)

Table 2.7

Measurements of *Homo erectus pekinensis* skull

Key No.	*Measurement*	*Skull Number*					
		II	*III*	*V*	*X*	*XI*	*XII*
1 (2)	Maximum cranium length g-op (i)	194?	188	213	199	192	195.5
1a	Horizontal projective	—	184	—	197.5	190.5	193.5
1b	Ophyron-occipital: on-op (i)	—	176	—	190	181	183
1d	Nasion-opisthocranion (inion): n-op (i)	—	184	—	194	185	192
3	Glabella-lambda line: g-l	183?					
		180?B*	172	—	186	169	174
3a	Nasion-lambda line: n-l	180					
		177?B	170	—	184	166	175
4	Inner skull	—	156	—	173	167	168
			157				
5	Nasion-basion line: n-ba	—	—	—	—	rest. (105.5)	—
5 (1)	Nasion-opisthion line: n-o	—	144?	—	—	145	147
6 (2)	Horizontal occipital	—	48?	—	—	47	50
	Bregma position projected to g-op	81?	74	—	74	79.5	78
	Lambda position projected to g-op	178?	167	—	181	165	169
	Bregma position projected to n-o	—	52	—	—	56	62
	Lambda position projected to n-o	—	148	—	—	142.5	154
	Inion position projected to n-o	—	178	—	—	181	182
	Opisthocranion position projected to n-o	—	178	—	—	181	182

B*= After Davidson Black

Breadth

8c	Temporo-parietal	—	133	—	138?	135	139?
	Torus angularis	127	131	—	137	136	140
	Maximum intercristal	—	144?	—	150?	145	147?
	Average "maximum"	—	137.2	—	143	139.8	141
8 (2)	Inner skull	—	122	—	128?	128	129
9	Least frontal: ft-ft	84?	81.5	91	89	84	91
9 (1)	Postorbital	106?	88	—	98?	93	95
10	Greatest frontal: co-co	108?	101.5	112	110?	106	108
10b	Stephanion: st-st	—	78?	—	—	81	103
11	Biauricular: au-au	—	141	148.5	147	143	151
12	Biasterionic: ast-ast	103	117	124	111?	113	115

(*continued*)

Table 2.7 (continued)

Measurements of *Homo erectus pekinensis* skull

Key No.	*Measurement*	*II*	*III*	*V*	*X*	*XI*	*XII*
13	Bimastoid: ms-ms	—	106?	—	—	103?	—
	Interporial: po-po	—	122.6	—	124?	120	128?
	Lateral interglenoidal	—	131?	—	—	126?	130?
	Medial interglenoidal	—	82?	—	—	84	92?
	Stylo-mastoid	—	91.4	—	100?	88	92?
	Distance between temporal lines	—	90	—	94	86	104?
Height							
17	Basi-bregmatic	—	—	—	—	rest. 115	—
19	Opisthion above FH**	—	105?	—	106.5?	104	103
20	Auricular-bregmatic: po-b:	— (93.5B)	96.5	—	106	94	101.5
21	Auricular above FH	100 (97?B)	95 (92.7B)	99.5	105	93.5	100
	Auriculo-lambda: po-l	—	91	—	100	93	93.5
	Inner skull	—	105	—	—	102	110
	Bregma above g-op (i)	78?	68	—	80	66	72.5
22a	Calvarial above g-op (i)	79?	71	—	82	67	74.5
22	Calvarial above n-op (i)	83?	75	—	85	71.5	81
22b	Calvarial above g-l	57?	47	—	57.5	45	49.5
	Bregma above n-o	—	89	—	—	90	95
	Vertex above n-o	—	103	—	—	100	101
	Lambda above n-o	—	83	—	—	81	83
	Opisthocranion above n-o	—	47	—	—	53	41
	Inion above n-o	—	47	—	—	53	41
	Circumference, Arcs and Chords						
23a	Horizontal cir.: on⌒on (op)	—	504?	—	520	507	518
23	Max. horiz. cir.g⌒g (op)	—	557?	—	582?	556?	560?
24	Auriculo-bregmatic arc: po⌒po (b)	—	277	—	310	280	280
25a	Median sagittal arc I: n⌒op	—	263	—	293	271	277
25	Median sagittal arc II: n⌒o	—	321	—	—	332	337
26	Nasion-bregma arc: n⌒b	123 (120B)	115	—	129	122	124
	Glabella-bregma arc: g b	115	110	—	120	112	110
29	Nasion-bregma chord: n-b	113 (109B)	102	—	115	106	113
	Glabella-bregma chord: g-b	110	100	—	112	104	107

(continued)

Table 2.7 (continued)

Measurements of *Homo erectus pekinensis* skull

Key No.	*Measurement*	*Skull Number*					
		II	*III*	*V*	*X*	*XI*	*XII*
27	Bregma-lambda arc: b⌒l	112 (110B)	100	—	113	92	102.5
30	Bregma-lambda chord: b-l	104	94	—	106	86	91
28	Lambda-opisthion arc: l⌒o	—	106?	—	—	118	118
31	Lambda-opisthion chord: l-o	—	80?	82?	—	86	86
26 (1)	Sagittal arc--pars glabellaris front: n⌒sg	28	25	—	28	26	32
29 (1)	Chord--pars glabellaris front: n⌒sg	22	22	—	25	21	28
26 (2)	Sagittal arc--pars cerebralis front: sg⌒b	93	88	—	96	97	91
29 (2)	Chord--pars cerebralis front: sg⌒b	82.5	83	—	94	89.5	88
27 (1)	Sphenion-asterion: arc sphn⌒ast	l 107?	r 109? l 106?	—	—	r 108? l 103?	107
30 (2)	Sphenion-asterion chord: sphn-ast	l 100	r 99? l 96?	—	r 99? l 99?	l 101	
27 (3)	Lambda-asterion arc: l⌒ast	r 90? l 90?	r 90 l 88	— —	r 93 l 88	r 99? l 85?	r 92 l 100
30 (3)	Lambda-asterion chord: l-ast	r 83? l 83?	r 81 l 77	— —	r 85 l 78	r 84? l 77?	r 87 r 87
28 (1)	Lambda-opisthocranion (inion) arc: l⌒op(i)	—	49	—	51	50	55
31 (1)	Lambda-opisthocranion (inion) chord: l-op (i)	—	47	—	49	48	52.5
28 (2)	Opisthocranion(inion)-opisthion arc: op(i)⌒o	—	60?	—	—	67	60
31 (2)	Opisthocranion(inion)-opisthion chord: op(i)-o	—	58?	—	—	63	57
Cranial capacity		1030	915	1140	1225	1015	1030
	Angles (in degrees)						
32a	Frontal profile: m-g∠g-(op)i	—	62	53	63	61	56
32(1)	Frontal inclination I: b-n∠ n-op(i)	45?	44	—	46.5	42	44
32(2)	Inclination of frontal squama: b-g ∠ g-op(i)	45?	42	—	45	38	42.5

(continued)

Table 2.7 (continued)

Measurements of *Homo erectus pekinensis* skull

Key No.	*Measurement*	*Skull Number* II	III	V	X	XI	XII
32(3)	Inclination of pars glabellaris: sg-n ∠ n-op(i)	—	73	—	70	55	65
32(4)	Inclination of pars cerebralis: sg-b∠n-op(i)	—	39	—	41	38	37
	Frontal curvature (angle of frontal bone flatness m-n∠n-b	18	22	—	21	24	16
33(a)	Occipital inclination(upper scale)I: l-op(i) ∠ op(i)-n	—	67	—	70	60	64
33(b)	Occipital inclination II: l-op(i) ∠ op(i)-g	—	65	—	68	57	61
33(2a)	Opisthion-opisthocranion (inion)(lower scale): o-op(i) ∠ op(i)-n	39?	36?	—	43	34	38
33(4)	Occipital curvature: lo∠op(i)∠op(i)-o	—	106	101?	104?	105	98
	Frontal inclination II: b-n ∠ n-o	—	59	—	58?	59	57
	Inclination of whole occipital: l-o∠o-n	—	93.5	—	99?	88	92
	Calvarian base: n-op (i) ∠FH	—	7	—	3	3	4
37	Cranial base: ba-n-o	—	—	—	—	5	—
	Inclination of occipital foramen: n-ba-o	—	—	—	—	rest. 159	—

**FH refers to Frankfurt Plane

Key No. According to Martin's *Lehrbuch der Anthropolgie (1928),* Stuttgart: G.F. Verlag

After Weidenreich (1943) and Qiu et al. (1973)

If the thickness of the bones of the skull vault is expressed by an average index giving the mean of the measurements of the four chief bones of the vault—the frontal, parietal, temporal, and occipital—this index is 9.7 mm. The thickness of the cranial bones is due to a peculiar thickening of the outer and inner tables (Table 2.10). The small frontal sinus is confined to the region of the glabellar torus. Even the large frontal sinus in juvenile Skull no. III does not extend beyond the interorbital region. Both the air cell in the temporal and the maxillary sinus are larger than in modern humans (Table 2.11).

There are several peculiarities on the *H. erectus pekinensis* endocast (Table 2.12). The digital impressions and cerebral juga are very clearly visible on the roof of the frontal part, especially in endocast II from Locus D. The inferior part of the parietal lobe and the posterior part of the temporal lobe strongly expand laterally in endocast I from Locus E but not in endocast III, a combined reconstruction of pieces unearthed at different loci. There is a larger space between the temporal lobe and the cerebellum in endocasts I and III.

Table 2.8
Cranial indices of *Homo erectus pekinensis*

			Skull number		
Indices	*II*	*III*	X	*XI*	*XII*
Length-breadth	—	72.3	71.4	72.4	72.6
Inner length-breadth	—	78.2	74.1	76.7	76.2
Length-height	—	—	—	rest. 59.6	—
Length-auricular height	51.5	50.5	52.8	48.7	51.2
Length-opisthion height	—	56.0	53.6	54.2	52.7
Breadth-height	—	—	—	rest. 75.6	—
Breadth auricular height	—	69.6	74.0	67.2	70.3
Bregma height above g-op(i)	40.2	36.2	40.2	34.4	37.1
Calvarial height above g-op(i)	40.7	37.8	41.2	34.8	38.0
Calvarial height above n-op(i)	—	40.7	43.7	38.6	42.2
Bregma height I above n-o	—	61.8	—	62.0	64.6
Vertex height above n-o	—	71.6	—	69.0	67.7
Lambda height above n-o	—	57.6	—	55.8	56.4
Opisthocranion height above n-o	—	32.7	—	36.6	29.7
Inion height above n-o	—	32.7	—	36.6	29.7
Bregma position above n-o	—	36.1	—	38.6	42.2
Vertex position above n-o	—	64.5	—	66.8	67.7
Lambda position above n-o	—	102.8	—	98.3	104.8
Occipital length I	—	26.1	—	25.2	25.9
Occipital length II (op(i) behind o in relation to n-o)	—	23.6	—	24.8	23.8
Sagittal cranial curvature	—	44.8	—	43.6	43.6
Transverse cranial curvature	—	50.9	47.4	51.1	54.8
Frontal breadth	—	80.3	80.9	79.3	84.2
Transverse fronto-parietal	—	59.7	62.6	60.4	64.1
Transverse parieto-occipital	—	86.2	78.3	81.3	81.0
Transverse postorbital-biauricular	—	62.4	66.7	65.0	62.9
Lower parietal breadth	—	94.4	93.8	94.4	92.2
Upper parietal breadth	—	63.8	63.8	60.2	68.9
Fronto-parietal (arc)	91.3	88.6	87.6	75.4	76.5
Fronto-occipital (arc)	—	95.7	—	96.0	95.3
Parieto-occipital (arc)	—	108	—	127	124
Fronto-sagittal (arc)	—	34.7	—	36.7	36.8
Parieto-sagittal (arc)	—	30.7	—	27.6	28.2
Occipito-sagittal (arc)	—	33.1	—	35.2	35.1
Frontal curvature	91.8	90.5	89.2	86.9	91.2
Flatness of frontal bone (subtense (st) to n-b/n-b	16.7	18.6	14.8	16.1	15.9
Glabellar curvature	78.7	88.0	89.3	81.0	87.6
Cerebral curvature	88.8	94.3	98.0	92.3	96.7

(*continued*)

Table 2.8 (continued)

Cranial indices of *Homo erectus pekinensis*

	Skull number				
Indices	*II*	*III*	*X*	*XI*	*XII*
Glabello-cerebral	26.7	26.5	26.6	23.5	31.8
Parietal curvature	93.1	94.0	94.0	93.5	95.7
Occipital curvature	—	74.2	—	74.5	72.8
Upper scale curvature	—	96.0	96.0	96.0	95.4
Lower scale curvature	—	96.6	—	94.0	95.0
Nasion-basion length	—	—	—	rest. 72.6	—

1. After Weidenreich (1943) .

2. There are some inconsistencies in the figures presented by Weidenreich. Some of the figures presented here are inconsistent with figures attained from the calculation based on figures presented by Weidenreich in a table of linear measurements in the same book. These inconsistencies may be due either to printing errors or to errors of calculation. Since the original specimens are lost, we are unable to clarify the cause of the inconsistencies. They are maintained in this table.

Table 2.9

Position of the porion in relation to the nasion-opisthion line in *Homo erectus pekinensis*

	Minimum-maximum	*Average*
Nasion-opisthion line	144 – 148	146
Foot position of the porion	96 – 106	102
Height of the porion	+1 – +6	+4.3
Porion position length index	66 – 71.6	69.6
Porion position height index	+ 0.7 – 4.1	+3.0

After Weidenreich (1943)

Despite massive facial bones, the face is relatively small and only slightly prognathous if the upper facial triangle is the basis of measurement. The ratio between the size of the face (represented by the area of the nasion-basion-prosthion triangle) and the calvaria (represented by the area on a similarly simplified midsagittal section) is 28.

The skull exhibits several distinctive features. The contours of the glabellar torus and the supraorbital tori show rather sharply defined surfaces, rather than rounded surfaces. A well-developed edge is particularly noticeable where the superior and anterior surfaces meet. The course of the supraorbital margin is interrupted by a wide and shallow notch. The notch, which occupies the medial section, is bordered laterally by a distinctly prominent supraorbital tubercle (process). Other than in juvenile Skull no. III, there is no lacrimal groove. A metopic suture is preserved in adult female Skull no. XI. The foramen caecum seems to be absent in all skulls, but the frontal crest is well developed. Endocranial relief is very pronounced and its extent, laterally and upwards, exceeds modern humans. It comes to a rather sudden stop just anterior to the coronal suture, where it seems to be barred by a thresholdlike transverse swelling.

The parietal is smaller, flatter, thicker, and more rectangular in its outlines than

Table 2.10

Thickness of the Cranial Bones of *Homo erectus pekinensis*

Specimen No.	*I*	*II*	*III*	*V*	*VI*	*VII*	*VIII*	*IX*	*X*	*XI*	*XII*
Frontal											
Glabella	—	20.0	23.0	—	—	—	—	—	23.0	18.7	22.0
Torus supra-orbitalis											
Med. part	—	14.2	13.5	13	—	—	—	—	12.6	14.0	17.0
Mid. part	—	14.0	11.5	12	—	—	—	—	13.0	14.0	16.0
Center of squama	13.0	10.0	10.0	6.0	(9.5)	—	—	(7.1)	7.0	11.0	7.0
Facies tempor.	—	6.5	4.8	—	4.6	—	—	(5.6)	(5.8)	4.6	5.5
Parietal											
Near bregma	—	9.0	9.6	—	(9.9)	—	—	—	7.5	7.0	9.7
Tuberosity	5.0?	11.0	11.0	—	11.2	—	—	—	12.5	16.0	9.0
Angulus mastoid	14.0	13.5	17.2	14.0	—	17.4	—	—	14.0	13.5	14.5
Occipital											
Center of occipit. planum	—	(10.7)	10.0	(7.0)	—	—	(5.0)	—	10.0	9.0	9.0
Center of occipit. torus	—	—	20.4	15.0	—	—	7.1	—	15.0	12.0	15.0
Fossa cerebella.	—	—	6.8	4.5	—	—	3.8	—	(5.0)	2.8	2.5
Temporal											
Center of squama	—	10.0	9.3	10.0	7.7	—	—	—	(5.2)	6.0	7.0
Sutura parieto-mastoid behind parietal notch	—	—	15(r)	15(r)	—	—	—	—	—	—	15
	—	—	18(l)	17(l)							
Sutura occipito mastoid.medial to mastoid	—	—	6.5 (r)	7.5 (r)	—	—	—	—	—	—	—
process	—	—	7 (l)	8 (l)	—	—	—	—	—	—	—

1. After Weidenreich (1943) and Qiu et al.(1973)

2. Figures followed by a ? are uncertain because of fractures or compression of the bone. Figures within brackets are taken next to the landmark concerned.

Table 2.11

Dimensions of the frontal sinus of *Homo erectus pekinensis*

Skull	*Max. Breadth*	*Depth (Sagit. Dim.)*	*Author*
III	24.5	14 – 15	Black, Weidenreich
X	17	11	Weidenreich
XI	8	7	Weidenreich
XII	20	4	Weidenreich

1. After Weidenreich (1943).

Table 2.12

Measurements of *Homo erectus pekinensis* endocasts

	Endocast I	*Endocast II*	*Endocast III*
Max. length	157	161.5	—
Max. breadth	121	118	118 (?)
Max. height	97.5	103 (?)	100 (?)
Percent length related to total length:			
frontal part	33.7	25.4	—
temporal part	37.6	41.5	42.8
cerebellum	28.3	—	30.4 (?)
Height-length index:			
fontal lobe	42.4	—	—
temporal lobe	15.9	—	18.5
parietal lobe	39.9	—	—
cerebellum	18.3	—	(18.5)
Real volume	950	1025 (?)	1000 (?)

Data according to Weidentreich (1936b)

in modern humans. The greater flatness is restricted to the two longitudinal margins (Table 2.13). The transverse margins are curved more than in modern humans. There is a circumscribed obelion depression in infantile Skull no. III. The pteryon is a narrow H-type. A pronounced Sylvian crest that occupies the entire sphenoid angle extends obliquely upwards and backwards, and disappears before reaching the center of the bone.

The length-breadth index for the occipital squama is very low. The chord of the upper scale is considerably shorter than that of the lower scale. The occipital squama is curved much more than in modern humans (Table 2.14). In four of the six preserved skullcaps, the apex of the occipital squama is separated from the rest of the occipital plane by accessory, transverse sutures. In two cases a sutura mendosa separates almost the entire upper scale. The two other skulls have a large interparietal bone, the height of which is about half the height of the upper scale. The cerebellar fossae are about half the size of the cerebral ones. The distance between the internal occipital protuberance and the inion is longer in "Peking Man" than in later hominids (Table 2.15).

The temporal squama resembles a right-angled triangle. The squama is very low in proportion to its length, and its length-height index averages 49.6 (Table 2.16). A deep and distinct parietal notch of acute angularity separates the squama from the mastoid portion. The zygomatic process continues as a strong and projecting supramastoid crest. The process is held away from the squama by a wide, shallow sulcus processus zygomaticus. There is a planum preglenoidale instead of a true articular tubercle. The mandibular fossa is very deep, with a short distance extending in the sagittal direction (Table 2.17); it develops a medial recess that extends between the tympanic plate and the entoglenoidal process. The medial wall of the fossa is convex and bulges into the fossa. The center of the mandibular fossa coincides with the cerebral surface of the cranial wall. There is no true postglenoid process, as appears in a gorilla or an orang-utan. A low, broad-based, transverse elevation marks the boundary between the lateral part of the mandibular fossa and the tegmen pori acustici. The zygomatic process forms a roof overhanging the porus

Table 2.13

Length of chords and arcs of the margins of the parietal of *Homo erectus pekinensis*

	II	*III*	*X*	*XI*	*XII*	*Average*
Margo Sag.						
Chord r	99	94	106	86	91	95.2
Arc l	105	99	113	92	95	101
Margo Coro.						
Chord r	—	79	—	82	93	
Chord l	90?	87	—	86	90	86.7
Arc r	—	111	—	101	105	
Arc l	97?	106	—	102	102	103.4
Margo Lamb.						
Chord r	80	86	87	81	86	
Chord l	82	81	82	80	90	83.5
Arc r	96	95	98	95	93	
Arc l	92	94	90	103	102	95.8
Margo Temp.						
Chord r	—	98	—	87	98	
Chord l	87	91	—	93	99	93.3
Arc r	—	100	—	96	102	
Arc l	92	93	—	94	106	97.6

1. After Weidenreich (1943)

Table 2.14

Length of the mid-sagittal chord and chord index of the upper and lower scales of the *Homo erectus pekinensis* occipital

	Skull III	*Skull X*	*Skull XI*	*Skull XII*
Upper scale	46	48	45	56
Lower scale	57?	—	65	56
Index	124	—	144	100

After Weidenreich (1943)

tegmen pori acustici. The entoglenoid process forming the medial wall of the mandibular fossa is not a true process, but is rather a steep slope entirely built by the squama. The cerebral surface of the squama is relatively long and relatively low, largely due to the extraordinary breadth of the squamosal suture. The squamosal suture is 22 mm at its widest point and 14 mm at the narrowest. The tympanic plate is very thick, and with reference to the auriculare, the plate lies much more medially.

The tympanic plate is oriented more horizontally than it is in modern populations. Rather than an acute angle, the axis of the tympanic plate forms almost a right angle to the midsagittal line. The part of the tympanic plate that forms the posterior wall of the mandibular fossa is either flat or slightly convex. It is concave in modern humans. The transverse slit that divides the porus edge and the entire lateral portion of the tympanic into an anterior and a posterior portion is only seen in Skull no. III and is an infantile character.

Table 2.15

Distance between inion (center of the occipital torus) and internal protuberance of *Homo erectus pekinensis*

Skull III	27.5
Skull V	29.5
Skull VIII (child)	17.0?
Skull X	38.0
Skull XI	34.0
Skull XII	35.0

After Weidenreich (1943) and Qiu, et. al. (1973)

Table 2.16

Length and height and length-height index of temporal squama of *H. erectus pekinensis*

Specimen	*Side*	*Length*	*Height*	*Index*	
III	r	72.5	33	45.2	(45.5)
III	l	74.0	33.5	45.5	(45.3)
V	l	71.5	35.0	49.3	(49.0)
X	r	70.5	39.0	55.3	
XI	r	63	29	46.0	
XI	l	62	35.5	57.3	
XII	l	74	36.5	49.3	
Average	69.4	34.5	49.7		

After Weidenreich (1943).

The figures in parentheses are recalculations by Xinzhi Wu.

The auditory aperture is relatively wide (Table 2.18), and despite some variation, a horizontally oriented, elliptic form prevails. In Skull no. X a circumscribed ear exostosis located on the anterior border of the tympanic extends from the porus 14.5 mm inward into the meatus. Because of its more horizontal position, the tympanic plate covers the lateral portion of the pyramid base to a greater extent than in modern humans. The pyramid is solid and its surface is smooth and even. An axial line drawn through the base of the tympanic plate runs in a transverse direction, while that of the uncovered part of the pyramid curves obliquely forward. The two lines form an obtuse angle, the corner of which falls on the entrance to the carotid canal. The lower margin of the tympanic plate and the spinelike projection on it are quite massive. There is neither a styloid nor a vaginal process. A distinct swelling, a processus supratubalis, marks the medial-anterior end of the tympanic plate and marks from above the entrance to the canalis musculo-tubarius. The carotid canal seems smaller than in modern populations. The jugular fossa is flatter, narrower, and shorter than in modern humans.

The anterior portion of the mastoid process extends further back than in the average modern human and appears more as a lateral bulge of the cranial wall than as an independent process with a downward trend. The mammillary process of the mastoid is relatively small. The supramastoid and mastoid crests, and the intervening sulcus, are well developed. The suprameatal spine is missing. In three out of eight cases (6 skullcaps and two additional pieces) there is an unevenness to the

Table 2.17

Dimensions of the mandibular fossa of *Homo erectus pekinensis*

Specimen	Side	Sex	*Length*	*Breadth*	*Depth*	*Breadth-Length Index*	*Depth-Length Index*	*Depth-Breadth Index*
III	r	m	18	25	11.5	72.0	63.8	46.0
III	l	m	16	?	13.0?	—	81.3	—
V	l	m	21	?	15.0	—	71.4	—
XI	l	f	21	27?	15.0	77.7	71.4	55.5
XII	l	m	18 ?	23?	15.0	78.3	83.2	65.2
Average			18.8	25.0	13.9	75.2	74.0	55.6

After Weidenreich (1943)

Table 2.18

Dimensions and morphology of the external auditory meatus in *Homo erectus pekinensis*

Specimen	*Side*	*Length*	*Height*	*Form and Character*
III	r	10.5	8.0	horizontally elliptic
	l	11.5	8.0	
V	r	9.5	12.0	vertically elliptic
	l	9.0	14.0	
XI	l	10.0	10.0	round
XII	l	11.0	9.0	horizontally elliptic

After Weidenreich (1943)

surface of the bone occupying the usual site of the spine. The occipito-mastoid suture runs along a relatively high crest, which marks the margins of the temporal and occipital bones.

Viewed from inside the skull, the pyramid appears as a stout, relatively flat, and large structure. The superior margin presents an obtuse or rounded corner, and the anterior and posterior superior surfaces are pressed down against the floor. Measured at its base, the height of the posterior surface of the pyramid amounts to not more than 18 mm. The sulcus sigmoideus is shallower and narrower than that of modern humans.

The lateral section of the medial middle cerebral fossa is relatively longer but located much lower than in modern humans. The facies temporalis of the sphenoid bone passes gradually over the facies infratemporalis, and both surfaces present a continuous curve without a sharp bend or strongly developed infratemporal crest. The facies temporalis of the sphenoid bone borders the maxilla so closely that the orbital fissure is reduced to a mere slit and the orbit is only accessible from below. The facies cerebralis of the sphenoid bone is very small in transverse direction. The greater wing of the sphenoid is very robust, and the edge of the squamosal margin between the facies temporalis and cerebralis is very thick (about 8 mm).

The naso-frontal suture, together with the fronto-maxillary suture, makes a continuous horizontal curve. The nasal bones are much wider than those of modern humans. The nasal index is 57.2 so the nose is considered to be chamaerrhine and is very close to being hyperchamaerrhine. A vertical line drawn from the rhinion falls in front of the nasospinale, as in most modern humans. The nasal floor is even and separated from the clivus nasoalveolaris by a margo limitans. There is neither a crestlike sill nor a prenasal groove, and the anterior nasal spine is missing. The anterior surface of the alveolar process of the maxilla is convex in sagittal cross section. A broad and prominent jugum alveolare at the canine extends far upward, borders the pyriform aperture, and is separated from the broad and prominent root of the zygomatic process by a narrow infraorbital sulcus. There is a broadly based and rounded malar pillar. The lateral contour forms a narrow arch, the incisura malaris, before reaching the zygomatic bone. The clivus nasoalveolaris, as a whole and apart from the alveolar juga, is convex. Buccal maxillary exostoses present in all specimens extend over the entire malar region.

The *H. erectus pekinensis* skull (Fig. 2.8) is dolichouranic (the cranial index falls between 70 and 74.9), and the maxilloalveolar index is 107.6. The roughness of the palatal surface is caused by irregular crests, depressed areoles, and fine, more or less longitudinal ridges. The skull does not differ, in this respect, from modern humans. The palatal orifice of the incisive canal lies farther back from the orale than in modern humans, and the direction of the canal is more oblique than in the latter. Because the platal index is 75.0, the skull is leptostaphyline.

The zygomatic bone is very high; the height on the male specimens is about 65 mm. The frontal orientation of the malar surface includes the fronto-sphenoidal process of the zygomatic. The angle of inclination of the malar facies of the process toward the midsagittal plane is 50°. The malar surface is oriented in a rather vertical direction. The angle of the malar profile is 106°. There is a well-developed marginal process at the temporal edge of the fronto-sphenoidal process of the zygomatic bone. The zygomatic arch runs below the level of the Frankfurt Horizontal.

Homo erectus pekinensis has an orbital index that amounts to 81.9 when the orbital breadth is mf-ek, and 90.0 when la-ek is the orbital breadth. Thus, it is considered to be chamaeconch or even hypsiconch despite the heavy supraorbitals. The cephalo-orbital index is 7.3. The angle of the orbital axis is 52°. The angle of vertical inclination of the orbit is 110°. The angle of frontal inclination is 11°, and the angle of horizontal inclination of the orbit is 10°. The supraorbital margin is a thick, rounded structure, the anterior surface of which continues without a distinct demarcation into the slightly vaulted orbital roof. The infraorbital margin is rounded and level with the even floor of the orbit. The superior fissure is reduced to a small, almost circular hole not much larger than the optic foramen. The inferior fissure was probably a narrow slit. The foramen zygomatico-orbitale inferius is a relatively wide, funnel-like orifice situated where the floor ascends toward the lateral wall and is in immediate continuation of the lower fissure. The saggital suture closes after the coronal suture.

The mandible of *H. erectus pekinensis* is large and bulky. (This description relies on Weidenreichs' 1936 report.) This is especially true for the male specimen. The main measurements and indices of the mandible are presented in Table 2.19

The robustness index (thickness of the mandibular body/height of the mandibular body) of the female specimen is very high. That of the male specimen is lower because it has a higher body than the female. There is obvious sexual dimorphism in the *H. erectus pekinensis* mandible. The inclination angle of the symphysis is rather low. There is no mentum osseum, and the mental trigon has just begun to develop. Multiple mental foramina are a rather striking characteristic of *H. erectus pekinensis*. There are two to five mental foramina on each side of the mandible (Table 2.20). The diagastric fossa is rather long, both absolutely and relatively compared to its breadth. It occupies the basal (lower) margin of the mandible in-

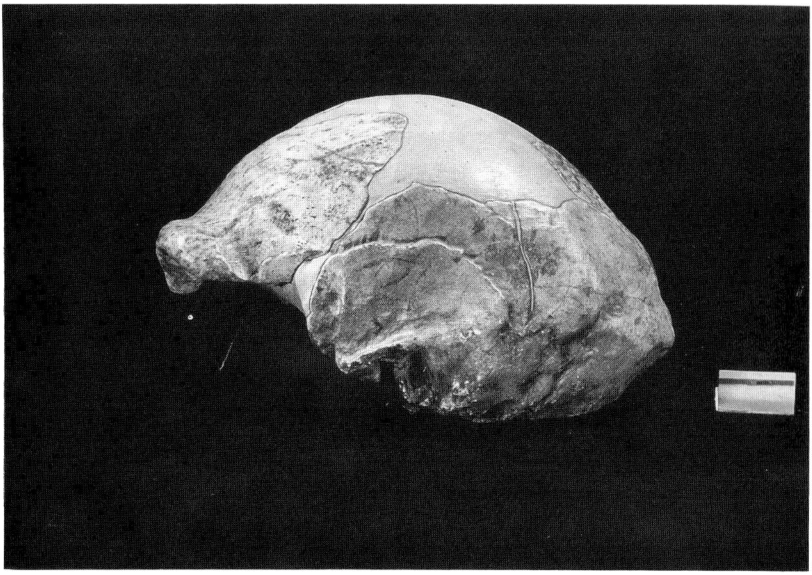

a

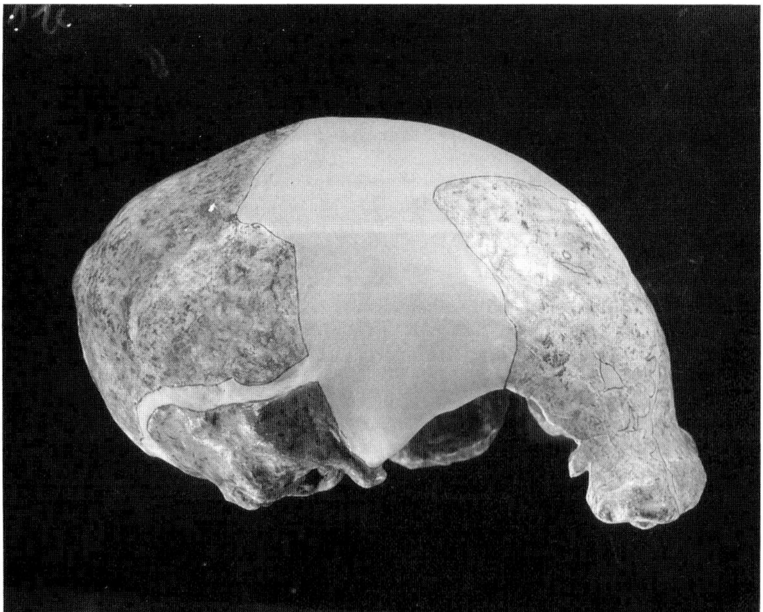

b

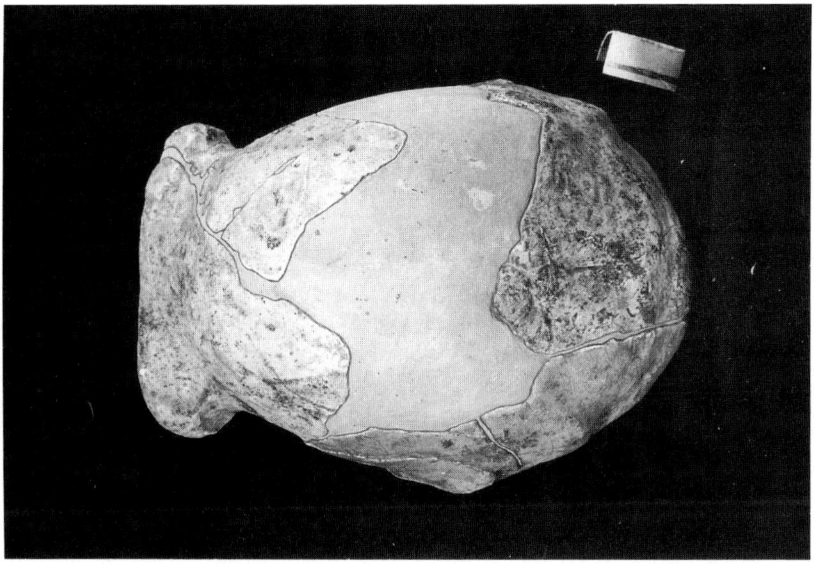

c

FIG. 2.8 Three views of the reconstructed *Homo erectus* skull H, Zhoukoudian: The anterior and posterior parts were found in 1966. a = right side; b = left side; c = superior view. (Courtesy of IVPP.)

stead of the posterior (inner) surface of the mandible. The ramus is steep and large, and has remarkable breadth, even in female specimens. Strong muscular markings characterize the medial surface of the ramus. The border of the region of the mandibular angle is everted like a hook, thus forming a true fossa masseterica. The angle becomes a prominent crest, which is thicker and more projecting in the male mandibular specimen GI than in the female specimen HI. There is a strongly developed crista ectocondyloidea with a distinct neck tubercle below the condyle. Despite the great morphological variability, the areas for the attachment of the two M. pterygoideus in *H. erectus pekinensis* reveal stronger markings in the male than in the female specimen.

Figures 2.9 and 2.10 show mandibles from males and females. A mandible found in 1959 was originally described as belonging to a female (J. Woo and Chao, 1959; J. Woo et al., 1959).

The coronoid process is triangular in cross section. It is flat on the lateral side but shows a distinct ridge on the medial side, the crista endocoronoidea. The endocoronoid ridge plus the thickening of the anterior border of the coronoid process combine to give the corocoid process a strong and robust appearance. The most striking feature of the condyloid process of *H. erectus pekinensis* is its relative slender appearance when compared with the lower jaw of La Chapelle-aux-Saints, for example.

The alveolar arch is a horseshoe-shaped, long, and relatively narrow curve. The front of the arch is equally rounded in the region of the canines and incisors, and is not flattened or retracted, as in modern humans. The alveolar arch of *H. erectus pekinensis* is less prognathic than in apes. The female *H. erectus pekinensis* condition is similar to that of modern males.

Table 2.19
Measurements of the mandible of *Homo erectus pekinensis*

			Adult		
Mandible	GI	HI	AII	PA86	KI
Sex	m	f	f	f	f#
Body of Mandible					
Height of symphysis	40*	31.5	—	—	32.5
Height of the body at level of mental foramen	34	26	25.6	26.2 (r) 27.1 (1)	27.1 (l)
Bimental breadth	46.1*	53.6*	—	—	—
Thickness of the body	16.4	15.4	15.2	16.7 (r) 16.5 (l)	15.5 (l)
Index of robustness	48.3	58.4	59.4	63.7 (r) 60.9 (l)	57.2 (l)
Index of the position of foramen supraspinosum	—	30.2	—	—	—
Angle of inclination+ (a)	59°*	60.5°*	—	63°	58°
Angle of inclination++ (b)	123°*	114°	—	—	—
Rami					
Height of the ramus	66.7	59	—	—	—
Breadth of the ramus	40.7	39.7	—	—	—
Breadth of the incisura semilunaris	33*	34*	—	—	—
Depth of the incisura semilunaris	12*	17.2*	—	—	—
Index of the ramus	60.8*	67.3	—	—	—
Depth-length index of the incisura semilunaris	36.4	50.6	—	—	—
Index of the position of the deepest point of the incisura semilunaris	49.9	36.9*	—	—	—
The Mandible as a Whole					
Length	103*	94*	—	—	—
Bicondylar breadth	146.4*	101.8*	—	—	—
Bigonial breadth	108.6*	97.8*	—	—	—
Length of the alveolar arch	65*	54*	54*	62.7	59.6
Breadth of the alveolar arch	63*	54*	57*	56.3	59.4
Length of the anterior alveolar arch	28*	23.5*	22*	32.5	26.2
Breadth of the anterior alveolar arch	47*	47*	43*	55	47.0
Length of the basal arch	46.5*	42*	—	—	44.8
Breadth of the basal arch	68.2*	67*	—	—	79.0
Length-breadth mandibular index	70.3*	92.4*	—	—	—
Breadth index	74.1*	95.5*	—	—	—
Height index of body and ramus	44*	36.2*	—	—	—
Index of the alveolar arch	103.3*	100*	94.8*	111.4	100.0
Index of the anterior alveolar arch	58.3*	50.0*	51.2*	59.1	55.7
Index of the basal arch	68.2*	62.7*	—	—	56.7
Mandibular angle	97	108	—	—	—
Angle between alveolar and basal line	-1.0°*	0.5°*	—	—	—
Angle of molar row	24.0°	13.0°	23.0°	16.0°	23.0°

1. Measurements marked with an asterisk are taken on restored specimens.

2. After Weidenreich (1936) and Wood (1964a, b).

This specimen was attributed to a male by Weidenreich (1936) on the basis of its tooth size. Woo (1964) attributed it to a female on the basis of its overall measurements.

+ The angle of inclination (a) is formed by the incision-gnathion line and the alveolar line.

++ The angle of inclination (b) is formed by the incision-gnathion line and the basal margin of the mandible.

(*continued*)

Table 2.19 (continued)

Measurements of the mandible of *Homo erectus pekinensis*

			Juvenile	
Mandible	BI	BIV	BV	FI
Sex	f	f	m	m
Body of Mandible				
Height of symphysis	—	—	—	—
Height of the body at level of mental foramen	—	—	—	—
Bimental breadth	—	—	—	—
Thickness of the body	—	—	—	—
Index of robustness	—	—	—	—
Index of the position of foramen supraspinosum	31.3	—	—	—
Angle of inclination+ (a)	63.5°	63°	59.5°	—
Angle of inclination++ (b)	123°	—	—	—
Rami				
Height of the ramus	—	—	—	—
Breadth of the ramus	—	—	—	—
Breadth of the incisura semilunaris	—	—	27	—
Depth of the incisura semilunaris	—	—	8.3	—
Index of the ramus	69.8*	—	—	—
Depth-length index of the incisura semilunaris	—	—	30.8	—
Index of the position of the deepest point of the incisura semilunaris	—	—	51.2	—
The Mandible as a Whole				
Length	80*	—	—	—
Bicondylar breadth	108.6*	—	—	—
Bigonial breadth	—	—	—	—
Length of the alveolar arch	—	—	—	—
Breadth of the alveolar arch	—	—	—	—
Length of the anterior alveolar arch	26*	32.5	28	—
Breadth of the anterior alveolar arch	49*	46	48	—
Length of the basal arch	14*	—	—	—
Breadth of the basal arch	44*	—	—	—
Length-breadth mandibular index	73.8*	—	—	—
Breadth index	81.6*	—	—	—
Height index of body and ramus	—	—	—	—
Index of the alveolar arch	—	—	—	—
Index of the anterior alveolar arch	53.2*	70.5	58.3	—
Index of the basal arch	31.8*	—	—	—
Mandibular angle	107	—	112	—
Angle between alveolar and basal line	4.0°*	—	—	—
Angle of molar row	—	—	—	—

1. Measurements marked with an asterisk are taken on restored specimens.

2. After Weidenreich (1936) and Wood (1964a, b).

This specimen was attributed to a male by Weidenreich (1936) on the basis of its tooth size. Woo (1964) attributed it to a female on the basis of its overall measurements.

\+ The angle of inclination (a) is formed by the incision-gnathion line and the alveolar line.

++ The angle of inclination (b) is formed by the incision-gnathion line and the basal margin of the mandible.

Table 2.20

Position and number of mental foramen in *Homo erectus pekinensis*

Specimen	*Sex*	*Position*	*Number*
Adult			
GI	m	P1 P2	5
KI	f	P1 P2(l)	4
AII	f	P2 M1	4
HI	f	P2 M1	3
PA 86f	P1 P2(l)	2	
		P1 C-P2(r)	4
Juvenile			
BI	?	m1 m2	at least 2
BIII	?	m1 m2	3
BIV	?	m1	at least 2

After Weidenreich (1936) and Wood (1964a, b)

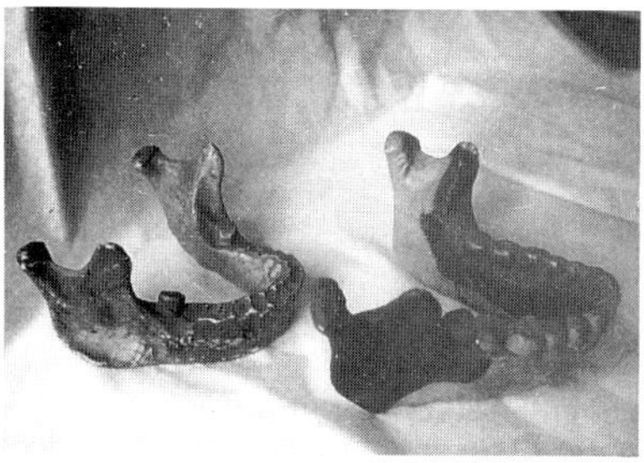

FIG. 2.9 Male mandible (G,left) and female mandible (H,right) of *Homo erectus from Zhoukoudian. (Courtesy of IVPP.)*

There are two types of mandibular torus in *H. erectus pekinensis*. The tubercle type, characterized by large or small distinct rounded swellings, may exist anterior to M1. The torus located at the posterior part of the alveolar arch is usually larger than that located at the anterior. The striation type of mandibular torus is represented by striations with intermediate or moderate elevations. These elevations are located in the upper margin of the mandibular body in the region of the molars, and the striations run in an oblique direction down and backward. Reconstruction of the physiognomy is illustrated in Figures 2.11 and 2.12.

M. M. Gerasimov (1949) reconstructed the face and skull of *H. erectus pekinensis* in 1939. His reconstruction, based on Skull no. II and some facial fragments, looks very prognathous and rugged. F. Weidenreich's reconstruction made in 1943 was based on Skull no. XI; one of the most complete calvariae; Maxilla no. III, which probably belonged to the skull; and Mandible HI (Weidenreich, 1943). Jukang Woo, Xinzhi Wu, and Cunyi Wang made a new reconstruction in 1959. Using Weidenreich's reconstructed skull as the skeletal basis, they made some changes in the soft tissue. For example, the upper and lower parts of the face are broader and there is a thicker masseter muscle. The external corner of the eyelids slant slightly downward instead of upward, as in Weidenreich's configuration. The nose, with a

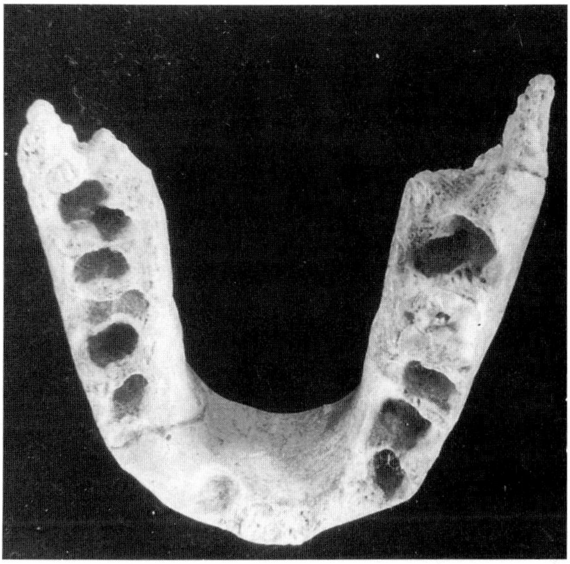

FIG. 2.10 *Homo erectus* mandible found at Zhoukoudian in 1959. (Courtesy of IVPP.)

more distinct nasal tip, is broader and more flattened at the alar region, and the nasal root is narrower. The lips are thinner. Weidenreich's reconstruction looks like a young woman, while the 1959 version of Woo and his colleagues gives an impression of a middle-aged woman that is more concordant with the estimated skeletal age.

Extremities

Unlike the head and face, the extremities of *H. erectus pekinensis* are morphologically more similar to those of modern humans. This is an example of mosaic evolution, a happenstance that often confused our earlier interpretations of human evolution. The major characteristics of the extermities studied by Weidenreich (1941) are given below.

Clavicle. Only a left clavicular fragment of *H. erectus pekinensis* has been found. Although the sternal epiphyseal articular portion and the acromial fourth of the bone are missing, it is sufficiently intact to allow a discussion of its main morphological characteristics. The clavicle belongs to a male. The estimated maximum length is 145 mm; the circumference at the middle of the maximum length is 34 mm. The degree of curvature is rather high. The clavicle is rather flat. The clavicle of *H. erectus pekinensis* does not display any specific difference from clavicles of modern humans. The variability in clavicular shape in modern humans is so great that it covers any special features recorded for *H. erectus pekinensis*.

Humerus (Fig. 2.13). There are three humeral fragments belonging to *H. erectus pekinensis*. Humerus no. I (B II) is represented by a small splinter of the distal part of the shaft of a left humerus of an adult male. Humerus no. II (J III) is from the left side and consists only of the shaft without either end. The size, proportion, shape, and muscular markings of the shaft are identical to those of modern humans, but the *H. erectus* humerus has a thicker wall and a narrower medullary

FIG. 2.11 Bust of *Homo erectus* from Zhoukoudian. (Courtesy of IVPP.)

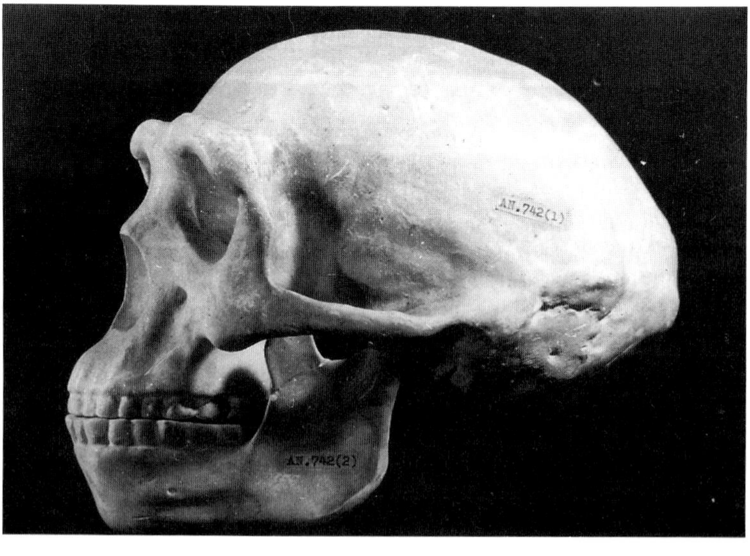

FIG. 2.12 Reconstructed face and skull of *Homo erectus* from Zhoukoudian. (Courtesy of IVPP.)

canal. The narrowest diameter of the canal amounts to only 22% of the entire transverse diameter of the shaft. Toward the proximal end of the shaft, the canal's diameter increases to 64% of that of the shaft. However, it is still narrower than in modern humans, where the diameter equals 71% of the shaft.

Humerus III was found in 1951 among broken bones collected in previous years. The specimen is the middle third of the shaft. The diameter of the canal at the proximal end occupies 61% of that of the shaft. The color of the bone suggests that it was burned (J. Woo and Chia, 1954). The measurements of the humerus are given in Table 2.21.

Lunate. The radio-ulnar diameter of this reconstructed bone is 14.4 mm. The restored proximo-distal and dorso-volar (height) diameters are 14.5 and 16.5 mm, respectively. Weidenreich (1941: 68,69) wrote:

> *The Os lunatum of* Sinanthropus *shows all the characteristics of the semilunar wrist bone of modern man. It is smaller than the average size of most races, but equals that of the bones of the North Chinese population. The two most conspicuous peculiarities are, firstly, in greater breadth (radio-ulnar diameter) its height-breadth and length-breadth indices exceeding considerably the average for man, and, secondly, the lowness of the radial surface. The bone differs from the os lunatum of the anthropoids in size and some minor details but resembles it in proportions, especially that of chimpanzee, more than the average form in modern man.*

FIG. 2.13 Anterior aspect of the right humerus of *Homo erectus* found at Zhoukoudian in 1951. (Courtesy of IVPP.)

Femur. Five femoral fragments were found at Zhoukoudian. The femur is rather short, slender, and straight, and has a slight forward bend. Its shortest transverse diameter is approximately at the point slightly above the border between the third and the distal quarters of the shaft. A slight distinct broadening at midshaft is due to the slight outward curvature of the medial border. The distinctness of these features varies in different femora. The shaft of Femur no. IV shows a distinct median convexity of the medial border, the vertex of which is found between the subtrochanteric region and the midpoint of the shaft. The lateral border is perfectly straight; there are medial and lateral ridges. The lateral border is more pronounced than the medial border in Femur no. I, while the reverse is true in Femur no. IV. Table 2.22 provides the measurements of the femur.

At the base of the greater trochanter in Femur no. I is a low and oval swelling, the subtrochanteric tubercle. Another such swelling, approximately 19 mm long and 9 mm wide, situated about 25 mm distal to the subtrochanteric tubercle, forms a third trochanter. Just distal to this is the beginning of a narrow, rugged groove that is 55 mm long. This "subtrochanteric fossa" is imprinted on the posterior surface of the lateral bulge of the shaft. Femur no. IV has a third trochanter, but no subtrochanteric fossa. In both femora, the intertrochanteric and pectineal lines are absent. One of the most striking features of the *H. erectus pekinensis* femur is the extraordinary thickness of the wall and the proportional narrowness of the medullary canal. This is particularly evident at midshaft, where the opening of the canal is always smallest. Figures in Table 2.23 show that the canal is only about one-third the transverse diameter of that of the whole shaft. The entire space occupied by the canal is not more than one tenth of the surface of the transverse section. The thickness begins to diminish noticeably in the subtrochanteric region. There is a diffuse arrangement of the trajectorial systems at the upper end of the femur. This arrangment is combined with a tight-meshed and coarse cancellous substance.

Using Manouvrier's and Pearson's formulae, Weidenreich (1941) estimated a stature of 156 cm on the basis of male Femora nos. IV and I. He (1941: 71) wrote: "Provided the same ratio in stature as between present-day male and female individuals existed in *Sinanthropus,* the average stature of the *Sinanthropus* female would have been about 1440 mm."

Tibia (Fig. 2.14). In 1951 a 83 mm long fragment of the middle shaft of a left tibia was found in the laboratory among broken bones previously collected from the "Peking Man" site. This fragment is described in Woo and Chia (1954). The medial border is well preserved, smooth, and rounded. The lateral border shows signs of having been chewed by a carnivore. There is no distinct interosseus crest. The sagittal diameter at midshaft is 27 mm. The transverse diameter at the same level is 21 mm, and the circumference is 178 mm.

Table 2.21

Measurements of the humerus of *Homo erectus pekinensis*

SPECIMEN	*I*	*II*
Maximum length	rest. 324	—
Sagittal diameter of the shaft at the middle	20.7	21.6
Transverse diameter of the shaft at the middle	15.4	17.8
Circumference of the shaft at the middle	59	62
Torsion angle	137°	—

Measurements according to Weidenreich (1941) and Woo and Chia (1954)

Table 2.22

Measurements of the femur of Homo erectus pekinensis

Specimen	*I* *male*	*II* *female*	*IV* *male*	*V* *male*	*VI* *male*
Max. length	c.400	—	rest.407	—	—
Supratrochanter level (greater breadth)					
Tranverse diameter	34.3	—	34.3	—	—
Sagittal diameter	23.2	—	22.7	—	—
Circumference	90	—	94	—	—
Mid-shaft level					
Transverse diameter	29.7	26.4	29.3	29.5	29.2
Sagittal diameter	27.1	22.8	25.0	23.7	26.1
Circumference	98	77	85	83 ?	85
Supracondylar level					
Transverse diameter	—	—	34.7	—	—
Sagittal diameter	—	—	27.8	—	—
Circumference	—	—	104	—	—
Length of the base of the bending of the femur shaft	—	—	259	—	—
Subtense of the arch of the bending of the femur shaft	—	—	5.8	—	—
Curvature index	—	—	2.24	—	—
Torsion angle	—	—	35° rest.36°	—	—
Vertical diameter of the neck	32.5	—	rest. 35	—	—
Transverse diameter of the neck	23.8	—	23.8?	—	—
Vertical distance between distal basis of lesser trochanter and upper margin of the neck (parallel to axis of diaphysis)	54	—	50	—	—
Circumference of the neck	97	—	93	—	—
Collar angle	rest. 119°	—	rest. 127°	—	—
Collar index	73.3°	—	—	—	—

After Weidenreich (1941).

Compared to modern humans, the anterior border of the *H. erectus pekinensis* tibia is blunter, the contour of the cross section is more rounded, the compact bone is blunt, and there is almost no medullary canal at midshaft.

Teeth. Weidenreich (1937) analyzed 147 *H. erectus pekinensis* teeth, among which 64 are isolated and 83 are either embedded in their respective jaws or found in immediate connection with them or with skulls. One hundred and thirty-four teeth belong to the permanent dentition, of which 52 are attributed to the upper and 82 to the lower jaw. Thirteen teeth belong to the mandibular deciduous dentition. The 147 teeth studied by Weidenreich were attributed to 32 individuals.

Besides Weidenreich's monograph, Rukang Wu and Zikui Chao (1959), and Zhonglang Qui et al. (1973) described other teeth of *H. erectus pekinensis* found in the same cave. Dental specimens found after 1949 include one median upper incisor, one upper first premolar, one upper second premolar, one lower first premolar, two lower first molars, and one lower second molar.

Table 2.23

Indices of robusticity of the femur wall of *Homo erectus pekinensis*

Specimen	*I*	*II*	*IV*	*V*	*VI*
Transverse diameter of the canal	9.7	6.6	11.1	8.7	10.4
Transverse diameter of the entire shaft	29.9	27.4	27.8	27.8	28.4
Sagittal diameter of the canal	9.9	6.9	9.3	7.5	9.8
Sagittal diameter of the entire shaft	26.7	23.0	23.1	23.6	25.5?
Index of the transverse diameter	32.4	24.2	39.9	31.3	36.7
Index of the sagittal diameter	37.5	29.9	40.3	31.7	38.4?

After Weidenreich (1941)

FIG. 2.14 Anterior view of the left tibia of *Homo erectus* found at Zhoukoudian in 1951. (Courtesy of IVPP.)

Dental measurements are given in Table 2.24. Figures 2.15 and 2.16 show various *H. erectus pekinensis* teeth.

The teeth are represented by large and small specimens, presumably belonging to males and females, respectively. The specimens represent 10 adult males, 6 young males, 10 adult females, and 6 young females. *Homo erectus pekinensis'* dentition shows great variability. Teeth with a very primitive appearance may be found together with rather advanced types, occasionally in the same jaw.

The crowns and the roots are very large and robust. The roots of the incisors and the canines are especially high. The crowns of the lower dentition, except the incisors, are long relative to their height. In this regard they differ from other extant hominid teeth. Abundant accessory ridges (wrinkles) are a characteristic feature,

Table 2.24

Measurements of the teeth of *Homo erectus pekinensis**

			Male			*Female*		
			Mean	*[no.]*	*Range*	*Mean*	*[no.]*	*Range*
I1 (upper)	Crown	Height	(13.3)	[1]	[(9.5) – 13.3]	(7.7)	[3]	[(4.3) – (10.4)]
		Length	10.7	[3]	[10.7 – 10.8]	9.9	[2]	[(7.2)9.8 – 9.9][3]
		Width	7.9	[3]	[7.5 – 8.1]	7.7	[3]	[7.5 – 7.9]
	Root	Height	17.8	[3]	[11.5 – 24.0]	19.5	[2]	[18.3 – 20.7]
		Length	8.2	[3]	[8.0 – 8.3]	6.7	[3]	[5.1 – 7.6]
		Width	7.1	[3]	[6.4 – 7.7]	7.3	[3]	[7.0 – 7.6]
I2 (upper)	Crown	Height	11.4	[1]	—	11.9	[1]	[(2.7) – 11.9] [2]
		Length	8.2	[1]	—	8.3	[1]	[(6.0) – 8.3] [2]
		Width	8.1	[1]	—	8.1	[2]	[8.0 – 8.2]
	Root	Height	(17.3)	[1]	—	19.0	[1]	
		Length	6.2	[1]	—	5.7	[2]	[5.6 – 5.7]
		Width	8.1	[1]	—	7.8	[2]	[7.6 – 8.0]
I1 (lower)	Crown	Height	(8.3)	[4]	[(6.8) – (9.1)]	(7.5)	[3]	[(6.9) – (7.8)]
		Length	6.5	[4]	[6.2 – 6.7]	6.3	[3]	[6.0 – 6.8]
		Width	6.4	[4]	[5.8 – 6.8]	6.3	[3]	[6.1 – 6.5]
	Root	Height	18.0	[2]	[(14.2)–18.0][3]	(6.9)	[3]	[(16.8) – (17.2)]
		Length	4.0	[2]	[3.9 – 4.0]	3.9	[1]	
		Width	5.8	[2]	[5.8 – 5.8]	5.8	[1]	
I2 (lower)	**Specimens of Both Sexes Lumped Together**							
	Crown	Height	10.4	[3]	[(4.1)10.2–10.7][10]			
		Length	6.9	[9]	[6.3 – 7.2]			
		Width	7.0	[10]	[6.4 – 7.3]			
	Root	Height	18.9	[4]	[(5.8)18.0–19.4][9]			
		Length	4.8	[6]	[3.7 – 5.6]			
		Width	6.8	[6]	[6.0 – 7.2]			
C (upper)	Crown	Height	14.2	[1]	[(11.3)–14.2][3]	13.6	[1]	[(17.9) – 13.6] [3]
		Length	9.9	[3]	[9.6 – 10.5]	9.0	[3]	[8.5 – 9.3]
		Width	10.5	[3]	[10.4 – 10.6]	9.8	[3]	[9.8 – 9.9]
	Root	Height	23.0	[2]	[22.8 – 23.2]	22.1	[2]	[21.8 – 22.4]
		Length	8.5	[3]	[7.8 – 9.3]	7.4	[3]	[7.2 – 7.7]
		Width	10.8	[3]	[10.5 – 11.1]	9.2	[3]	[8.9 – 9.5]
C (lower)	Crown	Height	11.5	[2]	[(5.6)11.3–11.7]	11.7	[1]	[(4.9) – 11.7] [3]
		Length	8.8	[4]	[8.4 – 9.0]	8.2	[3]	[8.1 – 8.5]
		Width	9.8	[4]	[9.3 – 10.4]	8.4	[3]	[8.2 – 8.7]
	Root	Height	23.8	[3]	[21 – 26.2]	—	—	—
		Length	7.8	[3]	[7.0 – 8.5]	7.1	[3]	[7.2 – 7.8]
		Width	10.0	[3]	[9.0 – 10.38]	8.7	[3]	[8.2 – 9.3]

(continued)

Table 2.24 (continued)

Measurements of the teeth of *Homo erectus pekinensis**

			Male			*Female*		
			Mean	*[no.]*	*Range*	*Mean*	*[no.]*	*Range*
P1 (upper)	Crown	Height	9.7	[1]	[(7.2) – 9.7] [2]	(7.1)	[2]	[(5.2) – (9.0)]
		Length	9.0	[2]	[8.7 – 9.2]	8.0	[3]	[7.4 – 8.7]
		Width	12.7	[2]	[12.6 – 12.8]	11.3	[3]	[10.5 – 11.9]
	Root	Height	20.4	[1]	[(13.1) – 20.4]	18.3	[2]	[16.6 – 20.0]
		Length	7.1	[2]	[6.9 – 7.2]	5.9	[3]	[5.5 – 6.2]
		Width	12.8	[2]	[12.7 – 12.8]	11.1	[3]	[10.4 – 12.0]
P2 (upper)	Crown	Height	(8.0)	[1]	—	(8.3)	[2]	[(4.6) 8.2 – 8.3] [9]
		Length	8.9	[1]	—	7.9	[9]	[7.3 – 8.9]
		Width	12.5	[1]	—	11.3	[9]	[10.3 – 12.1]
	Root	Height	—	—	—	15.8	[4]	[13.3 – 17.7]
		Length	—	—	—	5.7	[6]	[5.3 – 5.8]
		Width	—	—	—	10.8	[5]	[10.5 – 11.5]
P1(lower)	Crown	Height	8.8	[2]	[(6.0) 8.5 – 9.0] [7]	7.6	[2]	[(3.0) 6.5 – 8.7] [8]
		Length	8.9	[7]	[8.2 – 9.8]	8.3	[8]	[7.9 – 9.0]
		Width	10.4	[7]	[10.0 – 10.8]	9.2	[7]	[8.2 – 10.2]
	Root	Height	18.1	[5]	[16.8 – 19.1]	15.1	[3]	[14.4 – 15.7]
		Length	6.4	[4]	[6.1 – 6.6]	6.1	[6]	[5.5 – 6.6]
		Width	9.4	[4]	[8.8 – 9.7]	8.5	[6]	[7.5 – 9.0]
P2 (lower)	Crown	Height	7.8	[2]	[(6.4) 7.3 – 8.3] [4]	6.5	[1]	[6.5 – (8.1)]
		Length	8.9	[4]	[8.5 – 9.2]	8.5	[2]	[8.2 – 8.7]
		Width	10.7	[4]	[9.8 – 11.1]	8.6	[3]	[8.0 – 9.6]
	Root	Height	18.3	[2]	[17.3 – 19.2]	—	—	—
		Length	6.3	[4]	[6.1 – 6.5]	5.7	[1]	—
		Width	9.5	[4]	[9.2 – 9.7]	7.8	[1]	—
M1 (upper)	Crown	Height	(6.8)	[1]	—	7.9	[2]	[(5.3) 7.5 – 8.3] [5]
		Length	12.6	[2]	[12.1 – 13.1]	10.6	[5]	[10.0 – 11.3]
		Width	13.4	[1]	—	12.4	[5]	[11.7 – 13.7]
	Root	Height	15.4	[1]	—	14.4	[3]	[13.7 – 15.4]
		Length	8.9	[1]	—	8.2	[4]	[8.0 – 8.4]
		Width	13.2	[1]	—	11.6	[4]	[11.3 – 12.1]
M2 (upper)	Crown	Height	—	—	—	8.2	[1]	[(5.6) – 8.2] [7]
		Length	—	—	—	10.9	[7]	[10.2 – 12.2]
		Width	—	—	—	12.7	[7]	[12.2 – 13.4]
	Root	Height	—	—	—	15.1	[5]	[13.5 – 17.7]
		Length	—	—	—	8.1	[5]	[7.3 – 8.9]
		Width	—	—	—	12.2	[5]	[11.3 – 12.8]
M3 (upper)	Crown	Height	– –			(5.8)	[7]	[(5.1) – (6.4)]
		Length	– –			9.7	[8]	[8.7 – 10.4]
		Width	– –			11.7	[8]	[10.4 – 12.5]
	Root	Height	– –			14.9	[6]	[10.3 – 16.5]
		Length	– –			7.8	[7]	[5.7 – 9.8]
		Width	– –			11.3	[7]	[10.3 – 12.0]

(continued)

Table 2.24 (continued)

Measurements of the teeth of *Homo erectus pekinensis**

			Male			*Female*			
			Mean	*[no.]*	*Range*	*Mean*	*[no.]*	*Range*	
M1 (lower)	Crown	Height	7.8	[4]	[(3.5)7.3 – 8.4][9]	7.2	[2]	[(3.8)7.1 – 7.2]	[4]
		Length	13.1	[9]	[12.3 – 14.1]	11.4	[5]	[9.9 – 12.2]	
		Width	12.3	[9]	[11.7–12.8] [10]	10.8	[5]	[10.1 – 11.2]	
	Trigonid	Width	12.4	[7]	[11.7 – 12.8]	10.7	[3]	[10.1 – 11.1]	
	Talonid	Width	12.1	[7]	[11.3 – 12.5] [8]	10.3	[3]	[9.6 – 10.7]	
	Root	Height	16.1	[4]	[(12.4)14.1–18.5] [6]	13.1	[1]	[(11.7) – 13.1]	[2]
		Length	10.6	[8]	[9.1 – 11.1]	9.4	[3]	[8.9 – 10.2]	
		Width	10.8	[8]	[9.9 – 11.9]	9.9	[3]	[9.7 – 10.2]	
M2 (lower)	Crown	Height	7.6	[3]	[(5.0)6.7–9.1?][7]	(6.3)	[4]	[(4.5) – (7.4)]	
		Length	12.5	[7]	[11.3? – 13.2?]	12.5	[4]	[11.9 – 13.1]	
		Width	12.5	[7]	[11.5? – 13.0?]	11.5	[4]	[11.1 – 12.0]	
	Trigonid	Width	12.3	[5]	[11.5? – 12.9]	11.5	[4]	[11.1 – 11.8]	
	Talonid	Width	12.3	[5]	[11.5? – 12.9]	11.3	[4]	[10.7 – 12.0]	
	Root	Height	16.5?	[2]	[16.0? – 17.0]	15.5	[1]		
		Length	10.1	[4]	[9.2 – 11.0]	9.8	[2]	[9.4 – 10.2]	
		Width	10.5	[4]	[9.7 – 11.3]	9.9	[2]	[9.5 – 10.3]	
M3 (lower)	Crown	Height	6.9	[2]	[(4.8)6.7–7.0?][7]	7.0	[1]	[(3.8) – 7.0]	[2]
		Length	12.2	[7]	[10.6 – 13.8]	10.7	[3]	[10.0 – 12.2]	
		Width	11.5	[7]	[10.8 – 12.4]	10.8	[3]	[10.0 – 12.1]	
	Trigonid	Width	11.4	[7]	[10.3 – 12.4]	11.1	[2]	[10.0 – 11.4]	
	Talonid	Width	11.2	[6]	[10.6 – 12.1]	10.5	[2]	[9.3 – 11.7]	
	Root	Height	14.0	[3]	[13.0 – 15.5]	11.6	[1]	—	
		Length	10.0	[5]	[8.1 – 11.4]	8.7	[1]	—	
		Width	9.7	[5]	[8.6 – 10.7]	8.6	[1]	—	

* The figures within the paranthesis designate those of partly damaged or immature roots or worn crowns as shown in the original papers. The means within the parantheses were calculated on the basis of data including such cases.

Notes:

1. The sexing of the teeth and the data are according to Weidenreich (1937), Woo and Chia (1951), and Qiu et al. (1973).
2. The figures within [ ] designate the number of the teeth involved or the range of variation.
3. The means without brackets were calculated from undamaged specimens.
4. The length and width of the root of the first and second lower molar is their total length and width.
5. Where the lengths of different branches of the root are different, we adopt the length of the longest branch as the length of the root.
6. The question mark is given by Weidenreich (1937).
7. As a further explanation of the table,the following illustrates the notation for root height of lower I2 of a female:

Mean *Range*

18.9 [4] [(5.8) 18.0 – 19.4] [9] indicates:

There are 9 total specimens: 5 are damaged or worn; the mean height is calculated on the basis of only 4 teeth, the shortest root is damaged and is 5.8 mm long (or high); the shortest intact root is 18.0mm. long.

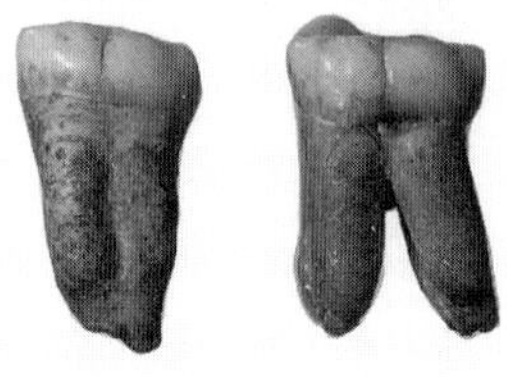

a b

FIG. 2.15 *Homo erectus* teeth found at Zhoukoudian in 1949 and 1951. Left, buccal view of left lower M1. Right, buccal view of left lower M2. (Courtesy of IVPP.)

a b c

FIG. 2.16 *Homo erectus* teeth found at Zhoukoudian in 1949 and 1951. Left to right: buccal view of right upper P2; buccal view of right upper P1; buccal view of left upper median incisor. (Courtesy of IVPP.)

even on deciduous molars. An additional widespread feature is a very pronounced cingulum on the upper and lower canines and on the premolars and molars. The molars also show traces of stylar cusps, and the canines have triangular differentiations on the mesial and distal surfaces.

The upper central incisor has a very strongly developed basal tubercle, which continues into several fingerlike prolongations. The lingual surface is shovel-shaped, with the border markedly thickened and folded around in the lingual direction. The basal tubercle on the upper lateral incisors forms part of the thickened folded border. The crown and root of the upper incisor form a straight line. The lower incisors have a moderately concave lingual surface and a relatively weak basal tubercle. The cutting edge is distinctly crenulated.

The upper canine is a very large and pointed tooth with a well-developed cingulum and a complicated lingual surface. The development of the cingulum on the smaller and less robust mandibular canine is approximately the same as on the upper canine. However, in the configuration of the lingual surface and the tendency to form a cutting edge instead of a tip, the mandibular canine bears a closer resemblance to the incisors than to the upper canine. The maxillary canines strongly resemble the lower canine of the fossil orang-utan.

The buccal cusp of the first upper premolar is larger and higher than the lingual cusps. The chewing surface of both cusps exhibits a wrinkled pattern. Because the mesial portion of the basal part of the buccal surface projects considerably, the outline of the crown is asymmetrical. The root of the first upper premolar is divided into two branches or at least shows a deep indentation, in both cases with traces of a third branch. The second upper premolar is smaller than the first one. Unlike the first premolar, the lingual and buccal cusps of the second upper premolar are about equal in size and height. There is no basal projection on the second upper premolar; therefore, the tooth crown is a fairly regular oval shape in the occlusal view.

The first lower premolar typically presents an asymmetrical, oblong shape with a well-developed cingulum and a distinct talonid, and both cusps are shifted towards the mesial border of the crown. The metaconid is lower than the protoconid and is represented by an isolated cusp situated buccally from the thickened lingual rim. Variation in the talonid is rather marked. The root is very robust and shows indications of a division into two or three branches. The second lower premolar typically reveals the same asymmetrical oblong form as the first. The cingulum is well developed; a large talonid covered by wrinkles radiates from its center. The two trigonid cusps are shifted towards the mesial border of the crown. The root is very robust and shows the same tendency to divide into two or three branches as observed in the first premolar.

The paracone of the first and second upper molars markedly projects mesially and buccally. Cingulum differentiations are very distinct on the buccal surface. There is no clearly discerned trigon pattern. The anterior trigon crest is represented by the mesial rim of the chewing surface, and the posterior crest is represented by the slopes of the metacone and the protocone. The fovea anterior is a triangular area between these cusps and the mesial rim. The sloping surface of the metacone is covered by wrinkles. This surface continues with a separated distal section of the protocone. The root of the first molar is very robust, and its lingual branch is straight and diverges considerably. The mesial portion of the buccal branches is much thicker than the lateral portion and is shifted buccally to correspond to the position of the paracone. The three branches of the root of the second molar have the tendency to fuse. The third upper molar is small and triangular; the apex is rounded off due to the reduction of the metacone and hypocone. The branches of the root are more or less fused. The Carabelli tubercle appears to be absent.

The second lower molar is slightly larger than the first. The crowns of both the molars typically show clear indications of a cingulum and stylar cusps on the buccal surface. The enamel reaches down to a lower level on the buccal than on the lingual surface. The anterior part of the crown is broader than the posterior segment. There are six or five cusps. The metaconid is the largest and longest cusp, and the hypoconid is occasionally very small. There is a well-developed *Dryopithecus* Y-5 pattern and a slight tendency to transform into a plus pattern by reducing the metaconid. The fovea anterior is triangularly shaped. The root is very robust, with the two branches straight and divergent. The mesial branch is shorter, lower, and thicker, and tends to subdivide the apical part into a buccal and a lingual portion. The branches of the root of the second molar tend to fuse. The third lower molar is generally the smallest. Like the case in the upper molars, the distal cusps are reduced in size. The branches of the root tend to fuse, and the lower molars exhibit taurodontism.

As regards the deciduous dentition, the incisor crowns are strong and robust, and the incisor's roots are relatively robust and high. The canine has a pyramidlike shape with a distinct point and a well-developed cingulum. The first molar is long and narrow with a pronounced cingulum. The tooth is clearly molarized. The talonid is slightly longer than the trigonid. The mesial branch of the root shows an oblique position and projects considerably mesially and buccally. The second molar is more molarized than the first, its cingulum is very pronounced, and the root and its branches are very large and stout. The second molar appears after the eruption of the permanent incisors and before the deciduous canine and molars are shed. The canine is the last deciduous tooth to erupt.

Judging from the wear facets on the incisors, there must have been overbite in the dentition of young individuals with a moderate degree of attrition. The dentition of old individuals indicates strong wear due to edge-to-edge bite. The upper permanent canine projects considerably beyond the general level of the chewing surface, exceeding the crown height of the first upper premolar. The dental arcade is a relatively narrow curve extending forward, so the incisors form a relatively equal curved line. There is no mandibular or maxillary diastema in the permanent dentition. The lower deciduous canine is separated from the lateral incisor and from the first molar by a relatively wide gap. The gap between the lateral incisor and canine is broader than the gap between the canine and first deciduous molar.

From *H. erectus pekinensis* to recent humans there has been a size reduction of the crowns and roots. The cingulum and wrinkles on the crown have mostly been lost; therefore, the tooth surface is smoother. The crowns of the lower premolar have become completely symmetrical due to the shortening of the talonid. The lower molars become more square in shape. The cusps, especially the metaconid, decrease in size; their number diminishes (four instead of six); and their arrangement alters (a plus pattern instead of the *Dryopithecus* Y-5 pattern). The lower canine loses its cutting edge and, by acquiring a tip, takes on the appearance of the upper canine. The size reduction, especially reduction of root size, of the lower front teeth, including the canine, leads to a reduced alveolar process. The reduced alveolar process leads to a relatively protuding basal part of the mandible, resulting in the formation of a chin (Daegling, 1993, discusses the evolution of the chin). A similar transformation happens in the upper teeth. The anterior nasal spine and "hollow cheeks" in modern humans may be the result of a reduced alveolar process.

Weidenreich (1935: 441) suggested that "the difference in size of *Sinanthropus* teeth is a sexual one" and "the large type characterizes the male and the small one the female." He concluded: "The morphological character of the *Sinanthropus* population of Locality 1, therefore, remained unchanged during the long period of time necessary for the filling-up of the cave." (Weidenreich, 1935: 45l).

When Zhang (1991) analyzed temporal variation in the Zhoukoudian hominid dental sample, he lumped the teeth from Layer 5 and those layers above it in a "Later group." Those teeth from Layer 7 and layers under it are referred to an "Early group." Dental samples from the lower to upper levels showed a distinct size increase in the incisors, and a decrease in the size of the canines and cheek teeth. Zhang concluded that an evolutionary change in the direction of early *H. sapiens* is suggested in the dental sample from Zhoukoudian Locality 1.

Zhang calculated the means of various measurements of the crowns of the teeth found in the upper and lower parts of the cave deposits. The means obtained are given in Table 2.25.

Geology

The *H. erectus pekinensis* cave site is on a small hill of Ordovician limestone located at the eastern border of the Western Hills. The cave deposits are estimated to be 140 m long from east to west and 20 m wide at the center, and narrow toward both the eastern and western ends. The 40 m of excavated deposit are divided into 13 layers. Test wells have been dug recently to investigate the deposits beneath Layer 13 (Pei and Zhang, 1985; Yang et al., 1985):

Layers 1–2: Fossil bearing breccia and sandy clay about 4 m thick, intercalated with stalactites.

Layer 3: Fossil bearing breccia from which human fossils, many stone artifacts, and evidence of fire have been found. Big blocks of angular limestone found in

Table 2.25

Comparisons of tooth size in early and late groups of *Homo erectus pekinensis*.

			Early Group			Later Group		
			Mean	*No.*	*Range*	*Mean*	*No.*	*Range*
Upper dentition								
Incisor	(median)	length	10.7	[3]	[10.7 – 10.8]	9.9	[1]	—
		width	7.8	[4]	[7.5 – 8.1]	7.8	[1]	—
	(lateral)	length	8.3	[2]	[8.2 – 8.3]	—	—	—
		width	8.1	[3]	[8.0 – 8.2]	—	—	—
Canine		length	9.4	[6]	[8.5 – 10.5]	—	—	—
		width	10.2	[6]	[9.8 – 10.6]	—	—	—
Premolar	(first)	length	8.4	[5]	[7.4 – 9.2]	—	—	—
		width	11.9	[5]	[10.5 – 12.8]	—	—	—
	(second)	length	8.0	[7]	[7.2 – 8.9]	8.1	[2]	[7.9 – 8.3]
		width	11.5	[7]	[10.3 – 12.5]	11.3	[2]	[11.2 – 11.3]
Molar	(first)	length	11.4	[6]	[10.1 – 13.1]	—	—	—
		width	12.6	[5]	[11.7 – 13.7]	—	—	—
	(second)	length	10.8	[3]	[10.3 – 11.4]	11.3	[3]	[10.5 – 12.2]
		width	12.9	[3]	[12.4 – 13.4]	12.6	[3]	[12.2–13.2]
	(third)	length	9.5	[5]	[8.7 – 10.3]	9.9	[2]	[9.8 – 9.9]
		width	11.7	[5]	[10.4 – 12.5]	12.0	[2]	[12.0 – 12.0]
Lower dentition								
Incisor	(median)	length	6.3	[3]	[6.0 – 6.7]	6.5	[2]	[6.5 – 6.5]
		width	6.3	[3]	[5.8 – 6.8]	6.4	[2]	[6.3 – 6.5]
	(lateral)	length	6.7	[4]	[6.3 – 7.2]	7.0	[3]	[6.9 – 7.2]
		width	6.9	[4]	[6.4 – 7.3]	7.1	[4]	[6.9 – 7.3]
Canine		length	8.9	[4]	[8.5 – 9.0]	8.2	[3]	[8.1 – 8.4]
		width	9.6	[4]	[8.7 – 10.4]	8.6	[3]	[8.2 – 9.5]
Premolar	(first)	length	8.8	[6]	[8.2 – 9.8]	8.4	[7]	[7.9 – 9.0]
		width	10.0	[6]	[9.2 – 10.7]	9.3	[6]	[8.2 – 10.2]
	(second)	length	8.8	[4]	[8.5 – 9.2]	8.6	[2]	[8.2 – 9.0]
		width	10.7	[4]	[9.6 – 11.1]	8.4	[3]	[7.1 – 9.8]
Molar	(first)	length	12.7	[7]	[9.9 – 14.1]	12.2	[6]	[11.3 – 13.6]
		width	11.9	[6]	[10.1 – 12.6]	11.7	[5]	[10.9 – 12.6]
	(second)	length	12.4	[7]	[11.3 – 13.0]	12.6	[3]	[11.9 – 13.1]
		width	12.1	[6]	[11.5 – 12.8]	11.3	[3]	[11.1 – 11.5]
	(third)	length	11.8	[6]	[10.6 – 12.7]	11.3	[3]	[10.0 – 13.8]
		width	11.4	[6]	[10.7 – 12.4]	10.1	[2]	[10.0 – 10.2]

Measurements according to Zhang (1991).

the basal part of this layer may represent the collapsed roof of the original cave. The layer is about 3 m thick.

Layer 4 (upper cultural layer): Breccia and clay silt layer containing human and many rodent fossils, stone artifacts, burnt bones, burnt stones, and a thick ash lens laminated by water action. The layer is about 6 m thick.

Layer 5: This layer of hard travertine with breccia and sand is about 1 m thick. The layer contains a small number of stone artifacts and fossils, including part of the right half of a female mandible.

Layer 6: Fossil bearing breccia and cemented sand layer about 5 m thick. This layer contains burnt bones, stone artifacts, human fossils, and a "hearth" with ash. The hearth is 1.2 m wide and 0.75 m thick.

Layer 7: A layer of laminated fine sands with breccia and clay balls that contains small amounts of artifacts and human fossils, as well as fossils of animals that lived along the water's edge. The layer is 6.5 m thick at the western part, becomes thinner eastward, and disappears at the western part of the Gezitang (Pigeon) cave. Gezitang is a large cave at the northeastern part of Locality 1, Zhoukoudian.

Layers 8–9 (Lower cultural layer): These fossiliferous breccia are about 6 m thick and are filled with sand, clay, and ashes. Many human fossils, including three skullcaps, numerous stone artifacts, and mammalian specimens, were found in these layers.

Layer 10: This layer of breccia, sandy clay, and ashes is approximately 2 m thick and yielded some artifacts and human fossils.

Layer 11: Breccia, about 2 m thick, with a very few stone artifacts were unearthed from the upper part of this layer. The lower fissure, where the first "Peking Man" skull was found, corresponds with the layer.

Layer 12: This layer of dark brown coarse sands with breccia and small pebbles is about 2 m thick. It yielded a few broken animal fossils, but no human bones or stone artifacts.

Layer 13: This layer of sandy clay has a thick layer of hyaena coprolites and fossils of *Megaloceros pachyosteus,* a species of deer with a thick jaw bone. No human fossil was found in this layer, but a dubious stone artifact has been reported. The layer is about 2 m thick.

———erosion?———

Layer 14: This layer of brownish red clay and sands with gravels is about 2.0–3.85 m thick.

Layer 15: This brownish-red gravel layer is about 1.75–2.8 m thick.

———erosion———

Layer 16: This laminated grey silty layer is about 2.5 m thick.

Layer 17: This brownish-yellow gravel with coarse sands was dug to 0.7 m in depth without reaching the base.

Recently obtained chronometric dates for the different layers of the cave deposits are presented in Table 2.26.

The rich faunal assemblage locality includes the following (Hu, 1985; Qi, 1989):

Mammals

Ninety-seven mammalian fossil species were recovered in Locality 1 according to Hu (1985). There are slight differences between Hu's list and a list by Qi (1989).

Table 2.26
Chronometric data of Locality 1, Zhoukoudian

Layer	*Thousand years*	*Method*	*Author*
1-3	256 + 62/-40	Uranium series	Zhao et al. (1985)
	230 + 30/-23	Uranium series	Zhao et al .(1985)
	260 + 80/-50	Uranium series (carbonate)	Zhao et al. (1985)
	420 +11/-5	Uranium series (travertine)	Shen and Jin (1991)
3-4	282	ESR	Huang et al. (1991)
4	292 ± 26	TL	Pei (1985)
	312 ± 28	TL	Pei (1985)
	299 ± 55	Fission track	Guo et al (1991)
7	370 to 400	Paleomagnetism	Qian et al (1985)
8-9	older than 300	Uranium series (bones)	Zhao et al (1985)
	418	ESR	Huang et al. (1991)
10	462 ± 45	Fission track	Liu et al. (1985)
	417 to 592	TL	Pei (1985)
11	578	ESR	Huang et al. (1991)

Those forms marked with an * appear on Qi's list and those marked with a # are only found on Hu's list.

Primates
Homo erectus pekinensis
Macaca robustus
Insectivora
Scaptochirus primitivus
Neomys bohlini
Neomys sinensis
Crocidura sp.
Erinaceus olgai
Sorex sp.
Chiroptera
Rhinolophus pleistocanicus
Miniopteris cf. *schrebersi*
Ia io

? *Pipistrellus* sp.
Myotis sp.*
Rodentia
Citellus cf. *mongolicus*
Tamias wimani
Petaurista brachyodous
Marmota bobak
Marmota complicidens
Marmota sp.
? *Castor* sp.
Trogontherium cuvieri
Cricetinus varians
Cricetulus cf. *griseus*
Cricetulus cf. *obscurus*
*Cricetulus longicaudatus**
Cricetulus cf. *migratorius**
*Apodemys sylvaticus**
Mus sylvaricus#
Mus musculus
Micromys cf. *minutus*
Rattus rattus
Gerbillus roborowskii
Clethrionomys rofucanus
? *Eothenomys* sp.
Alticola sp.
Pitymys simplicidens
Microtus brandtioides
Microtus epiraticeps
? *Phaiomys* sp.
Myospalax wongi
Myospalax sp.
Myospalax epitingi
Hystrix cf. *subcristata*
Lagomorpha
Ochotona koslowi
Ochotona sp. A.
Ochotona sp. B.
Lepus cf. *wongi*
Lepus sp. A.
Lepus sp. B.
Carnivora
Canis lupus

Canis lupus variabilis
Canis cyonoides
Nyctereutes sinensis
Cuon antiquus#
Cuon cf. *alpinus**
Vulpes cf. *vulpes*
Vulpes cf. *corsac*
Canidae gen. et sp. indet.
Ursus thibetanus kokeni
Ursus arctos
Ursus cf. *spelaeus*
? *Ailuropoda* sp.
Meles cf. *leucurus*
Lutra melina
Gulo sp.#
*Gulo schlosseri**
Mustela cf. *sibirica*
Mustela sp.
Martes sp.
Hyaena brevirostris sinensis
Crocuta ultima
Machairodus inexpectatus
Panthera youngi
Panthera cf. *pardus*
Panthera cf. *tigris*
Felis teilhardi
Felis sp. A.
Felis sp. B.
Felis cf. *microtis**
Felis cf. *chinensis**
Cynailurus sp.#
Acinonyx sp.*

Proboscidea
Palaeoloxodon cf. *namadicus*

Perissodactyla
Dicerorhinus choukoutienensis
Coelodonta antiquitatis
Equus sanmeniensis

Arctiodactyla
Sus lydekkeri
Paracamelus gigas
Camelidae gen. et sp. indet.

Moschus moschiferus pekinensis
? *Hydropetes* sp.
Capreolus sp.#
Pseudaxis grayi
Megaloceros pachyosteus
*Megaloceros flabellatus**
Cervus sp.
Gazella sp.
Spirocerus peii
Spirocerus cf. *wongi*
Ovis cf. *ammon*
Ovis sp.
Ovibovinae gen. et sp. indet.
Bubalus teilhardi
Bison sp.
? *Naemorhedus* sp.
Bovidae gen. et sp. indet.

The mammalian faunal assemblage relatively dates Locality 1 to the Middle Pleistocene. Kahlke and Hu (1957) correlated this site with the European Holsteinian Interglacial. Most scholars agree that Layers 3–10 were deposited in the interglacial period.

The composition of the mammalian fauna changes in different layers (Hu, 1985). Layers 10 and 11 contain *Homo erectus pekinensis, Scaptochirus primitivus, Neomys sinensis, Neomys bohlini, Cricetulus* cf. *obscurus, Mus sylvaticus, Mus musculus, Gerbillus roborowskii, Microtus epiratticops, Ochotona koslowi, Canis lupus variabilis, Nyctereutes sinensis, Vulpes* cf. *vulpes, Ursus arctos, Ursus* cf. *spelaeus, Hyaena brevirostris sinensis, Machairodus inexpectatus, Panthera* cf. *pardus, Felis teilhardi, Dicerorhinus choukoutienensis, Equus sanmeniensis, Sus lydekkeri, Pseudaxis grayi,* and *Megaceros pachyosteus.* These animals usually inhabit forests and grasslands.

Homo erectus pekinensis, Macaca robustus, Marmota sp., *Trogontherium cuvieri, Hystrix* cf. *subcristata, Canis lupus, Canis lupus variabilis, Canis cyonids, Nyctereutes sinensis, Cuon antiquus, Vulpes* cf. *corsac, Ursus thibetanus kokeni, Ursus arctos, Ursus* cf. *spelaeus, Meles* cf. *lecurus, Gulo* sp. (?), *Hyaena brevirostris sinensis, Machairodus inexpectatus, Panthera* cf. *tigris, Felis teilhardi, Felis* sp. A., *Palaeoloxodon* cf. *namadicus, Dicerorhinus choukoutienensis, Coelodonta antiquitatis yenshanensis, Pseudaxis grayi, Megaceros pachyosteus, Spirocerus peii, Ovis* cf. *ammon,* and *Bubalus teilhardi* have been unearthed from Layers 8–9. Carnivores predominate and most of them are typical forest dwellers. Grassland animals are less important elements in this assemblage.

Layer 7, mainly composed of sands, contains a much smaller sample of animal remains than other layers. Insectivora, chiroptera, and human remains are not found in this layer. A list of mammals of this layer includes: *Macaca robustus, Marmota complicidens, Trogontherium cuvieri, Cricetulus cf. griseus, Cricetulus* cf. *obscurus, Micromys* cf. *minutus, Canis lupus, Nyctereutes sinensis, Lutra melina, Hyaena brevirostris sinensis, Panthera* cf. *tigris, Felis* sp. B, *Dicerorhinus choukou-*

tienensis, Equus sanmeniensis, Sus lydekkeri, Moschus moschiferus, Pseudaxis grayi, Megaceros pachyosteus, Ovis cf.*ammon*, and *Bubalus teilhardi*. Rather complete skulls of *Sus lydekkeri* and *Bubalus teilhardi,* and complete antlers of *Pseudaxis grayi,* have been found. The fossils show traces of water abrasion and probably were water transported into this cave. Stone artifacts are rare.

Layer 6 yields mammals similar to those inhabiting the temperate zone of the Palaeoarctic Region. They are *Homo erectus pekinensis, Macaca robustus, Erinaceus olgai? Castor* sp., *Trogontherium cuvieri, Cricetinus varians, Cricetulus* cf. *obscurus, Micromys* cf. *minutus, Canis lupus variabilis, Nyctereutes sinensis, Vulpes* cf. *corsac, Ursus thibetanus kokeni, Ursus arctos, Meles* cf. *leucurus, Hyaena brevirostris, Hyaena sinensis, Panthera* cf. *tigris, Dicerorhinus choukoutienensis, Equus* cf. *sanmeniensis, Moschus moschiferus pekinensis, Pseudaxis grayi, Megaceros pachyosteus, Gazella* sp., and *Ovis* cf. *ammons.*

Layer 5 contains *Homo erectus pekinensis, Macaca robustus, Scaptochirus primitivus, Erinaceus olgai, Trogontherium cuvieri, Cricetinus varians, Cricetulus obscurus, Micromys* cf. *minutus, Rattus rattus, Microtus brandtioides, Microtus epiratticeps, Canis lupus, Canis lupus variabilis, Vulpes* cf. *corsac, Ursus thibetanus kokeni, Ursus arctos?*, *Ailuropoda* sp., *Meles* cf. *leucurus, Mustela* cf. *sibirica, Hyaena brevirostris sinensis, Machairodus inexpectatus, Panthera* cf. *tigris, Panthera* cf. *pardus, Felis teilhardi, Felis* sp. A, *Felis* sp. B, *Felis* cf. *microtis, Cynailurus* sp., *Moschus moschiferus pekinensis, Pseudaxis grayi, Megaceros pachyosteus, Gazella* sp., and *Ovis* cf. *ammon.* Carnivores predominate; when the deposits were forming, the cave was probably occupied by carnivores such as *Hyaena.*

Homo erectus pekinensis, Macaca robustus, Scaptochirus primitivus, Neomys sinensis, Neomys bohlini, Crocidura sp., *Erinaceus olgai, Rhinolophus pleistocaenicus, Miniopterus schreiberii, Ia io, Cricetinus varians, Cricetulus* cf. *griseus, Cricetulus* cf. *obscurus, Micromys* cf. *minutus, Rattus rattus, Gerbillus roborowskii, Microtus brandtioides, Microtus epiratticeps, Ochotona koslowi, Canis lupus variabilis, Ursus thibetanus kokeni, Ursus arctos, Palaeoloxodon* cf. *namadicus, Dicerorhinus choukoutienensis, Equus sanmeniensis, Moschus moschiferus pekinensis, Megaceros pachyosteus,* and *Ovis* sp. are found in Layer 4. The predominance of rodents in this layer may imply that grasslands are the major regional environmental feature.

Homo erectus pekinensis, Macaca robustus, Neomys sinensis, Neomys bohlini, Crocidura sp., *Erinaceus olgai, Miniopterus schreiberii, Cricetinus varians, Cricetulus* cf. *grisen, Cricetulus* cf. *obscurus, Micromys* cf. *minutus, Canis lupus, Vulpes* cf. *corsac, Ursus thibetanus kokeni, Ursus arctos, Meles* cf. *lucurus, Hyaena brevirostris sinensis, Crocuta ultima, Panthera* cf. *tigris, Dicerorhinus choukoutienensis, Equus sanmeniensis, Pseudaxis grayi, Megaceros pachyosteus,* and *Cervus elaphus* appear in Layers 1–3. The composition of these layers is similar to that of Layer 6. According to Qiu et al. (1973) the excavation in 1966 yielded a fragmentary maxilla and a left half of a mandible of *Hyaena sinensis,* and a skull and mandible of *Crocuta ultima* in Layer 3. These two species coexisted in part of the deposit. A basal segment of an antler of *Cervus elaphus* was also found during the 1966 excavation. Some primitive carnivores like *Machairodus inexpectatus* and *Felis teilhardi* disappeared in the upper part of the deposit.

Shenglong Lin (1985) studied the hunting behavior of *H. erectus pekinensis* by analyzing the profile of the large mammals discovered at Zhoukoudian. The numbers of each species varied considerably. *Hyaena brevirostris sinensis, Megaceros pachyosteus*, *Pseudaxis grayi*, *Dicerorhinus choukoutienensis*, and *Equus sanmeniensis* are frequently distributed in many layers. Coprolites showed that the hyaena probably inhabited this cave. During the period corresponding to the middle and lower part of the deposits, the cave was frequently occupied by carnivores. The rapid decrease in carnivore numbers toward the upper cave layers suggests that these layers were more frequently occupied by *Homo erectus*. This alternative occupation of the cave by hominids and carnivores indicates that the fossils of large mammals might have been deposited by both *H. erectus* and carnivores. In fact, these two predators may have competed with one another for food and shelter. The rarity of fossils in some *H. erectus* living floors (e.g., Quartz II of Gezitang and Layer 4) suggests that carnivores were not the main prey of *H. erectus*.

Herbivore remains were brought into the cave by *H. erectus* and/or carnivores. Many herbivore bones found in the cultural layers, such as Quartz II of Gezitang, are broken into small pieces and were burnt. Among the herbivores, *Megaloceros pachyosteus* and *Pseudaxis grayi* predominate, and their fossils are distributed in many layers of the deposit. Young (1932, cited by Lin, 1985) counted the faunal specimens obtained from the 1927 to 1931 excavations. There are at least 50 well-preserved *Megaceros pachyosteus* antlers, more than 100 skulls, more than 1000 maxillas, and fragments of several thousand mandibles and other bony parts. The mandibles represent at least 2000 individuals. The mandibles of *Pseudaxis grayi* represent more than 1000 individuals. The next most common animals are *Sus lydekkeri* (at least 200 individuals), *Bubalus teilhardi* (more than 80 individuals), and *Moschus moschiferus pekinensis* (more than 30 individuals). Most mammalian fossils in Layers 7 and 12 were probably transported into the cave by water. Some skulls of *Sus lydekkeri* and *Bubalus teilhardi* as well as antlers of *Pseudaxis grayi* are rather complete.

Lanpo Jia (1978) attempted to establish climatic changes during the time of "Peking Man" on the basis of mammalian fossils unearthed from the cave (see Appendix II). In Layers 11 and 10, mammals inhabiting a cold environment are more abundant than those inhabiting warmer locales. In Layers 8–9 mammals indicative of warm environments increased in number and are equal in representation to cold climate animals. Jia, therefore, suggested that Layers 8–9 represented a transition from a cold to a warmer phase. The incidence of cold climate fauna decreased in Layer 7, and warm climate mammals predominate in Layer 6. In Layer 5 cold climate animals again increased in number. Cold climate animals were rare in Layer 4, which represents one of the warmest periods during which *H. erectus* occupied Zhoukoudian. In Layers 1–3 warmer climate animals predominate.

Jia concluded that from Layer 11 through Layers 1–3 the climate fluctuated. It was rather cold and dry in Layers 11–10, and became warmer and wetter in Layers 6–9. It tended to be cooler in Layer 5, and slightly warmer and wetter again in the upper four layers.

Birds

The 62 avian species recovered from Zhoukoudian belonged to 9 orders, 19 families, or 48 genera (Hou, 1985) as follows:

Struthioniformes
  Struthio anderssoni
Gruiformes

Rallus sp.
Rallus aquaticus
Gallinula chloropus
Falconiformes
Buteo buteo
Accipiter sp.
Accipiter nisus
Galliformes
Alectoris graeca
Perdix dauuricae
Coturnix coturnix
Coturnix sp.
Pucrasia macrolopha
Phasianus colchicus
Columbiformes
Syrrhaptes paradoxus
Columba livia
Streptopelia chinensis
Streptopelia sp.
Strigiformes
Asio flammeus
Ninox scutulata
Apodiformes
Apus apus
Apus pacificus Latham
Piciformes
Dendrocopos major
Picus canus
Dryocopus martius
Passeriformes
Alauda arvensis
Melanocorypha mongolica
Eremophila alpestris
Calandrella cinerea
Galerida cristata
Hirundo daurica
Hirundo rustica
Riparia riparia
Motacilla flava
Anthus spinoletta
Lanius cristatus
Sturnus cineraceus

Cyanopica cyana
Pyrrhocorax pyrrhocorax
Corvus monedula
Corvus torquatus
Cissa erythrorhyncha
Luscinia calliope
Luscinia cyane
Hodgsonius phoenicuroides
Zoothera dixoni
Rhyacornis fuliginosus
Paradoxornis webbianus
Phragamaticola aedon
Prinia polychroa
Parus sp.
Parus major
Passer domesticus
Fringilla montifringilla
Carpodacus roseus
Loxia curvirostra
Coccothraustes coccothraustes
Emberiza sp.
Emberiza spodocephala
Emberiza leucocephala
Emberiza pusilla
Emberiza chrysophrys
Emberiza cia.

Avian fossils suggest that although the Middle Pleistocene climate at Zhoukoudian was rather dry, there was a rich vegetation.

Flora

The results of Kong's (1985) palynological analysis of Locality 1 follow:

Trees
Abies
Pinus
Cupressaceae
Salix
Juglans
J. mandshurica
Pterocarya
Alnus
Betula

Carpinus
Corylus
Ulmus
Celtis
Zelkova
Quercus
Castanea
Rhus
Aralia
Bushes and grasses
Ephedra
Compositae
Artemisia
Chenopodiaceae
Humulus
Polygonum
Leguminosae
Rosaceae
Ranuculaceae
Ramnus
Scabiosa
Plantago
Gardenia
Ligustrum
Jasminum
Potamogeton
Sparganium
Typha
Cryptogram
Selaginella sinensis
S. senguindenta
S. unicinata
S. delicatula
Selaginella sp.
Osmunda
Lygodium
Polypodium
Physcomitrium
Bryum
Riccia
Peltolepis
Reboulia hemisphaerical

The flora indicated that in the 13th layer the climate was warm and dry. In the 12th layer it became drier, the forest shrank, and the grassland expanded. In the later period of Layer 12, the climate became wetter. Palynology indicated that in the early and later parts of Layers 8–9 the climate was warm and dry, but in the middle part of the layers the climate was wetter. The vegetation in Layer 6 was rather temperate in nature. From Layer 4 upwards drought-enduring plants increased in number. In the later period of Layer 4, Chenopodiaceae and *Polygonum* predominate, indicating that vegetation in the region had become temperate forest and grassland. Spores from Layer 1 suggested that the climate probably became cooler and drier. Kong (1985) emphasized that despite the existing fluctuation, the climate during the long period of hominid occupation was more or less stable, and similar to today's temperate zone.

Archaeology

The thousands of stone artifacts found at Locality 1 have been analyzed by many scholars. Here we summarize the work of Wenzhong Pei (Wenchung Pei) and Senshui Zhang (1985), who analyzed 17,091 pieces of worked stone, flakes, and nuclei of various sizes. Although *H. erectus pekinensis* used 44 kinds of rocks to manufacture their stone tools, most artifacts (88.8%) were made of quartz, and primarily from vein quartz. Artifacts made of rock crystal, sandstone, and flint (chert) constitute 4.7%, 2.6%, and 2.4%, respectively, of the total assemblage. The quartz and rock crystal may have been introduced from nearby, the other materials were taken from a nearby river bed. Because many flakes, nuclei, and debris were found in the cave, perhaps this was a tool manufacturing spot. Most flakes were made by a bipolar technique, direct hammering, and anvil percussion. The bipolar technique was mainly used for producing quartz flakes, the direct hammering technique was applied to many kinds of materials, and anvil processing was basically used on sandstone and was of lesser importance. Flakes produced by the bipolar technique are smaller than those made by anvil processing.

The specimens analyzed by Pei and Zhang (1985) include 445 nuclei, 5160 flakes, 61 hammers, 19 anvils, 2228 scrapers, 406 points, 47 awls, 113 burins, 160 choppers, 8 bolas, and 8444 substandard products. Pei and Zhang divided the stone implements of "Peking Man" into two categories. The first group includes hammers and anvils; the second includes scrapers, points, awls, burins, choppers, and bolas. Most stone implements of *H. erectus* from Zhoukoudian are scrapers, which constitute 75.2% of the 2962 tool specimens in the second category. They are generally 30–50 mm in length, 20–40 mm in width, and lighter than 20 g. Points constitute 13.7% (406 pieces) of the tools in the second category; most are 20–40 mm long and lighter than 10 g. Awls were only derived from the upper part (Layers 1–5) of the cave deposit and constitute 1.59% (47 pieces) of the tools in the second category. Except for a few specimens, the awls are shorter than 30 mm in length. Gravers, which came from almost all segments of the uppermost 11 layers of the deposit, constitute 3.81% (113 pieces) of the tool assemblage of the second category. Choppers, the largest and crudest implements, constitute 5.4% of the total tool assemblage of the second category, and were found in all of the upper 11 layers. Most were unifacially flaked, and bifacially flaked choppers were made by alternating chipping and inverse trimming. In addition, there are eight stone bolas from Layers 1–5, 8, and 9.

There are also utilized flakes and utilized raw materials. Large flakes were used as choppers and smaller flakes were generally used as scrapers. About 10% of the flakes had no secondary preparation. The utilized raw materials include pitted (anvils and hammers) and flaked gravels (stone hammers). The elongated hammers were used to produce flakes.

About 70% of the assemblage is flake tools; 30% of the tools are made from a nucleus. Small tools (shorter than 40 mm) and micro-tools (shorter than 20 mm) constitute one of the most important elements of the so-called Sinanthropus Industry. No handaxes have been found. Most tools were made by direct hammering and were retouched on the dorsal side. Because most tools were crudely retouched on the edge, the worked margin varies in shape. The *H. erectus* tool assemblage at Zhoukoudian contains a large number of scrapers, points, and choppers, that is, tools normally found in the early Paleolithic. However, there are blades, microblades, and awls, which seem to be resemble artifacts of the later Paleolithic.

The Sinanthropus Industry covered a period of thousands of years. It can be divided into three stages: an early stage represented in Layers 11–8, a middle stage represented by Layers 6–7 and Quartz II of Gezitang [Kotzetang] Cave, and a late stage represented in Layers 4–5 and Layers 1–3. Dates for each layer are given in Table 2.26.

Early on, sandstone was the main material for making important stone tools, but it became less prevalent in later stages of tool manufacture. Quartz, seldom used in the early stage of tool manufacture, was used at Zhoukoudian increasingly in later stages. Flint use increased from the middle to later stages when the quality of quartz used became better. The use of milky quartz especially increased.

Both bipolar and direct hammering methods were the main manufacturing techniques in the early stage. The bipolar technique predominated in later stages. Anvil use characterized only the early stage of tool making. It was only occasionally used thereafter, and tended to disappear in the late stage. Direct hammering was developing from the early to the late stage. During the late stage flakes made by the direct hammering and bipolar techniques were more regular in shape, and more of them have the long and thin shape of blades and microblades.

Retouching became progressively more advanced from the early to the late stage. Points from the early stage of tool making were usually only retouched finely on one side and crudely on the other side. In the middle stage the points were smaller and usually finely worked on both sides.

In the early stage flake tools were less common than those made of nuclei, small stones, etc. Flake tools predominated in the middle stage and were even more common in the late stage. Choppers constituted about half the tool assemblage of the second category in the early stages, while points and burins rarely occurred. The percentage of choppers and scrapers alternatively decreased and increased in later stages. Burins rapidly increased in number from the middle to the late stage, during which a new kind of implement—the awl—appeared.

Stone tools became smaller from the lower part of the deposit through Layers 4–5. In Layers 1–3 tool size slightly increased once again. Multipurpose artifacts are more common in the early stage than in later stages. Tools became more specialized through time.

During early excavations fragmentary bones of black, grey, green, and pale yellow were recovered. Fine cracks and/or deformations are frequently seen on the bones. In 1930 and 1931, chemical analysis of a blackened deer antler and other broken bones showed that the color change was due to burning. A second chemical analysis was done on black earth from the Quartz II layer at the basal part of the deposit in Gezitang Cave. (This layer was so named because of the abundance of quartz flakes.) In this layer there is isolated carbon without iron and manganese, and burnt seeds of *Celtis* (hackberry) have been collected. The ash and burnt material was distributed in a limited area.

Two large hearths were discovered on a big block of limestone in the basal part of the third layer. Rather concentrated ash layers were also found in Layers 4–5, Quartz II, Layers 8–9, and the basal part of Layer 10. The thickness of these ash concentrations ranges from 6 m to several dozens of centimeters. The ash layer contains many burnt bones, burnt stones, and burnt seeds of *Celtis,* as well as carbon particles. The color of the burnt bones is white or green. The surfaces of the burnt stones are replete with cracks; some limestone blocks were burnt into lime; and on some areas of the surface of the living floor the earth was hardened and was a red or reddish brown color due to the burning.

HEXIAN (118°12′ E, 31°53′ N)

The Hexian *H. erectus* fossils were found in Longtandong Cave, located on the north slope of Wanjiashan mountain ("hill of Wang's family"). Wanjiashan is located south of the town of Taodian in Hexian County, Anhui Province, in eastern China. The cave and the mammalian fauna were found in 1973 when local farmers dynamited the north slope of Wanjiashan during construction of an irrigation canal. In July, 1980, Wanju Qing of the Bureau for Water Resources collected a left second upper human molar. The molar was sent to the Institute of Vertebrate Paleontology and Paleoanthropology, Academia Sinica, for study. In October an IVPP field team excavating the site in conjunction with local archaeologists recovered the following human fossils: an almost complete skullcap (PA 830), a fragmentary left mandibular body with an attached M2 and M3 (PA 831), a right upper second premolar (PA 832), a left upper second molar (PA 833), and a left lower first and second molar belonging to the same individual (PA 834 [1–2])

In 1981 fragments of the frontal (PA 840) and parietal bones (PA 841), as well as five teeth (PA 835–839), were recovered. The frontal fragment consisted of the right supraorbital torus and the adjacent part of the frontal. The parietal fragment is represented by the part covering the right parietal tubercle and the posterior segment of the temporal line. A number of researchers have analyzed the Hexian materials. We relied upon the work of R. Wu and Dong (1982), M. Wu (1983), W. Huang et al. (1982), Han and Xu (1989), and T. Chen et al. (1987).

Human fossils

Skullcap

The skullcap (PA 830) is rather complete (Fig. 2.17). It retains an almost complete frontal bone (missing the posterior part of the roof of the orbit and the lower portion of the nasal part of this bone), an almost complete left parietal and a right parietal (missing the sphenoidal angle, the parietal tubercle, and the anterior portion of the mastoid angle), an almost complete left temporal (lacking the zygomatic process and lower part of the mastoid process), a right temporal (missing the anterior portion and part of the upper portion of the squama, zygomatic process, and lower part of the mastoid process), an occipital squama, and the upper portion of the greater wing of the left sphenoid. The thick and robust calvaria has strong muscular mark-

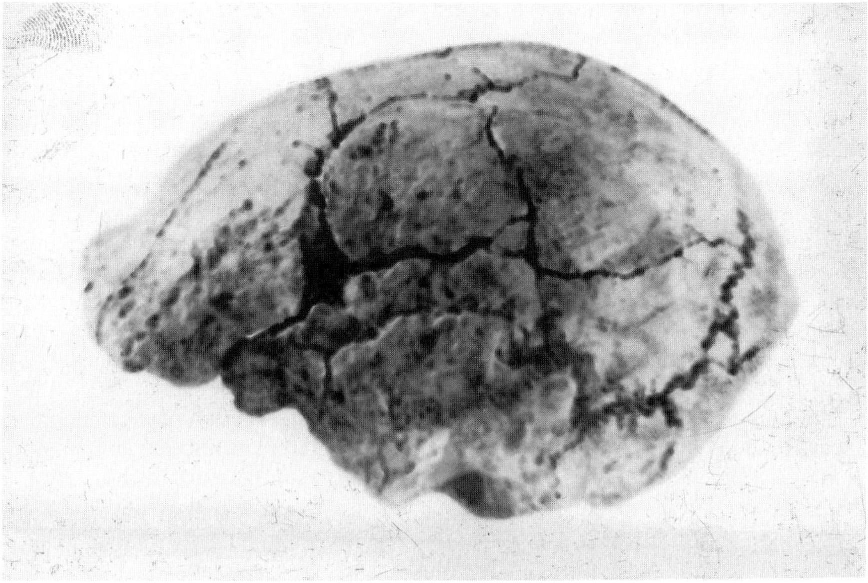

FIG. 2.17 *Homo erectus* skullcap from Hexian. (Courtesy of IVPP.)

ings. Both internal and external cranial sutures are visible, except for part of the left endocranial spheno-parietal suture. The calvaria probably belongs to a 20-year-old male. A flattened area between the bregma and vertex appears sagittally and between the temporal lines of both sides. This area was distorted by pressure exerted by the overlying deposit of earth.

The top of skull PA 830 is ovoid, its broadest part is at the level of the supramastoid ridge. The cranial vault is very low. The frontal bone is receding and the angle of inclination of the frontal squama (b-g-i) and the lower part of the frontal bone (m-g-i) is more acute than is true of early *H. sapiens* in China. The supraorbital tori are broad in the transverse direction and thick in the vertical direction. The supraorbital torus becomes thinner from the medial toward the lateral end in specimen PA 830, but is thickest at the midsection in specimen PA 840. There is a supraorbital notch instead of a foramen in both PA 830 and PA 840. A supraorbital process is more distinct on the right than on the left side in skull PA 830. The orbital roof is flat in PA 830. The frontal sinus is small and does not extend to the glabella and the roof of the orbit. The frontal squama posterior to the supraorbital tori is slightly bulging. The supraorbital sulcus of both PA 830 and 840 is shallower than appears in "Peking Man." Postorbital constriction in PA 830 and PA 840 is much less than characterizes *H. erectus pekinensis*. In PA 830 the frontal tubercle is slightly developed. The metopic suture is visible between the level of the frontal tubercle and bregma. There is no parasagittal depression. The parietal tubercle is well-developed. There is no obelion foramen. The well developed temporal lines connect with the angular torus at the mastoid angle of the parietal bone. An H-shaped pteryon appears on the left side.

The temporal squama is rather high and its upper border is convex instead of being straight. The parietal notch is rather deep between the squamosal and mastoid portion of the temporal bone. The posterior root of the zygomatic process of the temporal bone extends obliquely backwards and upwards, and connects with the supramastoid ridge. The latter connects with the angular torus. The sulcus of the zygomatic process is wide and shallow. The supramastoid ridge is well developed. The supramastoid sulcus is distinct; the mastoid notch is wide and shallow. The mastoid process is rather small, and its lower part turns medially.

The occipital torus forms a transverse smooth prominence. The central part of the occipital torus is most prominent, the lateral part turns downwards and terminates at the region of the asterion. The supratoral sulcus is distinct. The transition between the occipital and nuchal planes is angular. The lower margins of both sides of the occipital torus are curved. The two curved margins meet at the midsagittal plane and are continuous with the external occipital ridge. There is no external occipital protuberance or Inca bones. A Wormian bone (36 mm wide and 28 mm high) exists near the right asterion.

The mandibular fossa is narrow and deep. Instead of an arcuate eminence, there is a rather deep notch at the anterior border of the right mandibular fossa. The thick tympanic plate is neither as vertical as in modern humans nor as horizontal as in chimpanzees. The degree of the angle formed by the axis of the tympanic plate and the pyramid is intermediate between apes and humans.

There is neither a styloid process nor a vaginal process. The occipito-mastoid ridge is less developed than in "Peking Man". The stylomastoid foramen, styloid foramen, and mastoid notch are situated along a straight line, as in modern humans.

The sagittal sulcus of the internal cranial surface disappears at the junction between the upper and middle third of the frontal squama. There is no foramen caecum. Granular pits can be seen on the internal surface of the parietal bone close to its sagittal margin. The Sylvian crest is well developed, originating at the sphenoid angle of the parietal bone, extending obliquely upwards and backwards, and disappearing at the junction between the anterior and posterior halves of the parietal bone. The cerebellar fossa is obviously smaller than the cerebral fossa.

The left transverse sulcus is broader, deeper, and in a higher position than its right counterpart. The right sulcus is narrow and shallow, as is the sagittal sulcus. These two are continuous. Unlike in apes, the transverse sulcus does not reach the mastoid angle of the parietal bone. The pyramid of the temporal bone is robust. The distance between the upper border of the sigmoid sulcus and the sulcus for the superior petrosal sinus is greater than that in "Peking Man."

The sulcus for the common trunk of the left middle meningeal artery is about 10 mm long on the internal cranial surface. It divides into two branches. The fronto-parietal branch extends forward and upward and penetrates a bony tube at the sphenoidal angle of the parietal bone. After penetrating this tube, it divides into two branches and radiates on the fronto-parietal region. The superior temporal branch runs backward and then upwards at the posterior part of the temporal squama. Its terminal branches radiate on the posterior half of the parietal. The common trunk of the right middle meningeal artery is 19 mm long. The course of the fronto-parietal branch is not clear because of the lack of bone. After branching from the common trunk, the superior temporal branch divides into two branches, the inferior temporal branch and the terminal branch of the superior branch proper. The terminal branches of the superior temporal branch proper radiate at the posterior half of the parietal region. Those tributaries of the inferior temporal branch radiate at the mastoid angle of the parietal. In both the left and right sides the temporal branch of the middle mennigeal artery is thicker than the fronto-parietal branch. The small branches of both are as sparse as those of *H. erectus pekinensis.*

R. Wu and Dong (1982) compared the measurements and indices of *H. erectus* specimens from Hexian, Zhoukoudian, Trinil, and Ngandong from Java. They found that the *H. erectus* skulls from Hexian, Zhoukoudian, and Trinil are similar in such measurements as cranial length and height, the inclination angle of the frontal squama (b-g-i), their transverse curvature, the fronto-parietal width index, and curvatures of the parietal and occipital bones. In the following measurements the Hexian specimen is closer to *H. erectus* from Zhoukoudian than from Trinil: the interporion breadth, cranial circumference, transverse arc, sagittal arc, inclination angle of the inferior part of the frontal bone (m-g-i), the occipital curvature angle (l-i-o), and the cranial capacity. The Hexian skull is larger than that from Trinil in all these measurements, except the occipital curvature angle. In a superior view, the Hexian skull is broader and more rounded than *H. erectus* from Zhoukoudian but closer to that of Trinil. The Hexian skull is more modern in such features as its minimum frontal breadth, cranial sagittal curvature, frontal curvature, and its parieto-occipital index (arc). The Hexian skull is broader than either Zhoukoudian and Trinil. The cranial measurements of the Hexian specimen are given in Table 2.27.

Mandible

A partial mandibular body with an M2 and M3 in situ recovered from Longtandong belongs to a middle-aged individual according to the attrition on the molars. The height of the body (measured at the level between M1 and M2 or the most posterior mental foramen) is between that of the male and female mandibles of Zhoukoudian *H. erectus* specimens (measured at the level of the mental foramen). The thickness of the body (measured at the same level as the height) is much larger than the Zhoukoudian *H. erectus* mandibles. The index of robusticity is much higher for Hexian than in both sexes from Zhoukoudian. R. Wu and Dong (1982) suggested that the Hexian mandible belonged to a male. Three mental foramina are situated at the level of P2-M1.

Teeth

The following teeth, found in 1981, were described by Maolin Wu (1983).

Upper right median incisor (PA 835). The crown is nearly intact, except for the distal part of the slightly worn cutting edge. The lingual surface of the crown is shovel-shaped. A thick rim along the mesial and distal margins of the crown is more distinct than on the mesial margin. A well-developed and rounded basal tubercle appears on the lingual surface. Four fingerlike prolongations of different length radiate from the lower margin of the basal tubercle and terminate at the central fossa of the crown.

The conically shaped robust root has a broad and shallow vertical groove on both the mesial and distal surfaces. There is no constriction at the neck of the tooth. The borderline of the enamel is curved on both the lingual and labial surfaces. The top of this curve is at a similar level on both surfaces.

Upper left first molar (PA 836). The occlusal surface is square and the crown height is low because of attrition. Both mesio-buccal and disto-lingual corners of the crown are rounded. The hypocone is well developed. A vertical furrow extends from the occlusal surface to the neck of the tooth on both the buccal and lingual

Table 2.27
Cranial and mandibular measurements of *Homo erectus* from Hexian

Martin No.*	*Measurement*	
1(2)	Cranial length:g-op(i)	190
5(1)	Sagittal chord:n-o	131 (?)
8c	Max. width at level of upper margin temporal squama	145
8	Cranial width	160
9	Min. frontal width:ft-ft	93
11	Interauriculare width:au-au	144
12	Occipital width:ast-ast	141.8
20	Height from bregma to the line between porions:po-b	95
23	Circumference:g g(op)	571
24	Transverse arc:po po(b)	291
25	Sagittal arc:n⌒o	340?
26	Frontal sagittal arc:n⌒b	120?
27	Parietal sagittal arc:b⌒l	110
28	Occipital sagittal arc:l⌒o	110
29	Frontal sagittal chord:n-b	99?
30	Parietal sagittal arc:b⌒l	103
31	Occipital sagittal arc:l⌒o	83
32a	Inclination angle of lower part of frontal:m-g-i	58°
32(2)	Inclination angle of the frontal squama:b-g-i	41°
33(4)	Curvature angle of occipital:l-i-o	101°
8/1(2)	Cranial index	84.2
20/1(2)	Length-height index	50
20/8	Breadth-height index	59.4
5(1)/25	Sagittal curvature	38.5?
11/24	Transverse curvature	49.5
9/8	Fronto-parietal index	58.1
12/8	Parieto-occipital breadth index	88.6
27/26	Index of fronto-parietal (arc)	91.7?
28/27	Index of parieto-occipital (arc)	100.0
29/26	Frontal curvature	82.5?
30/27	Parietal curvature	93.6
31/28	Occipital curvature	75.5
	Cranial capacity	~ 1025 c.c.
Frontal bone:		
	Minimum distance of lateral surfaces of frontal squama	101
	Distance between extremities of both sides of brow ridges	111
	Index of post-orbital constriction	91
	Height of supraorbital torus:	
	PA 830: Left: medial	19
	middle	16
	lateral	12
	Right: medial	18
	middle	17
	lateral	13

(*continued*)

Table 2.27 (continued)

Cranial and mandibular measurements of *Homo erectus* from Hexian

	PA 840: Right:	medial	13
		middle	16
		lateral	12
	Diameter of frontal sinus:		
	Left:	sagittal	7.5
		transverse	14
	Right:	sagittal	11.5
		transverse	11
	Temporal bone:		
	Length of squama	(left)	70
	Height of squama	(left)	42
	Index of squama		60
	Angle formed by longitudinal axis of the tympanic plate and sagittal plane of the skull		
	Left		90°
	Right		95°
	Angle formed by longitudinal axes of the tympanic plate and the pyramid		
	Left		30°
	Right		35°
	Angle formed by longitudinal axis of the pyramid and sagittal plane of the skull		60°
	Distance between temporal lines of both sides		77
	Distance between sagittal planes passing through the exterior auditory orifice and auriculare		
	Left		15
	Right		14
Occipital bone			
	Distance between internal and external occipital protuberance		22
	Width of foramen magnum		at least 32
Thickness of cranial bones			
	Center of frontal squama		7.0
	Parietal tubercle:	Left: (PA 830)	13.5
		Right: (PA 840)	11.0
Thickness of cranial bones			
	Mastoid angle of parietal bone		18.0°
	Center of occipital torus		18.0
	Cerebellar fossa		6.0
	Center of temporal squama		10.0
Mandible:			
	Thickness of the body (between Ml and M2)		20.7
	Height of the body (between Ml and M2)		32.0
	Index of robustness		64.7

1. According to Wu Dong (1982) and Wu (1984)

2. * From Martin and Saller's *Lehrbuch der Anthropologie*, 1957, Stuttgart: G.F. Verlag

surfaces. A trace of a cingulum remains at the base of the mesio-buccal surface.

The buccal branches of the root are lost. The single lingual root is very robust and straight. Taurodontism is indicated by the preserved remaining part of the root.

Upper right second molar (PA 837). The occlusal surface is quadrangular. A vertical furrow, similar to that exhibited on specimen PA 836, is found on both the buccal and lingual surfaces. The tooth is slightly worn and has a small area of dental exposure at the paracone. Because of the attrition, most of the grooves between the main cusps are not clearly shown. A transverse groove between the paracone and metacone, and an oblique groove between the protocone and hypocone, are discernible. On the slopes of the paracone, protocone, and metacone is a rounded and thick ridge, lateral to which are small accesssory ridges. There is a distinct cingulum on the lingual and distal surfaces. Whether the cingulum exists on other surfaces cannot be judged because the basal part of these surfaces is broken. The root is missing.

Lower left second molar (PA 838). The cusp pattern is a Y-5. The mesial breadth of the crown is much longer than the distal breadth. A trace of a cingulum appears on the buccal surface.

The anterior and posterior branches of the very robust root fuse at the upper part. The fused area covers about half of the height of the root on the buccal surface and about the upper one third on the lingual surface. The anterior branch of the root is shorter and broader than the posterior branch. The anterior branch bifurcates at the tip, while the posterior branch has only a single tip. The shallow groove on the middle of the anterior surface of the anterior branch of the root is broader than that of the posterior branch.

Lower left second molar (PA 839). Although worn, the tooth has a discernible Y-5 cusp pattern, in which the five main cusps are nearly equal in size. The length and breadth of the trigonid are much greater than those of the talonid. The cingulum is visible on the mesial, buccal, and lingual surfaces, and the preserved part of the distal surface.

Teeth found in 1980—specimens PA 83l, 832, and 834—have not been reported in detail. The preserved molars from Hexian found in 1980 have abundant wrinkles and, except for the third molar, which is attached to the mandible, are generally larger than the averages of those of "Peking Man." The cingulum is most developed on upper M2 and lower M1. The third lower molar is the shortest. This is similar to the situation in the Zhoukoudian specimens, but different from that of the Trinil molars. Specimen PA 834, including a first and second molar, is heavily worn and belongs to a different individual than the skull and the other teeth found in the same year. Measurements of the tooth crowns of Hexian are given in Table 2.28.

Despite differences between the Hexian and Zhoukoudian specimens, R. Wu and Dong (1982) opted not to create a new subspecies name for the Hexian specimens.

Geology

Longtandong Cave is situated 23 m above sea level. Its Quaternary deposits were divided into five layers by Huang et al. (1982). From top to bottom they are a brownish-red, brownish black clay about 0.2–0.4 m thick; a brownish red clay

Table 2.28

Measurements of the tooth crowns of *Homo erectus* from Hexian

		Length	*Breadth*
Upper	median incisor(PA 835)	11.7	9.4
	second premolar(PA 832)	9.0	13.4
	first molar(PA 836)	12.3	13.7
	second molar(PA 833)	12.0	14.0
	second molar(PA 837)	12.5	15.5
Lower	first molar(PA 834	12.5	13.1
	second molar(PA 834)	13.3	13.6
	second molar(PA 838)	13.6	13.9
	second molar(PA 839)	14.3	13.4
	third molar(PA 831)	11.3	10.7

Data according to Wu and Dong (1982) and Wu (1983)

about 2.3 m thick; a yellowish-green silty clay about 0.1–0.3 m thick; a yellowish-brown sandy clay about 0.7–1.4 m thick, which yielded human and other vertebrate fossils; and a yellowish-grey sandy clay about 1.5 m thick.

Fauna

The vertebrate fauna include the following (mammals, Han and Xu (1989); birds and reptiles, W. Huang et al. (1982):

Mammals

Primates
- *Homo erectus*
- *Macaca robustus*

Insectivora
- *Blarinella quadraticauda*
- *Chodsigoa youngi*
- *Anourosorex squamipes*
- *Sorex* sp.

Chiroptera
- ? *Scaptochirus* sp.
- *Rhinolophus* cf. *ferrum-equinum*
- *? Myotis* sp.
- *Hipposideros* sp.
- *Miniopterus schrebersii*

Lagomorpha
- *Lepus* sp.

Rodentia
- *Tamias cf. wimani*
- *Trogontherium cuvieri*
- *Cricetulus varians*

Microtus brandtioides
Eothenomys melanogaster
Eothenomys eva alcinous
Eothenomys proditor
Eothenomys inez
Apodemus agrarius
Adodemus sylvaticus
Rattus rattus
Rattus norvegicus
Rattus edwardsi
Carnivora
Canis sp.
Cuon alpinus
Vulpes sp.
Arctonyx collaris
Lutra sp.
Hyaena sinensis
Megantereon sp.
Felis chinensis
Panthera pardus
Ailuropoda sp.
Ursus arctos
Ursus thibetanus kokeni
Proboscidea
Stegodon orientalis
Perissodactyla
Equus sp.
Tapirus sinensis
Megatapirus sp.
Dicerorhinus sp.
Artiodactyla
Sus lydekkeri
Sus cf. *xiaozhu*
Cervus (Pseudaxis) grayi
Megaloceros pachyosteus
Hydropotes inermis
Elaphurus davidianus
Bison sp.

Birds

Crossoptilon sp.

Reptiles

Ocadia sp.
Amyda sp.
Alligator cf. sinensis

Several chronometric dates are available. Li and Mei ([1983], cited in Wu et al., 1989), provided a TL date of 195,000 ± 16,000 years. According to Chen et al. (1987), most uranium dated samples are in the range of 150,000–190,000 years ago. However, the possibility of a date of 270,000–200,000 years ago is not excluded.

Archaeology

No archaeological remains have been reported.

TANGSHAN (119°03′ E, 32°03′ N)

In 1990 mammalian fossils were found in a limestone cave named Huludong (Calabash cave), on Tangshan hill (Warm Spring Hill) near Tangshan town, some 26 km east of Nanjing, capital of Jiangsu Province, eastern China. Huludong cave is less than 100 km east of the Hexian *H. erectus* site and separated by the Changjiang River (Yangtze River). The cave was exposed during quarrying by local farmers, who reported the mammalian fossils to the Nanjing Institute of Paleontology and Geology, Academia Sincia. Hanqui Xu and colleagues from this institute and Qinqi Xu of the IVPP began a short period of excavation in January 1993. In March 1993, local workers found a human skull while clearing the cave in preparation for making it a tourist attraction.

The specimen is an almost complete anterior part of a human skull, including an almost complete frontal bone, the left half of the face, a portion of the parietal bones, a part of the occipital, and a small part of the right upper face. The frontal bone is thicker than 10 mm and is very constricted behind the orbit. The brow ridges are thick.

The mammalian fauna from the cave were identified by Qinqi Xu as *Cricetulus* sp., *Ellobius* sp., Mustelidae, *Ursus arctos, Ursus* sp., *Hyaena sinensis, Megaloceros pachyosteus, Pseudaxis grayi, Sus* sp., and *Hydropotes* sp. Based on faunal correlations, Xu estimated that the fossils belonged the later Middle Pleistocene.

YUNXIAN (MEIPU) (111°10′ E, 33°00′ N)

In the summer of 1975, a field team from the IVPP, Academia Sinica, found a human incisor among fossils collected by a farmer from Dujiagou Village, Meipu Commune, Yunxian County, Hubei Province, in central China. This incisor was said to have been unearthed from Longgudong cave. From May to July and September to November, Chunhua Xu and others from the IVPP excavated in the cave and found three more human teeth (Fig. 2.18).

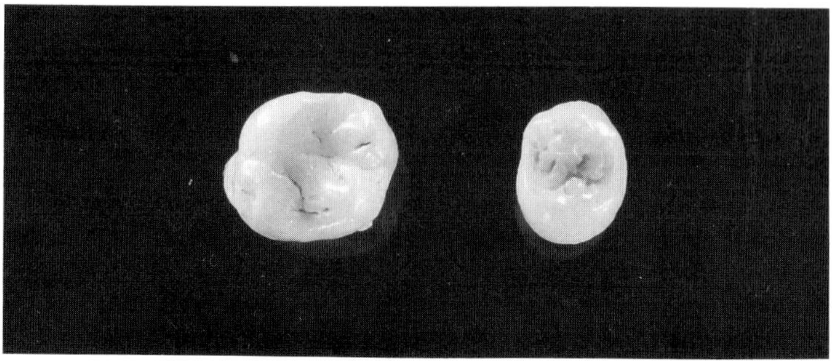

FIG. 2.18 *Homo erectus* teeth from Yunxian (Meipu) showing the left upper first molar (left) and the left upper second premolar (right). (Courtesy of IVPP.)

Human fossils

The following descriptions are from R. Wu and Dong (1980).

Teeth

Left upper median incisor (PA 634). The crown is almost complete but only the cervical part of the root is preserved. The slightly worn cutting edge exposes the dentine. The labial surface of the crown bulges vertically and horizontally, especially the area close to the cervix. The enamel extends more on the labial and lingual surfaces than on the mesial and distal ones. The lingual surface is shovel-shaped. The rim of the distal margin of the crown is more developed than the rim of the mesial margin of the crown. The basal tubercle is prominent. The fingerlike process, or as Weidenreich called it, the fingerlike prolongation of the basal tubercle, is not well developed near the cutting edge. The length of the incisor's crown is close to the average for *H. erectus* from Zhoukoudian.

Left lower lateral incisor (PA 635). This tooth is heavily worn and only two thirds of the crown height are preserved. The lingual surface is weakly concave. The basal tubercle appears as a smooth eminence at the basal part of the lingual surface. The cervical line is obviously curved on the mesial and distal sides, but is more or less straight on the lingual and labial surfaces. The enamel extends more on the lingual and buccal surfaces than on the mesial and distal surfaces.

Much of the robust root has been preserved. There is a longitudinal sulcus on both the proximal and distal surfaces of the root. The root turns slightly toward the lingual direction. The cervix is not constricted.

Left upper second premolar (PA 636). The crown is complete and slightly worn; the root is lost. The buccal cusp is larger and higher than the lingual cusp. The buccal half of the crown is slightly wider than the lingual half. The enamel extends more upward on the buccal than on the lingual surface. A trace of the cingulum appears as a triangular eminence situated at the basal part of the buccal surface, and as two small ridges located lateral to small fovea that are on both sides of the triangular eminence. The lingual surface of the crown looks like a spherical smooth eminence. Transverse accessory small ridges occupy the occlusal surface of both cusps.

Left upper first molar (PA 637). The crown is rather complete and slightly worn. The labial branches of the root are lost. The occlusal surface is rhombdoidal in shape. The disto-lingual angle is more rounded than the mesio-buccal angle. The paracone and protocone are slightly higher than the distal two cusps. A trace of a cingulum can be discerned.

Buccal, lingual, mesial, and distal ridges appear on the occlusal surface. The transverse sulcus between the paracone and the metacone extends upward and across the buccal ridge to the middle of the buccal surface. An oblique sulcus between the metacone, protocone, and the hypocone extends across the lingual ridge to the middle of the lingual surface. On the mesial half of the occlusal surface there is a V-shaped sulcus. Because of the amount of attrition, the lingual wing of the sulcus is not as distinct as the buccal wing. After joining, the two wings of the V-shaped sulcus reach the transverse sulcus. On the distal half of the occlusal surface, a transverse fovea connects with the oblique sulcus.

The occlusal surface of each cusp possesses several fine ridges. The intermediately located ridge is larger than those located to either of its sides. The middle larger fine ridge of the metacone connects with an accessory ridge of the protocone to form a well-developed oblique ridge.

The protocone is the largest cusp. The metacone is larger than the paracone and the hypocone is the smallest cusp. There is no Carabelli cusp, but there is a very small fovea on the lingual surface near the marginal ridge.

The lingual branch of the root is simple and does not bifurcate. The lingual deviates from the buccal branch of the root. The root inclines slightly to the distal side. A longitudinal deep sulcus exists towards the middle of the lingual surface of the root. The measurements of the Yunxian teeth are given in Table 2.29.

Geology

Xu (1978) described the geology of Longgudong cave. Longgudong is a big horizontal cave situated about 40 m above the water surface of the Taohe River. The entrance opens to the northwest. Although the cave deposits were dug by farmers, three layers are discernible in the remaining, undisturbed deposits. The upper layer is a hard travertine crust about 0.3 m thick. The middle layer, 0.5–2.5 m thick, is composed of a yellow sandy clay. Its central portion is loose, but its consistency is firmer close to the cave wall and cave floor. The human and other mammalian fossils were embedded in this layer. The lower layer, 0–0.6 m thick, is a hard yellow deposit containing small pieces of breccia.

Table 2.29

Measurements of *Homo erectus* teeth from Yunxian

		Incisor (upper 1st)	Incisor (lower 2nd)	Premolar (upper 2nd)	Molar (upper 1st)
Crown	Length	10.2	7.7	9.2	12.9
	Width	8.3	8.4	12.7	13.9
	Height	(12.5)	(6.6)	8.6	8.8
Root	Length	—	6.2	—	9.8
	Width	—	8.6	—	12.8
	Height	—	(23.2)	—	13.7*

According to Wu and Dong (1980)

* Height of the lingual branch

Figures in parenthesis are values of height of worn crown or damaged root.

Archaeology

From deposits transported to the cave from outside, a pebble nucleus showing traces of striking by a stone hammer was reported by Xu (1978).

YUNXIAN (QUYUAN RIVER MOUTH) (110°38′ E, 32°51′ N)

In 1989, a human skull was found near Mitousi Village, Yunxian County, Hubei Province in central China. Another skull was found in 1990. In addition to T. Li et al. (1991), Li and Etler (1992) provided some morphological details of these two skulls.

Human fossils

Both skulls are without mandibles. Because of postmortem compression, the vault of Skull 1 (EV 9001) is low, the right cranial wall was broken, and the palate was cracked. Skull II (EV 9002) was also compressed. The coronal and sagittal sutures are distinct and wavey in Skull I. These two sutures, as well as the lambdoidal suture, are even more distinct in Skull II. Li et al. (1991) suggested that the skulls belong to adults.

In Skull I dentine on the maxillary incisors was exposed by attrition. On the occlusal surface of the second upper molar, the exposed dentine of different cusps connect with each other. Dentine has not been exposed on the teeth of Skull II, which appears to be chronologically younger than Skull I.

Both skulls are big and robust (Figs. 2.19 and 2.20). The cranial capacity of Skull II has been estimated by Li et al. (1991) as probably 1100 mm^3. In both skulls the maximum cranial breadth is above the level of the external auditory meatus. The cranial side walls slope inwards toward the apex of the vault. This produces a bell-shaped paracoronal cranial profile. The EV 9002 skull is rather low and long, and the basicranial axis is poorly flexed. These two features may be due to the depression of the whole specimen. The brow ridges form a continuous line. Each of the ridges appears straight from an anterior and superior view, and swept back posterolaterally. The tori are thickest at the medial part. The frontal squama is receding and flat. The ophryonic region has a gently concave depression between the brow ridges and the frontal squama, rather than the prominent post-toral sulcus seen in *H. erectus* specimens from Zhoukoudian. Postorbital constriction is very distinct. There is no median sagittal ridge or bregmatic eminence. The parietal bone is short and flat. There is no angular torus.

In skull EV 9002 the occipital torus is distinct and smooth. There is an angular transition from the occipital plane of the occipital bone to its nuchal surface. The occipital plane is shorter than the nuchal plane. The temporal squama is rather high, and its superior margin is curved. The fossae for the insertion of the nuchal musculature are shallow and poorly defined. The mastoid process is moderately sized, the tympanic region is robust, and the mandibular fossa is narrow and long. There is no appreciable development of an articular tubercle. The sphenoid does not contribute to the medial wall of the glenoid fossa.

The upper and the midface are very flat, short, and broad, and the midface is not well pneumatized. The lower border of the zygomatic process of maxilla is retracted. The canine fossa may be present. The root of the zygomatic process is located high on the maxillary body. The lower nasal margin is rather sharply de-

FIG. 2.19 Face and skull of EV 90001 from Quyuan River Mouth, Yunxian. (Courtesy of Tianyuan Li.)

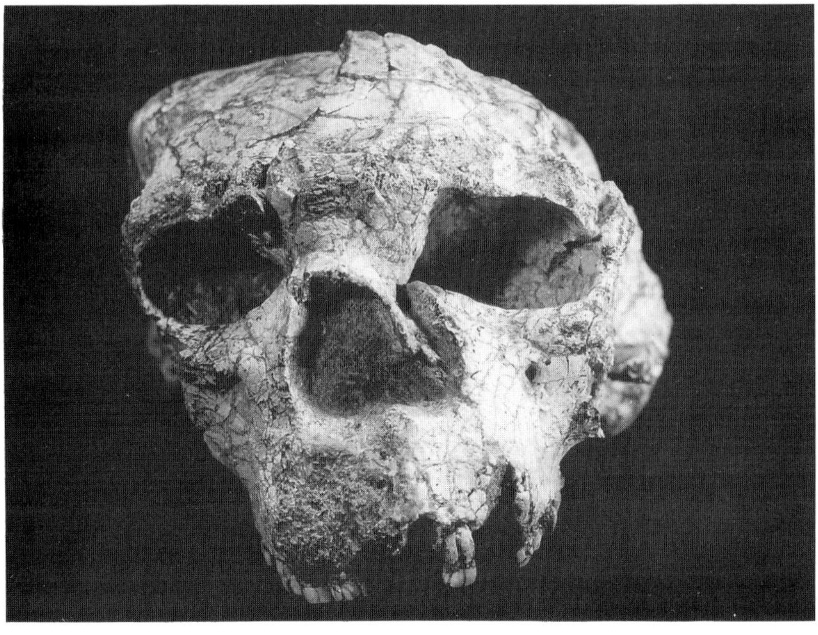

FIG. 2.20 Face and partial skull of EV 9002 from Quyuan River Mouth, Yunxian. (Courtesy of Tianyuan Li.)

fined. The interorbital distance is rather broad. The mastoid and supramastoid crests do not unite anteriorly. The zygomatic process of the temporal bone is robust and positioned away from the temporal squama by a broad sulcus. The auditory meatus is elliptical.

The palate is broad. The occlusal surfaces of the preserved teeth are somewhat more worn and fractured in Skull EV 9001. The teeth are big; in EV 9001 molar size increases from M1 to M3. The width to length ratio of the third molar exceeds 3:2. In Skull EV 9002 the left third molar is reduced in size compared to the second molar and is peg-shaped. Radiographs show an unerupted diminutive right third molar.

These skulls have features usually seen in *H. erectus,* such as a very distinct postorbital constriction and the low position of the maximum breadth of the brain case. However, there are other features concordant with archaic Chinese *H. sapiens*. The advanced features favor a *H. sapiens* attribution, but final taxonomic attribution must be delayed until further study.

Geology

The fossils are found in a terrace deposit at the juncture of the Quyuan and Hanjiang Rivers. T. Li et al. (1991) divided the deposit into five layers as follows:

1. Surface soil, 0.2 m thick
2. Brownish red clay, 0.3–1.5 m thick
3. Light brown sandy clay with calcified concretions within which the fossils are wrapped, 0.3–0.5 m thick
4. Light yellow sandy clay with small amounts of small concretions; fossils wrapped in slightly cemented sands, 0.5–0.7 m thick
5. Sandy loose clay of a lighter color than the overlying layers, thickness unknown.

The associated fauna includes: *Hyaena* sp., *Tapirus* sp., *Sus* cf. *scrofa, Cervus* sp., and *Bos* sp. T. Li et al. (1991) considered this site to date to the middle part of the Middle Pleistocene based on faunal analysis.

Archaeology

Some stone artifacts were surface collected and some were found in situ. They include a nucleus, flakes, scrapers, points, and choppers. The material of manufacture is not reported.

YUNXI (110°35′ E, 32°58′ N)

In 1977 a field team lead by Xinzhi Wu of the IVPP, Academia Sinica, excavated Bailongdong (White Dragon Cave), situated at the foot of Shenwuling (God Fog Hill), Yunxi County, Hubei Province, in central China. The fossils from this cave were first found by Jiazhen Wang of the Cultural Museum of Yunxi County in 1976. Six teeth, including two right upper P2, a lower left Pl, a right upper M2, a right lower M1 or M2, and a right lower M3, were found in the cave. The associated fauna contains more than 20 species of mammals, including *Rhizomys, Meg-*

antereon, and *Cervus*. The few stone artifacts from the sandy soils are still unstudied. The preliminary faunal correlation suggests a Middle Pleistocene dating.

YIYUAN (118°09′ E, 36°12′ N)

In 1981 a team from the County Library of Yiyuan County in Shandong Province of eastern China made a local survey of cultural relics. They found a fragment of a human cranial bone and other mammalian fossils in a fissure deposit on Qizianshan Hill, near Tumen town. In November, a team organized by the Bureau of Cultural Affairs of Shandong Province found five additional human teeth from the same site (Locality 1) as well as an adjacent area (Locality 3). Further work in 1982 found two more human teeth. Lu et al. (1989) have studied the fossils and geology of Yiyuan. Their work is summarized below.

Human fossils

Cranial fragments

The human remains consist of a broken skullcap (Sh.y. 001), including small fragments of the parietals, frontal, and the occipital bones. These fragments are separated from each other. The cranial fragment is 9 mm thick at bregma and is 13 mm thick near the left asterion. The internal cranial surface is smooth, and the sutures are obliterated here and on the external cranial surface. The preserved posterior segment of the sagittal suture is simple, while the preserved part of the lambdoidal suture is relatively complex. The bone surface is eroded, except for a small portion of a ridgelike structure existing in front and behind the bregma. The temporal line is indistinct.

Sh.y. 002.1 and Sh.y. 002. 2 represent two fragments of the supraorbital part of the frontal bone. Sh.y. 002.1 is the left part of the supraorbital torus and a part of the anterior wall of the temporal fossa. The torus is 13 mm, 12 mm, and 16.5 mm thick at the medial, middle, and lateral regions, respectively. There is a supraorbital notch instead of a foramen. The supraorbital sulcus is distinct and quite similar to *H. erectus* from Zhoukoudian and different from that of Hexian. Lu et al. (1989) suggested that the degree of postorbital constriction is intermediate between the Zhoukoudian and Hexian specimens.

Sh.y. 002.2 is a lateral section of the right supraorbital torus. Lu et al. (1989) suggested that this fragment and Sh.y. 001 belonged to the same individual. They consider Sh.y. 002.1 to represent another individual.

Teeth

Lower right canine (Sh.y. 005). The crown of this canine is heavily worn. Thick lateral ridges appear at both the mesial and distal sides of the lingual surface. The ridge on the distal side is especially thick and combines with a triangular enamel eminence on the distal crown surface. There is a distinct basal eminence on the lingual surface. A longitudinal ridge extends from the basal eminence to the biting edge. There is a shallow groove between the triangular enamel eminence of the distal surface and the labial surface. The terminal part of the robust root turns towards the distal direction.

Upper right first premolar (Sh.y. 003). Wrinkles on the occlusal surface of this well preserved tooth are strongly developed. The buccal cusp is slightly larger and higher than the lingual cusp. The axis of the main ridge of the buccal cusp divides the ridge into an anterior and a posterior triangular slope. This main ridge terminates at the longitudinal sulcus; here it divides into four small ridges.

The rounded lingual cusp is located at the mesial part of the occlusal surface. Near the longitudinal sulcus, the main ridge divides into three small ridges, which connect with the small ridges of the buccal cusp. The ridge along the margin of the crown is thick.

Small pits and fine grooves of different depths, sizes, and shapes appear on the occlusal surface. The pits and grooves are especially distinct on the disto-buccal ridge and the lingual and buccal cusps. Assumedly, these imperfections were caused by rough food.

Enamel extension is greater on the buccal than on the lingual surface. Rudiments of the cingulum exist as a triangular enamel eminence at the cervical part and as small lateral ridges bordering the grooves on the anterior and posterior sides of the triangular eminence.

The lingual surface of the crown is rounded. An enamel ridge parallel to the cervical line disappears on the mesial and distal surfaces. A depression on the basal part of the mesial surface of the crown connects with the longitudinal depression of the root.

At its middle the root divides into buccal and lingual branches. The buccal branch is straight and robust, and has a longitudinal groove on both the mesial and lingual surfaces of this branch.

Upper left first premolar (Sh.y. 004). This tooth is similar to premolar Sh.y. 003 in most features, including its size and degree of attrition. However, Sh.y. 004 has four small ridges extending from both the buccal and lingual cusps.

Upper left second premolar (Sh.y. 071). This tooth is larger than the other upper premolars. The main ridge of the buccal cusp is narrower than the upper first premolar and divides into two small accessory ridges. The lingual cusp is flatter and divides into four small accessory ridges. The triangular eminence is less distinct than that of the first premolars. Both mesial and distal surfaces are flat, while the mesial surface of the first premolar is concave.

Only the cervical part of the lingual branch of the root is preserved. Its buccolingual diameter is 5.8 mm. Judging from the size, color, degree of attrition, mineralization, general morphological characteristics, and the shape of the contact facets on both teeth, these three premolars probably belong to the same individual.

Upper left second premolar (Sh.y. 007). This tooth is so worn that the wrinkles are just barely discernable. The cervical part is less constricted than on the first premolar. According to Lu et al.'s (1989) original description, a slightly longer buccal root combines with a lingual one to form a single root that bifurcates near the tip.

Upper right first molar (Sh.y. 008). The crown of the maxillary right M1 is rhombdoidal in shape because of the protrusion of the paracone towards the mesial and buccal sides. The two mesial cusps are higher than the two distal cusps. The protocone is the largest cusp, followed in size by the metacone, paracone, and the hypocone. The oblique ridge is well developed. Viewed from the buccal side, the paracone is more protruding than is the metacone. A shallow buccal groove between the paracone and the metacone is limited to the lower half of the height of the crown. A rudimentary cingulum appears on the buccal surface of the paracone. The lingual surface is smooth and rounded, and has no grooves. The mesial surface is slightly concave; the distal one is rounded. The preserved lingual branch of the root inclines to the lingual side. Lu et al. (1989) suggested that this tooth belongs with the specimens represented by Sh.y. 003, 004, and 071.

Lower right first or second molar (Sh.y. 072). Lu et al. (1989) suggested that this heavily worn tooth is from a female. The order of size reduction of the cusps is metaconid, protoconid, hypoconid, entoconid, and hypoconulid. Although the occlusal surface is flat, the metaconid and entoconid are slightly higher than the protoconid and hypoconid. The trigonid and talonid are similar in width and length.

The buccal surface is rounded and swollen. The protoconid and hypoconid incline towards the lingual side. An indistinct trace of the cingulum exists as a transverse ridge at the basal part of the crown between the buccal and disto-buccal grooves. There is no groove on the lingual surface.

A large part of the root has been lost. The preserved part shows that the mesial branch of the root seems to incline to the distal side. There is no constriction between the crown and the root.

Lu et al. (1989) suggest that the Yiyuan cranial fragments are morphologically closer to Zhoukoudian than to Hexian. This assessment is based on robustness and morphological features, as well as the supratoral groove of the frontal bone. The Yiyuan teeth are similar to those of Zhoukoudian. However, the fragmentary nature of the Yiyuan specimens makes taxonomic ascription difficult. The remains have also been referred to as an early *H. sapiens*. As noted in Chapter 3, the early *H. sapiens* skull from Dali has thick brow ridges, a thick cranial wall, and is associated with thick jaw deer. Therefore, the possibility of an early *H. sapiens* attribution for the Yiyuan hominid cannot be dismissed. Lu et al.'s (1989) measurements of the Yiyuan teeth are cited in Table 2.30.

Table 2.30

Measurements of the *Homo erectus* teeth from Yiyuan

			Sh.y.005	*Sh.y.003*	*Sh.y.004*	*Sh.y.007*	*Sh.y.071*	*Sh.y.008*	*Sh.y.072*
Crown	Length		(8.2)*	(9.0)	(9.0)	(8.0)	(8.2)	(11.9)	(11.0)
			8.2	8.5	8.6	7.7	7.8	11.5	10.0
	Width		8.7	12.8	12.8	11.3	12.5	14.7	12.7
	Height	Buccal	6.7	8.6	8.5	6.2	8.0	8.1	5.1
		Lingual	6.0	6.8	6.0	4.0	6.0	5.1	5.1
Root	Length		6.7	6.6	—	5.3	—	—	—
	Width		8.8	13.0	—	10.5	—	—	—
	Height	Buccal	—	16.0	—	15.5	—	—	—
		Lingual	—	11.3	—	13.8	—	—	—

Data according to Lu et al. (1989).

Figures in parentheses were measured according to Weidenreich (1937) all others followed the methodology of M. Wolpoff (1971, *Trends in Hominid Dental Evolution.* Cleveland: Case Western Reserve University Press).

Geology

Lu et al. (1989) described the geology of the site. Three localities were excavated on the east slope of Qizianshan Hill; human fossils were found in two of the three mammalian fossil localities. Hominid specimens Sh.y. 001, 002, 005, and 007 are from Locality 1; Sh.y. 003, 004, 008, 071, and 072 are from Locality 3. The geological strata of Locality 1 from the surface downward include:

1. Loose brown silty clay, 0.5 to 0.6 m thick.
2. Cemented brownish red clay with a few concretions, 0.55 to 0.85 m thick.
3. Hard brownish-red silty clay with human bones and teeth, and other mammalian fossils, 0.15 to 0.40 m thick.
4. Brownish-red silty clay with limestone breccia, covered by a calcified crust of 10 cm thickness, with a few mammalian fossils, 0.30 to 0.55 m thick.
5. Coarse sands of different colors with a few mammalian fossils and gravels intercalated with a thick calcified crust. Exposed thickenss of 0.9 to 1.3 m.

The geological strata of Locality 3 from top to bottom include:

1. Loose brown silty clay, 0.15–0.5 m thick
2. Hard brownish-yellow clay, 0.35–0.45 m thick.
3. Cemented dark brown sandy clay with a few concretions and breccia containing a large amount of mammalian fossils, including five human teeth, 1.9 m thick.
4. Brown clay, 1 m thick.
5. Very hard brownish yellow sandy clay with concretions, exposed thickness is 1.2 m.

Locality 2 is 16 m south of Locality 1, while Locality 3 is 40 m south of Locality 2. Most of the mammalian fossils found in the three localities were embedded in the reddish-brown (Localities 1 and 2) or brown (Locality 3) sandy clay layer. Lu and his colleagues concluded that, although the color of the fossiliferous beds of these localities are slightly different, they are of the same age.

Most of the associated mammalian fossils are represented by isolated teeth. A few mandibles and many fragmentary limb bones were found. According to Lu et al. (1989) the fauna include:

Primates
Homo erectus
Macaca robustus
Rodentia
Trogontherium sp.
Carnivora
Ursus arctos
Ursus thibetanus

Hyaena sp.
Canis variabilis
Panthera tigris
Perissodactyla
Dicerorhinus mercki
Equus sanmeniensis
Artiodactyla
Sus lydekkeri
Megaloceros pachyosteus
Cervus sp.

The most abundant species unearthed from these sites, The Bovinae *Ursus arctors, Equus sanmeniensis, Sus lydekkeri* and *Megaloceros pachyosteus,* indicate a Middle Pleistocene date, and a warm and wet climate with a forest-grassland growth.

Archaeology

No artifacts were recovered.

NANZHAO (112°41′ E, 33°28′ N)

In September 1978, a human right lower second premolar (PA 684) was collected by Yun Liu and his colleagues at Xinhua Shan Hill (Apricot Flower Hill), about 3.5 km northwest of a town named Yunyang, in Nanzhao County, Henan Province, in central China. Qiu et al. (1982) described the site and the fossils.

Human fossil

The slightly worn cusp of this well preserved tooth indicates that it came from a young individual. The length, width, and height of the crown are 7.8 mm, 9.7 mm, and 8.0 mm, respectively. The tip of the root is broken. The length, width, and height of the preserved part of the root are 6.2 mm, 8.3 mm, and 14.6 mm, respectively.

Geology

In October, 1978, an IVPP field team and the local institutions for managing cultural relics started work in this region. They found mammalian fossils in a layer formed by a brownish yellow sandy clay, limestone breccia, and pebbles. The geological strata yielding the fossils include from the surface downward:

1. Disturbed earth, 0–1 m thick.
2. Dark brown sandy clay, with limestone breccia, 0–0.8 m thick.

3. Brown sandy clay, 0.5 m thick.
4. Yellow gravels in which the granules become coarser from top to bottom, 0.1–1 m thick.
5. Brownish-yellow sandy clay with limestone breccia and pebbles, intercalcated with a thin layer of greyish green sandy clay, rich in fossils. Exposed thickness is 1.8 m.

The fauna found at this site include (Qiu et al., 1982):

Primates
- *Homo erectus*

Carnivora
- *Hyaena sinensis*
- *Felis tigris*
- *Felis* sp.
- Machairodontinae gen. et sp. indet.
- *Arctonyx* sp.
- *Canis lupus variabilis*
- *Cuon alpinus*
- *Ursus arctos*

Proboscidea
- *Stegodon* sp.

Perissodactyla
- *Equus sanmeniensis*
- *Tapirus augustus*
- *Rhinoceros sinensis*

Artiodactyla
- *Sus* sp.
- *Megaloceros pachyosteus*
- *Cervus* sp. A
- *Cervus* sp. B
- Caprinae gen. et sp. indet.
- *Bibos* sp.

Rodentia
- Castoidae gen. et sp. indet.
- Sciuridae gen. et sp. indet.
- *Hystrix subcristata*

Based on faunal assemblages, Qiu et al. (1982) suggested that this site is closer in age to Zhoukoudian than to Gongwangling. The presence of Castoridae (living near water), horses (living in the grassland), hyaena (dwelling in the grassland and the hills), tiger, bear, jackal, giant tapir, and wild pig (a forest form) indicate a complex landscape. The existence of giant tapir, elephant, and hyaena may suggest a warm climate.

Archaeology

No artifacts were recovered.

JIANSHI (110°04′ E, 30°38′ N)

In the spring of 1968, a field team led by Chunhua Xu from the IVPP found a hominid lower molar in a drug store in Badong County, Hubei Province in central China. In 1970 they found *Gigantopithecus* teeth and other mammalian fossils during excavations in Longgudong Cave near Gaopin, Jianshi (see Chapter 6). From the same stratum yielding the *Gigantopithecus* teeth, they found three big hominid lower molars.

Human fossils

This site and the fossils were studied by Xu et al., who published a paper signed as Jian Gao in 1975. Their measurements of the crowns of the three hominid teeth from Jianshi and that from Badong appear in Table 2.31.

Teeth

Lower right first molar (PA 507). Attrition on this tooth that was collected from the drug store (herbal shop) at Badong suggests that it belongs to a young individual. The crown is quadrangular; the distal surface is slightly rounded, and the other surfaces are flatter. Two buccal grooves occupy half of the height of the crown. The mesial buccal groove is vertical and the distal one is curved. On the upper part of the buccal surface of the protoconid is a S-shaped groove, which extends obliquely from the anterior part of the protoconid to the middle part of the buccal surface of the protoconid. This structure also appears in *H. erectus pekinensis* (nos. 52, 98, 99) and has been called a *protoconid cingulum* by Weidenreich (1937). The lingual surface is more or less vertical, while the buccal surface inclines towards the lingual side. The lingual vertical groove occupies the upper two thirds of the crown. The mesial cusps are higher than the distal ones. The metaconid is the highest of all the cusps. A longitudinal deep groove on the middle part of the occlusal surface separates the buccal and lingual cusps into almost equal parts.

The protoconid and metaconid are almost equal in size to the metaconid, and are followed by the entoconid, hypoconid, and the hypoconulid in decreasing order of size. The hypoconulid looks like a thick ridge. The furrows of the occlusal surface

Table 2.31

Measurements of the molar crowns from Jianshi and Badong

	Jianshi			*Badong*
	PA 504	*PA 502*	*PA 503*	*PA 507*
Length	15.3	14.6	15.2	14.4
Breadth	13.6	14.1	14.0	12.2
Breadth(anterior)	13.6	14.0	14.0	11.8
Breadth(posterior)	13.3	14.4	13.9	12.2
Height	9.4	8.7	9.2	(8.6)

According to Gao (1975)

are arranged as a "plus pattern." A small cusp corresponding to the tuberculum sextum, and appearing in the middle of the distal margin of the crown, deviates slightly to the lingual part of the distal margin of the tooth. The longitudinal groove of the occlusal surface is slightly wavy. The anterior fovea communicates with the central fossa, which is surrounded by the five main cusps. The posterior part of the longitudinal sulcus bifurcates into two branches, one situated between the hypoconid and hypoconulid continues with the distal buccal groove. Another branch cuts through the cusp ridge between the entoconid and hypoconulid, and connects with the posterior fovea. Each main cusp has a thick main ridge, with one or two accessory ridges along its side. The ridges extend from the tip of the cusp to the furrow between the main cusps.

Lower second (?) molars (PA 502, 503, 504). All the second molars were obtained during the excavation of Longgudong from the same stratum yielding the *Gigantopithecus* teeth. PA 504 and PA 503 are from the right side; PA 502 is from the left side. Viewed superiorly, PA 504 is elliptical, while PA 502 and 503 are nearly quadrangular. The grooves on the buccal and lingual surfaces of PA 502 and PA 503 are shallow and short. A trace of a cingulum exists at the basal part of the buccal surface. There is a well-developed protoconid cingulum on PA 503 and PA 504. The protoconid cingulum of PA 502 is less developed.

The mesial cusps are slightly higher than the distal ones. The metaconid is the highest cusp and is similar to the protoconid in size. The hypoconid is slightly larger than the entoconid. The very small hypoconulids of PA 502 and PA 503 exist as a low ridge that diverges from the distal ridge. A tuberculum sextum on PA 504 occupies part of the posterior fovea. It deviates slightly to the lingual side from the median line of the occlusal surface of the crown. The tuberculum sextum on PA 502 and PA 503 is very small, and on PA 503 it is located on the median line of the crown. There is a small lingual accessory cusp located on the lingual side between the metaconid and the entoconid on all three molars. There is a small tubercle on the marginal ridge between the entoconid and the tuberculum sextum of PA 504 and PA 503. The two small tubercles at this position on PA 502 are slightly smaller than the hypoconulid and tuberculum sextum.

The main furrows on the occlusal surfaces of PA 502 and PA 503 form a reverse T pattern instead of a Y or plus pattern. The triangular ridge separating the anterior fovea, and the central fossa is bisected by the longitudinal sulcus of the occlusal surface. The distal segment of the longitudinal sulcus bifurcates into two accessory grooves after passing between the entoconid and the hypoconid. One accessory groove runs backward and medially, and terminates at the posterior fovea, while another passes between the hypoconid and the hypoconulid and then continues with the distal buccal groove. The posterior fovea is deeper and wider than the anterior fovea. Small ridges on the slope of the low cusps of specimens PA 502 and PA 503 are weakly defined.

The mesial and distal branches of the root of PA 502 are flattened. The cervical part of the root is not distinctly constricted. A vertical and shallow trough divides the anterior branch of the root into buccal and lingual segments. A broader vertical trough exists on the distal surface of the posterior branch of the root.

Jian Gao (1975) attributed these teeth to *Australopithecus;* Zhang (1984) disagrees and favors attribution to *H. erectus* because of dental similarities. (Note: The name Jian Gao is a pseudonym, the use of which occurred often during the Cultural Revolution. The authors were Chunhua Xu, Linghong Wang, and Kangin Hang).

Geology

Longgudong cave is southeast of Jintang village and about 1 km southwest of the town of Gaoping. Its entrance is about 85 m above the surface of the Longdong River. Hsu (Xu) et al. (1974) recorded three sections of deposits in different parts of the cave. The section on the east wall, from the surface down, includes:

1. Brownish-yellow loam, 40 cm thick
2. Calcium carbonate cemented layer, 12 cm thick
3. Brownish-yellow loam, 25 cm thick
4. Calcium carbonate cemented layer, 2 cm thick
5. Purplish-red clay, thickness uncertain

The section on the northwest wall, from the top down includes:

1. Brownish-yellow loam, 80 cm thick
2. Calcium carbonate cemented layer, 15 cm thick
3. Purplish-red clay, 35 cm thick
4. Calcium carbonate cemented yellow deposit, 20 cm thick
5. Brownish-yellow sandy clay (the first and the chief fossil bed), 110 cm thick

The section located approximately 7 m from the cave's entrance, from the surface down includes:

1. Brownish-yellow sandy clay (first fossil bed), thickness not reported in this area
2. Calcium carbonate cemented yellow deposit, 8 cm thick
3. Brownish-sandy clay, 16 cm thick
4. Calcium carbonate cemented yellow deposit, 4 cm thick
5. Brownish-yellow sandy clay, 2 cm thick
6. Calcium carbonate cemented yellow deposit, 6 cm thick
7. Yellow sandy clay, 23 cm thick
8. Calcium carbonate cemented yellow deposit, 28 cm thick
9. Brownish-yellow sandy clay (second fossil bed), 35 cm thick
10. Calcified yellow sands (third fossil bed), 30 cm thick
11. Calcified breccia, 5 cm thick
12. Light reddish-brown silty clay, 23 cm thick

The fossils unearthed from the cave were described by Hsu et al. (1974). The fauna suggests an early Pleistocene age and a climate warmer than that which characterizes the area today. Fauna from the first fossil bed include:

Mammals

Primates

Gigantopithecus blacki

Rodentia

Hystrix sp.

Carnivora
Cuon javanicus antiquus
Ursus sp.
Ailuropoda cf. *melanoleuca fovealis*
Arctonix sp.
Hyaena licenti
Felis sp
Machairodontinae gen. et sp. indet.
Proboscidea
Gomphotheriidae gen. et sp. indet.
Stegodon sp.
Perissodactyla
Equus yunnanensis
Tapirus sinensis
Rhinoceros sp.
Artiodactyla
Sus sp. A
Sus sp. B
Cervus sp
Muntiacus sp.
Bovinae gen. et sp. indet.
Ovinae gen. et sp. indet.

Reptile
Chelonia

Fauna from the second fossil bed include:

Primates
Gigantopithecus blacki
Carnivora
Ursus sp.
Hyaena sp.
Proboscidea
Stegodon sp.
Perissodactyla
Equus yunnanensis
Tapirus sinensis
Rhinoceros sp.
Artiodactyla
Cervus sp.
Bovinae gen. et sp. indet.

Fossils from the third bed include:

Primates
  Homo erectus
Rodentia
  Mus sp.
Carnivora
  Ailuropoda cf. *melanoleuca fovealis*
Proboscidea
  Stegodon sp.
Perissodactyla
  Equus yunnanensis
  Tapirus sinensis
  Rhinoceros sp.
Artiodactyla
  Cervus sp.
  Bovinae gen. et sp. indet.

LUONAN (110°08′ E, 34°06′ N)

Local farmers reported that they collected the human tooth from a small limestone cave situated near Donghe Village, Luonan County, Shaanxi Province in northwestern China.

Human fossil

Tooth

The tooth, an upper right first molar (Cat. no. 77 Ln001), was described by Xue (1987). The crown was well preserved and only slightly worn. The mesiodistal diameter is 12.4 mm. The bucco-lingual diameters are 13.4 mm and 12.4 mm at the mesial and distal parts, respectively. The protocone is the largest cusp, while the paracone is slightly larger than the metacone. Both the paracone and the metacone are slightly higher than the protocone. The transverse groove between the paracone and the metacone extends from the occlusal surface towards the cervix of the tooth. A small ridge connects the protocone and paracone at the anterior part of the occlusal surface. A low ridge extends from the peak of the metacone to the protocone. The hypocone is separated from the protocone and metacone by an oblique groove, which extends from the occlusal surface towards the cervix. Between 1 and 3 small ridges are distributed on the opposite surfaces of various cusps, but only those ridges on the protocone and metacone are distinct. On the central part of the sloping surface of the paracone is an elliptical deep hole, which is a dark brown color. A small elliptical pit appears on the apical part of the metacone. No cingulum was reported.

Based on its rather large size, Xue (1987) attributed the human molar to *H. erectus*. However, as we will see in the next chapter, the upper molar from Xujiayo (an early *H. sapiens*) is even longer than that from Luonan. The absence of a cingulum on the Luonan molar also challenges its attribution to *H. erectus*.

Geology

Local farmers also provided two nonhuman teeth: a panda's upper first molar and a lower molar from a tapir. All the teeth were supposedly collected from the same small cave, a fissure less than 1 m in width. Xue (1987) reported that the three teeth were cemented in a pale yellow conglutinated sandy clay.

Archaeology

No artifacts were recovered.

WUSHAN (109°36′ E, 30°48′ N)

According to Gu and Fang (1991a), two human fossils were unearthed from this site situated in Wushan County in the eastern part of Sichuan Province in central China. The specimens are a fragment of the mandibular body and an upper median incisor of *H. erectus*. Only the mandibular body between the anterior border of P2 and M2 is preserved. Two teeth, P2 and M1, are intact.

The height of the mandibular body at the level of M1 exceeds 21 mm. There is no mandibular torus. Both teeth are heavily worn and their crowns are rather low. The relatively large premolar is nearly spherical in horizontal section. The lingual cusp is slightly higher than the buccal cusp, on which the dentine was exposed. A sagittal sulcus lies between these two cusps. The talonid on the premolar is wide and long, and is relatively larger than in modern humans. The talonid has no cusp. The cervical part of the premolar is constricted. The exposed part of the premolar shows that the root has two branches. The root is not as short or stout as that found in "Peking Man." The metaconid is the highest of the five cusps. Molar cusp height decreases from the entoconid, to the protoconid, to the hypoconid. The breadth of the trigonid and talonid are similar. There are two molar roots.

The shovel-shaped upper incisor is a permanent erupting incisor with mammilary processes on the cutting edge. The mesial and distal marginal ridges run upward to meet a small basal tubercle. This makes the central part of the lingual surface concave. A central ridge leaves the small basal tubercle and bifurcates at the middle of the crown. Table 2.32 provides the measurements for the Wushan hominid teeth.

Geology

P. Huang et al. (1991) reported that the deposits in Wushan (Damiao) are composed of two deposition units. The lower unit has three parts; human and other mammalian fossils were unearthed from the middle part, which yielded the following:

Primates
- *Homo erectus*
- *Gigantopithecus blacki*

Rodentia

Table 2.32
Measurements of tooth crowns from Wushan

	Incisor	*Premolar*	*Molar*
Mesiodistal diameter	8.12	7.37	11.10
Labio(bucco)lingual diameter	7.02	9.12	10.10

Pteromys huananensis
Belomys parapearsoni
Chuanocricetus lii
Mimomys peii
Hystrix magna
Hapalomys eurycidens
Niviventer confucianus
Wushanomys brachyodus
Clethrionomys sebaldi
Sciurotamias teilhardi

Carnivora
Ailuropoda microta
Megaviverra pleistocaenica
Pachyrcrocuta licenti
Homotherium palanderi

Proboscidea
Sinomastodon yangziensis

Perissodactyla
Equus aff. *yunnanensis*
Tapirus sanyuanensis
Nestoritherium cf. *sinensis*

Artiodactyla
Cervavitus ultimus
Metacerculus capreolinus
Dicoryphoerus ultimus

The faunal assemblage suggests an early Pleistocene age. Based on their palynological analysis, Yingshuo Chen and Neiqiu Du (1991) suggested that during occupation by *H. erectus* and *Gigantopithecus*, the evergreens in the area increased and grass decreased. The palynological assemblage indicates a warm and wet environment. According to Chun Liu et al. (1991) paleomagnetic dates are from 2.01 to 2.04 mya. If correct, these are among the oldest dates obtained for hominids in China.

Archaeology

Two artifacts, a chopping tool and an extremely pitted stone hammer, were recovered.

XICHUAN (110°20′ E, 32°59′ N)

All the human teeth from Xichuan come from traditional Chinese medicine drug stores. In the summer of 1973, Zhanxiang Qiu of the IVPP found an upper premolar in a drug store in Nanyang Municipality, Henan Province, in central China. The fossil tooth was supposedly bought in Xichuan County. In September 1973, a team led by Rukang Wu of the IVPP surveyed drug stores in Nanyang and Xixia counties and found 12 human teeth, all of which supposedly came from Xichuan County.

According to Rukang Wu and Xinzhi Wu (1982), most of the teeth morphologically resembled those from *H. erectus*. Crown measurements of the human teeth from Xichuan are given in Table 2.33, and root measurements are given in Table 2.34.

Teeth

Left lower canine (PA 523). The buccal surface bulges with a horizontal riblike band. A triangular enamel prominence appears on the basal part of the distal surface and a shallow groove is situated between it and the buccal surface. An enamel prominence also exists on the mesial surface. There is no demarcation between this prominence and the buccal surface. The cingulum is only present on the distal side. A ridge on the central part of the lingual surface is bordered by mesial and distal ridges. The middle part of the cutting edge is horizontal without a pointed tip; the cervical part is constricted. There is no longitudinal groove on the surface of the root. The morphology and size of this tooth is close to late *H. erectus* or early *H. sapiens*.

Right upper second premolar (PA 525). The lingual cusp is much larger than the buccal cusp. There is a large talonid. Enamel extends further up on the lingual surface than on the buccal surface. There is an obvious cingulum at the base of the lingual surface. The robust root has shallow grooves on the anterior and posterior surfaces. The root abruptly narrows terminally to form a single tip. Based on the primitive features exhibited in the cingulum, talonid, and the root, this premolar is close to that characterizing early *H. sapiens* or late *H. erectus*.

Left upper second premolar (PA 524). The mesial and distal borders of the occlusal surface are more or less parallel lines. The occlusal surface is well worn, but the lingual cusp is evidently slightly smaller and situated closer to the mesial side than the buccal cusp. There is no longitudinal groove on the buccal surface. The lingual surface is lower than the buccal one. The indistinct cingulum is only discernible on the distal part of the buccal surface. The cervical part of the tooth is not very constricted. The mesial surface of the root is divided into two equal parts by a deep vertical groove. The morphology of the tooth more closely resembles that of *H. erectus* than *Australopithecus*.

Lower first premolar (PA 526 [left] and PA 527 [right]). These teeth are similar in their degree of attrition and general morphology, so they probably belong to the same individual. The crown's breadth is close to that of *H. erectus* and exceeds that of the Neandertals. The length of the crown is intermediate between *H. erectus* and the Neandertals.

On the occlusal surface, the longest diameter of the crown is from the mesio-buccal to the disto-lingual corner. The buccal cusp was more heavily worn than the

Table 2.33

Measurements of tooth crowns from Xichuan

		Length	*Breadth*
Lower canine	(PA 523)	8.3	8.4
Upper second premolar	(PA 524)	8.3	12.4
	(PA 525)	7.8	9.9
Lower first premolar	(PA 526)	8.1	9.8
	(PA 527)	7.9	9.7
Lower second premolar	(PA 528)	9.1	11.2
Upper first or second molar	(PA 529)	12.7	14.8
	(PA 530)	12.6	14.3
Lower first or second molar	(PA 531)	12.0	12.1
	(PA 532)	12.8	13.2
	(PA 533)	11.7	11.4
Lower third molar	(PA 535)	10.9	11.0
Deciduous lower second molar	(PA 534)	12.5	10.5

According to Wu and Wu (1982).

Table 2.34

Measurements of the roots of human teeth from Xichuan

		Length	*Breadth*	*Height*
Upper second premolar	(PA 524)	6.0	12.0	—
	(PA 525)	6.1	9.2	17.2
Lower first premolar	(PA 526)	6.4	9.5	15+
	(PA 527)	6.3	9.2	15+
Lower second molar	(PA 533)	10.8	10.7	18.0
Lower third molar	(PA 535)	9.7	10.3	16.3

According to Wu and Wu (1982)

lingual cusp. The anterior fovea is larger than the posterior fovea. A weak cingulum on the basal part of the buccal surface extends upward to the occlusal surface.

The root is straight and robust, and the middle and upper parts are almost equally thick. Longitudinal grooves appear on both the mesial and distal surfaces, but the groove on the mesial surface is deeper. The grooves divide the root into a larger buccal and a shallower lingual part. Judging from their general morphology, these teeth probably belonged to *H. erectus*.

Lower left second premolar (PA 528). Although the occlusal surface was worn smooth and flat, there is no dentine exposure. The mesio-buccal corner is especially protruding; therefore, the maximum crown diameter is from the mesio-buccal to the disto-lingual corner. The buccal are larger than the lingual cusps. Fine wrinkles exist in the posterior fovea, which is much larger than its anterior counterpart. The buccal and lingual cusps are situated almost at the middle of the occlusal surface. The buccal is larger than the lingual surface. The cingulum is just discernible. A groove extends forward and upward on the buccal surface. According to the size and morphology, this tooth probably belongs with *H. erectus*

Upper right first or second molar (PA 529). The dentine of this tooth is slightly exposed on the protocone and hypocone. The main grooves are very distinct and accessory ridges are very rare. While a vertical groove on the buccal surface extends from the cervical line to the occlusal surface, there is no groove on the lingual surface. A cingulum is discernible mesially on the buccal surface. The mesial surface is slightly concave; the distal surface is slightly convex. The root has three branches, but most of the root is missing. This extraordinarily large specimen is probably a *H. erectus* tooth.

Upper right first or second molar (PA 530). This tooth is most heavily worn on the antero-lateral part so the crown is very low. A vertical groove extends the whole height of the buccal surface. No sulcus exists on the lingual surface and no cingulum is visible. The remaining part of the root exhibits well developed taurodontism. This is probably a *H. erectus* specimen.

Lower first or second molars (PA 531 [right], PA 532 [left]). PA 531 has five cusps, fine wrinkles are sparse, and the anterior fovea is T-shaped. The larger talonid and smaller trigonid are of equal breadth. Neither the buccal nor the lingual surfaces are bulging. The lower part of the crown was damaged.

PA 532 has six cusps. A rather large accessory entoconid is situated between the metaconid and entoconid. The hypoconulid is well developed. The occlusal surface is nearly spherical, and the wrinkles are sparse. The trigonid is narrower than the talonid. There is no cingulum. These two teeth are close to those of *H. erectus* in size, while close to early *H. sapiens* in morphology.

Lower left second molar (PA 533). The central part of the occlusal surface is a fossa in which fine wrinkles divide the five cusps. The metaconid is the largest of the five cusps. The anterior fovea is Y-shaped, and there is no posterior fovea. The length and breadth of the trigonid is more or less similar to that of the talonid. There is a cingulum on the buccal and distal surfaces of the crown. The buccal surface is slightly bulging; the lingual surface is flatter. A groove between the protoconid and hypoconid terminates at the cingulum.

The root is robust and its anterior and posterior branches unite at the buccal surface. There is a shallow groove on the buccal surface and a deep one on the lingual surface. There are two grooves on the mesial surface but only one on the distal surface. Similarities between the root of this tooth and that of specimen PA 70 of *H. erectus pekinensis* suggest that PA 533 is a second molar.

Lower left third molar (PA 535). There are many fine wrinkles and cross-shaped grooves on the occlusal surface of the crown of PA 535. The trigonid is much broader and longer than the talonid. There are only four cusps without a hypoconulid. Both the buccal and lingual surfaces of the crown are bulging. The cingulum extends around the base of the crown.

The posterior branch of the robust root is more slender than the anterior one. The two branches unite at the buccal side. There are grooves on the lingual and mesial surfaces.

Lower right second deciduous molar (PA 534). Two vertical grooves dividing the three buccal cusps exist on the buccal surface of the crown. A groove on the lingual surface divides the metaconid and entoconid. The trigonid is higher than the talonid, especially on the lingual side. A very small paraconid has a fine ridge extending between it and the protoconid. A thick and high ridge between the protoconid and the metaconid is perpendicular to the long axis of the occlusal surface and divides the trigonid from the talonid. The former is smaller than the latter in length and breadth. This molar probably belongs to *H. erectus*.

Pulp cavity. Specimens PA 529, PA 533, PA 525, PA 526, and PA 527 all have big pulp cavities. The degree of mineralization and the amount of matrix is more or less similar, so these specimens are possibly collected from the same site with the same age. Several fossil sites of Xichuan County have been excavated, but no human material has been found. In the southern neighboring area of Yunxian County, two sites yielded human fossils, so perhaps the Xichuan teeth were transported from there.

SUMMARY

Although the earliest discovered and most famous *H. erectus* specimens in China come from Zhoukoudian, the oldest may come from Yuanmou in southern China. This site may be more than 1 million years old. However, because of the dating problems at Yuanmou, the Lantian specimen may prove to the oldest Chinese *H. erectus*. Other *H. erectus* materials date from perhaps 800,000 years ago to as recently as 150,000 years ago.

There has been a long-recognized general morphological similarity between Chinese *H. erectus* and subsequent *H. sapiens* in China. Such is the magnitude of the similarity that it questions the Out-of-Africa scenario. The case for regional evolution can probably best be made based on the Chinese fossils. Sohn and Wolpoff (1993) suggested that Chinese *H. erectus* may be ancestral to the Neandertals and may be the base of a pan-Asian dispersal.

The importance of achieving a clear understanding of the evolutionary relationships of the Chinese *H. erectus* cannot be overestimated. We hope that enough morphological data are given in this chapter for other researchers to formulate these evolutionary relationships.

3

Archaic *Homo Sapiens*

Early *Homo sapiens* in China may have descended directly from earlier *Homo erectus* populations, and many specimens are morphologically intermediate between *H. erectus* and *H. sapiens*. In fact, until recently some of the materials discussed here were referred to *H. erectus*. The sample discussed in this chapter is distributed the north, south, and central parts of China. Although the remains date from perhaps 300,000 to about 100,000 years ago, most are closer to a 150,000-year-old date.

DALI (109°40′ E, 34°52′ N)

In 1978, Shuntang Liu, a geologist from the Geological Bureau of Shaanxi Province, found an almost complete human skull at a loess terrace named Tianshuigou near Jiefang Village, Dali County, Shaanxi Province in northwestern China (Fig. 3.1).

Human fossils

The following description is summarized from X. Wu (1981). The Dali skull is rather complete, except that the posterior superior part of the skullcap, the left zygomatic arch, and the pterygoid process are missing. The lower part of the face was deformed by the upward depression of the alveolar process. This depression was caused by pressure exerted by the earth deposits.

The skull is rather robust and probably from a male slightly less than 30 years of age (Fig. 3.2). Its supraorbital torus is thick, and the temporal lines and the supramastoid ridges are well developed. The external cranial sutures are intact; however, the suture between the greater wing of the sphenoid and the temporal squama is obliterated.

In most measurements and indices the Dali skull falls within the range of variation of western early *H. sapiens* and is intermediate between *H. erectus* and late *H. sapiens*. Some measurements and indices are within the range of variation of the Zhoukoudian *H. erectus* skulls (Table 3.1).

In its nonmetric features, the Dali skull is more similar to early *H. sapiens* than to *H. erectus*. The broadest part of the skull vault is at the postero-superior margin of the temporal squama rather than near the cranial base. In a rear view, the upper part of the parietal is more horizontal than is true with *H. erectus*, while the lower part is more vertical than in *H. erectus*. Unlike the case with Neandertals, the posterior contour of the Dali skull is not spherical. The distance between the parietal tuberosities is close to the maximum breadth of the skull. The postorbital constriction is not as narrow as that in *H. erectus*.

FIG. 3.1 The early *Homo sapiens* site of Dali. (Courtesy of IVPP.)

A sagittal prominence on the frontal squama broadens at the middle and attenuates upward and downward. The upper end is slightly higher than the center of the frontal squama. The lower end is about 2 cm above the glabella. A weak eminence, reminiscent of the cruciate eminence, exists at the bregmatic region. Between the left parietal and occipital bones is a small bone, perhaps the remaining part of an Inca bone, the lower border of which is horizontal. The adjoining part of the right parietal and occipital bones is missing.

The left parietal is rather complete, except for a small supero-posterior part. Only the anterior portion of the right parietal is preserved. Both parietals are slightly elevated near the sagittal border. There is a shallow groove along the sagittal suture between these elevations. The parietal tuberosity is distinct and located slightly below the center of the parietal bone. Both the superior and inferior temporal lines are distinct and rather broad. The inferior temporal line passes the parietal tuberosity where the distance between both temporal lines is 1 cm. Both temporal lines are separated on the parietal bone. The inferior temporal line terminates at the angular torus, close to the asterion, and is thinner than that in *H. erectus pekinensis*. This torus in the Dali skull is circular and about 2 cm in diameter.

The superior border of the temporal squama is formed by a short anterior and a longer posterior part. The former is more horizontal. A small quadralateral process of the supero-anterior part of the squama inserts between the parietal and sphenoid bones, and connects with the frontal bone. The pteryon of both sides is of the I-type. The supramastoid ridge is robust and spindle-shaped and is almost in line with the zygomatic arch. Although it has a weak inclination in an upward direction, it is less inclined than that of the Neandertals. The small mastoid process points downward. The orifice of the auditory meatus is elliptically shaped with a vertically oriented longitudinal axis. A small, 4 cm in diameter, moundlike exostosis appears on the wall of the auditory meatus.

The upper border of the zygomatic process is lower than the Frankfurt Plane.

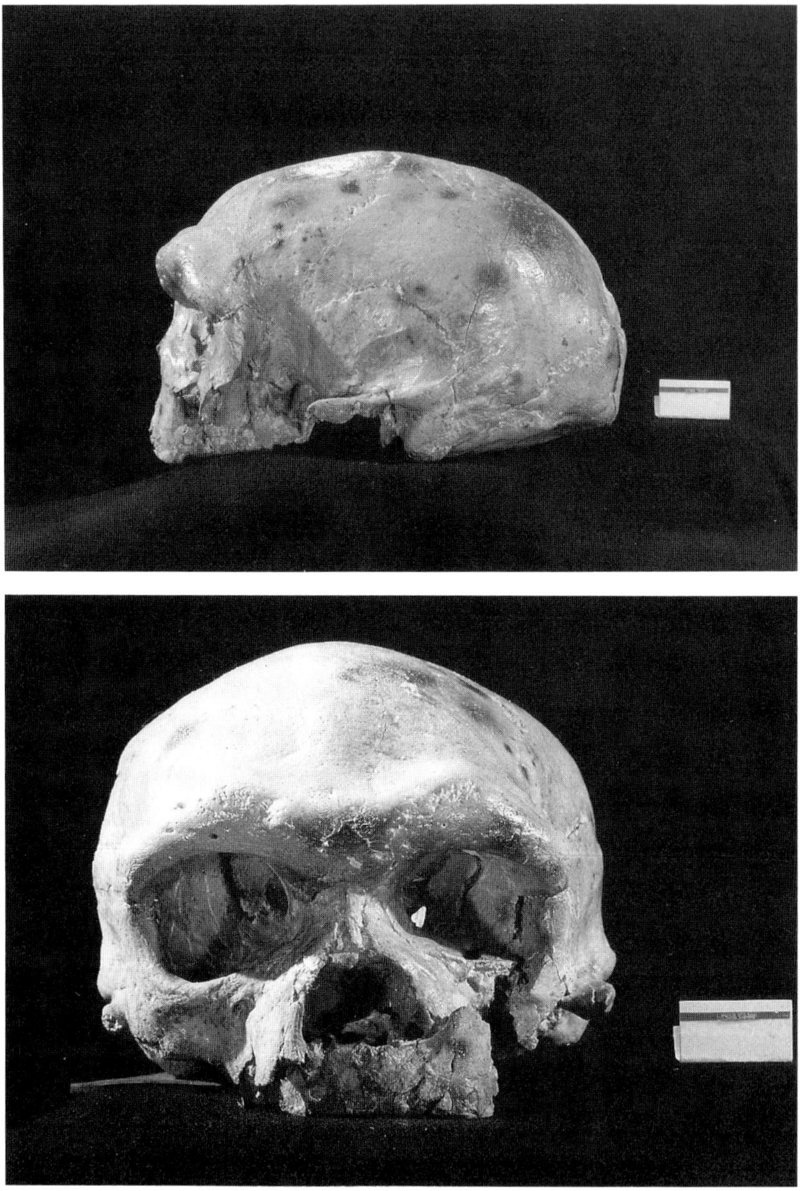

FIG. 3.2 Early *Homo sapiens* skull from Dali. (a) Lateral view. (b) Anterior view (Courtesy of IVPP.)

The longitudinal axis of the zygomatic arch forms a 10° angle with the Frankfurt Plane. The elongation of this axis passes through the upper half of the auditory orifice. The right zygomatic arch is thin, with its thinnest part 7 mm high and 4.3 mm wide. The left arch is similar judging from its remaining part. The postglenoid process is well developed. The concavity and thickness of the tympanic plate is intermediate between *H. erectus pekinensis* and modern humans. The long axis of the petrous part forms a 40° angle with the sagittal plane.

Table 3.1

Measurements of the Dali Skull

Cranial length(g-op)	207
Cranial width(eu-eu)	149
Cranial height(ba-b)	118
Auricular bregmatic height	102.5
Cranial sagittal arc(n-o arc)	379
Cranial transverse arc(po-b-po arc)	299
Distance between parietal tuberosities	136
Orbital height(r)	34
Orbital width(mf-ek, r)	45
Height of zygomatic	52.6
Minimum distance between postorbital surfaces of both sides of frontal bone	106.4
Distance between most lateral points of both sides of the brow ridges	125
Height of temporal squama	46.5
Length of temporal squama	72
Distance between inion and endinion	11
Nasomalar angle(fmo-n-fmo)	143
Inclination angle of frontal bone I(b-n-i)	54°
Inclination angle of frontal bone II(b-n-o)	49°
Inclination angle of frontal squama I (b-g-i)	50°
Inclination angle of frontal squama II(b-g-o)	45°
Inclination angle of lower frontal bone I(m-g-i)	72°
Inclination angle of lower frontal bone II(m-g-o)	67°
Thickness of cranial bones:	
center of frontal squama	9.0
center of parietal tuberosity	11.2
cerebellar fossa	3.9
center of temporal squama	7.0
Cranial capacity	1120
Cranial index	72.0
Length-height index	57.0
Width-height index	79.2
Transverse curvature	47.5
Index of postorbital constriction	85.1
Length-height index of temporal squama	64.6
Ratio of dimensions of cerebral fossa to cerebellar fossa	3:2

1. Cited from Wu (1989 in Wu et al.,1989)

No infratemporal ridge exists on the greater wing of the sphenoid, but the demarcation between the temporal and infratemporal surface is distinct. The occipital torus is thick in the middle but thins toward the lateral ends. There is a shallow supratoral sulcus. The glabellar region is receding. The infraglabellar notch is indented in lateral view.

The frontonasal and frontomaxillary sutures form an almost horizontal curve. The upper part of the nasal bone is much narrower than its lower part. The profile angle of the nasal bone is big, and the nasal saddle is flat. There is a median sagittal longitudinal narrow ridge on the nasal bone.

Both orbits are quadrangular. The medial part of the orbit is somewhat rounded, and the lateral part is slightly more angular. The right orbit is slightly higher than the left. The lacrimal depression of the orbit roof is shallower than that in modern humans and deeper than in "Peking Man." The superior orbital margin is evenly curved. There is no supraorbital notch, supraorbital foramen, or supraorbital tubercle. The infero-lateral part of the orbital margin is rounded. The superior orbital fissure is similar to that of "Peking Man." However, the inferior zygomatico-orbital foramen is located on the lateral wall of the orbit, not on the elongating line of the inferior orbital fissure.

The pyriform orifice is pear shaped. The nasal process of the maxilla bulges supero-laterally to the orifice, as is also true for Neandertals. The distance between the infraorbital foramen and the inferior orbital margin is 8 mm. A maxillary sulcus runs downward from the infraorbital foramen. A well-developed canine jugum is located lateral to the lower segment of this sulcus.

The anterior surface of the zygomatic process of the maxilla faces forward and forms a shallow groove with the anterior surface of the maxilla. The lower margin of the zygomatic process and the lateral surface of the maxillary body form a deepened curve, which is slightly weaker than that in "Peking Man," which possesses a very deep malar notch. The juncture of the lower margin of the zygomatic process and the body of maxilla is distant from the alveolar margin. This point is located midway between the alveolar margin and the inferior orbital margin. A forward protruding process at the upper end of the zygomatic process makes a sharp turn at the junction of this process and the zygomatic bone. The sharp turn is especially distinct when viewed inferiorly.

The zygomatic bone is only 52.6 mm high. The antero-lateral surface of the fronto-sphenoidal process of the zygomatic faces more forward than in Neandertals. A weak marginal process and an orbital tubercle appear on the posterior margin and the orbital margins of this process, respectively. There is a moderately developed tuberosity on the lateral surface of the bone.

There is a crista galli on the endocranial surface. A complete pattern of branching of the middle meningeal artery can be seen on the left side. The artery has three main branches. The fronto-parietal branch is slightly thicker than the superior temporal branch and terminates at the bregmatic and obelion regions. The superior temporal branch is the longest and is slightly thicker than the inferior temporal branch, which runs almost parallel to the superior branch. This pattern of branching is somewhat similar to that of the latest Peking skull (no. 5 from Locus H). The impressions of the branches of the meningeal artery on the endocranial surface are clearer than in "Peking Man."

Table 3.1 shows the important measurements and indices of the Dali skull. Many figures are within the range of early *H. sapiens* and intermediate between *H. erectus pekinensis* and modern humans. These measurements are the maximum length, maximum width, height, median sagittal arc, transverse arc, transverse curvature, inclination angle of the frontal bone, post-orbital constriction, distance between the external and internal occipital protuberance, and the ratio of the dimension of the cerebral fossa to that of the cerebellar fossa.

The Dali skull also has many *H. erectus* traits. These include the low cranial vault, thick cranial wall, well-developed brow ridges, small superior orbital fissure, low position of the zygomatic arch, angular turn from the occipital to the nuchal plane of the occipital bone, the supratoral sulcus of the occipital bone, and the maxillary sulcus, among others.

Progressive features of the Dali skull include a low zygomatic bone, thin zygomatic arch, the absence of a supraorbital tubercle, the short distance between the infraorbital foramen and the inferior orbital margin, the zygomatico-orbital foramen located on the lateral orbital wall, and the presence of a crista galli, among others.

Geology

The Dali human skull was unearthed from the third terrace of Lohe River. The deposits consist of 15 layers (X. Wu and You, 1979). They are as follows from the surface downward:

Strata dating to the upper part of the Middle Pleistocene:

13. Grayish-yellow silty clay, 7.6 m thick
12. Light brownish-red paleosol, 3 m thick
11. Grayish-yellow silty clay with small gravels and concretions, 7 m thick
10. Light brownish red paleosol rich with concretions, 2.1 m thick
9. Grayish yellow fine sands and silt, 1.8 m thick
8. Brownish-yellow, orange yellow coarse sands with cross-bedding, rich with molluscs, 1.2 m thick
7. Yellow green clay, 0.5 m thick
6. Grayish-yellow sandy clay rich with molluscs, 2.5 m thick
5. Grayish-yellow silt, fine sands, 6.2 m thick
4. Light purple moderate-sized and coarse sands, with cross bedding, cemented, 0.5–18 m thick
3. Brown grayish, brown gravels, yielding the human skull, stone artifacts, vertebrate fossils, molluscs, and carbon particles, 0.5–1.2 m thick

———unconformity———

Strata dating to the upper part of the Lower Pleistocene:

2. Grayish-green sandy clay, 1.8 m thick
1. Light gray sands with gravel, exposed thickness, 2 m

According to X. Wu and You (1979) the vertebrate fossils associated with the hominid remains are as follows:

Primates
Homo sapiens

Rodentia
- *Myospalax* sp.
- *Castor* sp.

Proboscidea
- *Palaeoloxodon* sp.

Perissodactyla
- *Equus* sp.
- *Rhinocerus* sp.
- *Coelodonta* sp.

Artiodactyla
- *Megaloceros pachyosteus*
- *Megaloceros* sp.
- *Pseudaxis* cf. *grayi*
- *Bubulus* sp.
- *Gazella przewalskyi*
- *Gazella* sp.

Bird
- *Struthio anderssoni*

Fishes
- Cypriniformes
- Siluroidea

Uranium series dates on ox teeth give an age of 209,000 ± 23,000 years ago (T. Chen et al., 1984).

Archaeology

Some small stone artifacts were found with the human fossils. Most of the stone tools are scrapers, others include points, a stone awl, and burins. Most tools were made by hammering, and the manufacture is simple and rough.

JINNIUSHAN (GOLD OX HILL) (122°27′ E, 40°35′ N)

In 1984, human fossils were unearthed from a limestone cave at Jinniushan, located west of Sitian Village, Yun'an Xiaong, 8 km southwest of the city of Yinkou, Liaoning Province (northeastern China).

Human fossils

The human remains include a skull (Fig. 3.3) without a mandible but having maxillary dentition, 6 vertebrae, 2 ribs, 1 left ulna, 1 left inominate, 1 left patella, 9 carpal bones (4 left, 5 right), 2 metacarpals (1 left, 1 right), 7 digits of the hand (4 proximal, 2 middle, 1 distal), 11 tarsals (5 left, 6 right), 2 metatarsals, and 13 digital bones of the foot (6 proximal, 1 middle, 6 distal). All these fossils, except one digital bone of the hand and one vertebra, were unearthed from a small area of

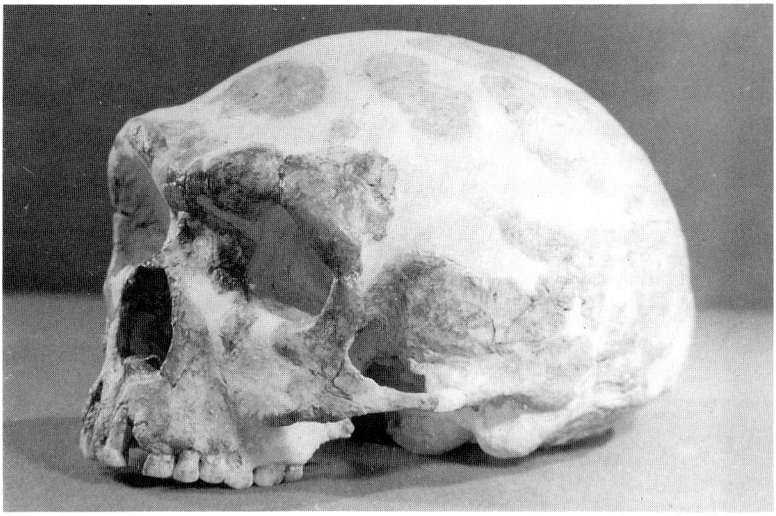

FIG. 3.3 Early *Homo sapiens* skull from Jinniushan. (Courtesy of IVPP.)

1.6 m^2 located about 2 m from the southern wall of the cave. On the basis of the proximity of the various human fossils, the color of the specimens, and the fact that the articular surfaces of the carpals and the tarsals can fit to make the corresponding joints comfortably articulated, Lu (1989) concluded that all the human bones belong to one individual.

The rather robust skull was broken into many pieces during excavation (Table 3.2). Rukang Wu (1988) and his assistant, Zongyi Zhao, restored the skull, which is rather complete. The following parts are missing: part of the lateral wall and a large part of the infratemporal surface of the maxillary body, part of the alveolar process behind the left first premolar, part of the lower margin of the right orbit, part of the lateral orbital wall, part of the frontal bone, the anterior part of the parietal bone, the left condyle and its adjacent part of the occipital bone, the upper part of the temporal squama, the left mastoid process, the external auditory meatus, the right mandibular fossa, part of the left greater wing of the sphenoid, a large part of the right pterygoid plate, part of the hard palate, and the vomer, among other bones. Part of the coronal suture and the bregmatic portion of the sagittal suture are closed. The other part of these sutures and the lambdoidal, parietomastoid, and squamosal sutures are distinct.

The incisors are very worn. Although there is no exposure of dentine on the occlusal surface of the right second maxillary premolar, streaks of dentine exposure occur on the occlusal surfaces of the other premolars. The dentine is exposed as a small point on the first molar. The enamel is weakly worn on the entocone of the second molar. The third molar is small and without any signs of attrition. Based on suture closure and dental attrition, Lu (1989) estimated that the Jinniushan individual is a male slightly more than 20 years old. According to Wu, the skull is from a male somewhat older than 30 years of age.

Rukang Wu (1988) and Zun'e Lu (1989) provide preliminary descriptions and measurements of the Jinniushan fossils. The brow ridges are less robust than in the Dali skull. The medial part of the brow ridge of the Jinniushan skull is more for-

Table 3.2

Measurements of the Jinniushan skull

Cranial length(g-i)	206.0
Cranial width(eu-eu)	148.0
Cranial height(ba-b)	123.0
Horizontal circumference(g-op)	603.0
Cranial sagittal arc(n-o)	362.0
Transverse arc(po-b-po)	308.0
Minimum frontal width*	114
Thickness of brow ridge:	
left side	
medial	14.3
middle	10.4
lateral	13.8
Thickness of cranial bone:	
center of frontal squama	5
bregma*	5.5
parietal tuberosity (left side)	6
lambda*	5.5
cerebellar fossa (l & r sides)	3
center of temporal squama (l & r)	4
Sagittal length of occipital surface of occipital (l-i)	59.2
Sagittal length of nuchal surface of occipital (i-o)	39.3
Length of foramen magnum*	50.7
Width of foramen magnum*	34.5
Upper facial height(n-pr)	74.2
Maximum facial width(zy-zy)	148.0
Nasal breadth*	31
Orbital height(l)	35.0
Orbital width(mf-ek, l)	52.0
Length of alveolar arch*(pr-alv)	61
Width of alveolar arch*(between distal surfaces of first molar of both sides)	67.8
Inclination angle of frontal bone(n-b/FH)	43°
Total facial angle(n-pr/FH)	89°
Nasal facial angle(n-ns/FH)	88°
Alveolar facial angle(ns-pr/FH)	86°
Cranial capacity	1390 cc.

1. Cited from Wu (1988).

2. Items marked with * are cited from Lu (1989)

ward than the lateral part. The medial part of the brow ridge is thickest, next thickest is the lateral part, while the middle part is the thinnest. The glabella region protrudes forward without any concavity. The supraorbital sulcus is shallower than seen in the Dali skull, but the Jinniushan skull is more constricted behind the orbits. A distinct median sagittal prominence on the preserved frontal bone extends from 45 mm above the glabella to 58 mm behind the bregma. It has been reported by Lu (1989) as weakly developed at the posterior of the parietal bone. The pteryon is H-shaped. There is a weak angular torus at the postero-inferior part of the parietal bone. The superior border of the temporal squama is weakly curved. The notch between the mastoid and squamosal portions of the temporal bone is as deep as in the Dali skull. The mastoid is small. The occipital plane turns to the nuchal plane with an angular turn marked by a weak occipital torus. This torus extends along the middle two thirds of the occipital squama.

The lateral cranial wall is as steep as in the Dali skull in rear view. The cranial wall is thinner than that of Dali. The fronto-nasal and fronto-maxillary sutures form a curve that is almost horizontal but is slightly convex upward. The nasal bones are intact and wide. The medial orbital margin is straight, the lateral margin is curved, and the inferior angle of the orbit is nearly a right angle. There is a supraorbital notch and the supraorbital process is distinct.

The zygomatic faces relatively forward as in other Chinese fossils. The zygomatic arch is thin. Its upper border is slightly lower than the Frankfurt Plane, and these two form an angle of about 10°. The two wings of the alveolar arch are slightly bulging toward the lateral side. Except for the lateral incisor and right first premolar, all the maxillary teeth are preserved. The teeth are big, there is no canine diastema, and the incisors are shovel-shaped. Wrinkles on the occlusal surface of premolar and molar teeth are more complex than seen in modern humans.

Lu (1989) gives a short description of the ulna. The ulna is intact and gracile in general appearance. Its length is 260 mm. The sagittal diameter of its upper part is greater than its transverse diameter. The surface of the olecranon is smooth. The shape of the coronoid process resembles that of modern humans. Its anterior margin is less curved and the medial part of the articular surface is shallower than in modern humans. The arc length of the sagittal ridge of the semilunar notch is 37.2 mm. This notch is shallower than in modern humans. The width of the upper, middle, and lower parts of this notch are 22 mm, 14.2 mm, and 21.4 mm, respectively. A deep fossa exists on the postero-medial surface of the coronoid process. Above this fossa is an ovoid fovea that is 4.2 mm × 3.0 mm wide and 3 mm deep. Another small fossa 15.5 mm × 6.5 mm wide and 4 mm deep is seen postero-lateral to the semilunar notch and postero-superior to the radial notch. Small nutrient foramina are scattered in a circle at the bottom of this fossa. The size and features of the radial notch approximate those of modern humans. The articular surface of the radial notch is everted toward the supero-lateral side. The tuberosity is a V-shaped eminence with a smooth surface. A very shallow elliptical fossa appears below the tuberosity.

The interosseus ridge is weak. Both anterior and posterior margins of the ulna are rounded. The tip of the intact styloid process turns forward and medially. The maximum transverse diameter of the bone occurs midway between the lower margin of the radial notch and the nutrient foramen. The bone wall is thick. The diameter of the marrow cavity is less than one third of that of the bone.

Geology

The geological sections from the surface down include (Lu, 1989):

1. Brown claylike silt containing a few fossils, 1.3 m thick
2. Brownish-yellow claylike silt with four thin layers of concretions, cemented, with fossils of deer, spiral-horn antelope, and cave bear, about 2 m thick
3. Hard breccia containing fragments of fossil deer, 0.5 m thick
4. Brownish-yellow claylike silt, big blocks of stone, many deer fossils belonging to *Cervus xanthopygus,* underlined by a calcified crust, 1.2 m thick
5. Breccia, big blocks of stone, filled with cemented and hard brownish-red sandy silt with fossils of rodents, deer, and rhinoceros, 2.5 m thick
6. Breccia, filled with brownish-red sandy silt, cemented and very hard, rich with fossils, including those of the thick jaw bone deer, robust macaque, badger, etc., 1.8 m thick
7. Brownish-red silty sands, thicker in the southern part, with horizontal lamination, rich with fossils, including rather complete skeletons of bear, pig, thick jaw bone deer, rhinoceros, etc.; human bones and broken animal limb bones unearthed at the bottom of this layer
8. Big stone blocks, yellow clay, silty sand fills in the spaces between the stones, only exists near the north and west walls; considerable fossils of rodent, bear, and rhinoceros

According to Lu (1989) the upper part of the cave is composed of brown and brownish-yellow clay soil and breccia. The uranium date of the upper part of the cave deposits is 160,000–200,000 years old, the lower part, is 200,000–310,000 years old, and Layer 7, where the human fossils were found, 280,000 years ago. W. Huang and You (1987) found that erosion of the earlier deposits produced spaces (secondary caves) for the later deposition of materials. They also found that deposits in the peripheral part are older than in the more central part of the cave. Human fossils are embedded in the more central part; therefore, they are later in time than the Middle Pleistocene deposits in peripheral parts of the cave. The very latest are ESR and uranium-series data from five animal teeth from the hominid locality. According to Chen et al. (1994) the results suggest a date of 200,000 years or somewhat older.

XUJIAYAO (113°59′ E, 40°06′ N)

This site, found in 1974, is located on the west bank of the Liyigou river, 1 km southwest of Xujiayao Village, Gucheng District, Yangao County, Shanxi Province, northern China. The site is about 60 km east of Datong city. Excavations in 1976 produced 9 human fossils; in 1977 8 pieces were recovered, and in 1979, 3 more pieces were added. The human fossils include a child's maxilla (no. 1), 12 parietal fragments (nos. 3 to 6, 8 to 14, and one not numbered) from individuals of different ages, 2 pieces of occipital bone (nos. 7 and 15), a temporal bone, a mandibular fragment, and 3 isolated teeth (2 left upper molars [nos. 2 and 16] and 1 lower molar (Jia et al., 1979, for specimens 1–9; Wu, 1980, for all other specimens).

Human fossils

The human fossils were embedded in the lacustrine deposits of the Datong Basin. Most fossils derive from 8 to 12 m below the surface.

Parietal bone

Among the 12 pieces of parietal bone, only 2 (nos. 6 and 10) are rather complete. Parietal no. 6 is the most complete and only the antero-superior and antero-inferior corners are missing. Jia et al. (1979) studied parietals nos. 3, 6, 8, and 9 and M. Wu (1980) studied parietals 10–14. Parietal no. 6, from the left side, is thinner than the other parietals, and probably belongs to a female or an adolescent. The postero-superior corner is naturally lacking, hinting at the possible existence of an Inca bone. According to Jia et al. (1979), the coronal lengths of the chord and arc measured at the middle point of the superior and the inferior border are 106 mm and 130 mm, respectively (index 81.5). The sagittal chord and arc of the middle part are 117 mm and 135 mm, respectively (index 86.7). A parietal tuberosity is distinctly visible.

Parietal no. 10, intact except for its anterior part (Fig. 3.4), is slightly larger and much thicker than parietal no. 6. The postero-superior corner is also lacking naturally. The area to which the temporal muscle attaches is distinctly bulging. The anterior part of the temporal line is not a curved rugged line, it appears as the upper margin of a bulging area to which the temporal muscle attaches. The posterior part of the temporal line is divided into two branches which are thick ridges. The parietal tuberosity is indistinct. Maolin Wu's (1980) measurements of his restoration of parietal no. 10 (Fig. 3.4) appear in Table 3.3.

Parietal no. 3 is a small fragment (probably from the left side) from around the parietal tuberosity. Parietal no. 4 is a fragment above the temporal line of the left parietal. Parietal no. 5 is a bregmatic part of the left parietal. Parietal no. 8, col-

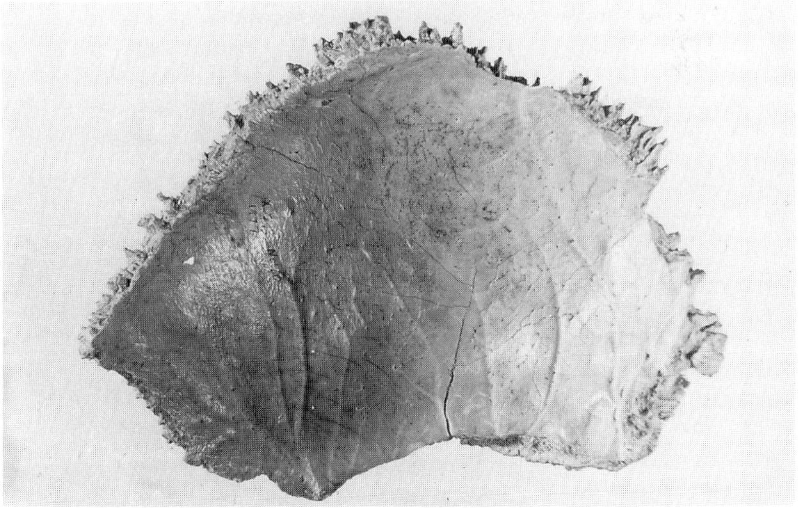

FIG. 3.4 Right parietal (no. 10) of a young male early *Homo sapiens* from Xujiayo. (Courtesy of IVPP.)

lected from the surface, is a small fragment from the left side. Parietal no. 9, about 17 cm^2 in dimension and 5 mm thick, is probably a fragment of the anterior-inferior part. Parietals nos. 11 and 12 are probably from the left side, Parietal no. 13 is a mastoid portion of the right parietal, and Parietal no. 14 is probably the anterior part of both sides with the sagittal suture fused.

No Xujiayao parietal possesses an angular torus, and none shows a sagittal ridge along the upper margin. The branches of the middle meningeal artery of both parietals are richer than in "Peking Man". A segment of the fronto-parietal branch is very thick on Parietal no. 6. The difference in size and thickness between the anterior and posterior branches is hard to judge. Excepting Parietal no. 10, the posterior branch of the middle meningeal artery is longer and thicker than its anterior branch. The branches are thinner than in *H. erectus pekinensis,* but thicker and less ramified than in modern humans.

Parietals nos. 3 and 4 are similar to no. 6 in the presence of straight grooves on the external surface. These grooves radiate forward, upward, and backward from the parietal tuberosity. The longest groove is 17 mm, and the widest is 1.5 mm. There seem to be traces of scraping before death, but the cause of this scraping is not determined. The posterior part of Parietal no. 14 possesses a small hole that seems to be the result of a healed wound. The bregmatic region of Parietal no. 5 is full of dense small foramina that are probably due to osteoporosis or represent cribra cranii externa.

The thicknesses of most parts of the parietals are within the range of variation exhibited by "Peking Man." The mastoid angle of Parietals nos. 6 and 10, as well as the bregmatic region of no. 6, are comparatively thinner than those in *H. erectus pekinensis*. They are still much thicker than in modern humans. Parietal no. 14 is extraordinary thick, with the thickest part being 14 mm thick (Table 3.4).

Table 3.3

Measurements of parietal No.10 from Xujiayao

Margin	*Chord*	*Arc*	*Index*
Sagittal	114.2	121.0	94.4
Coronal	(102.5)	(122.0)	84.1
Lambdoidal	105.0	120.0	87.5
Squamosal	(104.8)	(111.0)	94.4

1. Cited from Wu (1980)

Table 3.4

Thickness of the parietal from Xujiayao

Specimen	*Bregmatic region*	*Tuberosity*	*Mastoid angle*
No. 3	—	12.4	—
Nos. 4 and 5	9.0	10.8	—
No. 6	6.5	7.0	7.2
No. 10	8.5	12.6	13.0

1. Figures cited from Jia et al.,1979 except that of no. 10 which is cited from Wu (1980)

No. 14 is probably the anterior portion of parietals of both sides with an obliterated sagittal suture. This segment is very thick, the thickest part is 14 mm.

Occipital bones (Fig. 3.5)

Information on Occipital no. 7 is from Jia et al. (1979) and that for no. 15 is from M. Wu (1980). Occipital no. 7 is represented by the complete occipital plane and a part of the nuchal region of the squama. Occipital no. 15 includes the left half and a small part of the right half of the squama, except the apical part. Both bones are very thick; Occipitals nos. 7 and 15 are 19 mm and 18 mm thick at the center of the occipital torus, respectively (Table 3.5).

The shape of the occipital torus is similar in both bones, that is, it is broad in the middle and becomes narrower laterally. Below the middle part, there is a triangular rough surface, the apex of which connects with the external occipital ridge. The extending ridge of the torus terminates slightly below the asterion. A shallow supratoral sulcus exists on Occipital no. 15, but not on no. 7.

The cerebral fossa of Occipital no. 15 is bigger and deeper than its cerebellar fossa. The cerebral and cerebellar fossae of the right side are shallower than those of the left. In both nos. 7 and 15 the sagittal sinus continues with the right transverse sinus. The left transverse sinus begins slightly lower than that of the right. The internal occipital crest of Occipital no. 15 bifurcates near the posterior margin of the foramen magnum. In Occipital no. 7 the left cerebral fossa is much bigger than that of the right side.

Temporal bone (Table 3.6)

This specimen is fairly complete; only the anterior part of the zygomatic process and a small part of the superior margin of the squama are missing. Maolin Wu (1980) suggests that this bone probably belongs to a female because its mastoid process is small. Furthermore, the muscle markings are less robust than would be predicted based on male specimens from the same site.

On the other hand, the muscular markings on the outer surface of the temporal squama are strong, as is generally shown in males. If the small size of the mastoid process represents one of the primitive features of the Xujiayo bone, the possibility

FIG. 3.5 Occipital of early *Homo sapiens* from Xujiayo. (Courtesy of IVPP.)

Table 3.5

Measurements of the occipital squama from Xujiayao

Specimen	*Measurement*	*Author*
No. 7		
Sagittal arc	(109)	Jia et al (1979)
Sagittal chord	(86)	
Ast-ast arc	134	
Ast-ast chord	111	
Distance between inion and endinion	8	Wu-this book
No. 15		
Sagittal arc	(130)	Wu (1980)
Ast-ast chord	(102)	
Angle of curvature	(116°)	
Distance between inion and endinion	10	Wu-this book

Table 3.6

Measurements of the temporal from Xujiayao

Length of temporal squama	69
Height of temporal squama	44.5
Length of mandibular fossa	27.2
Breadth of mandibular fossa	30.0
Depth of mandibular fossa	16.5
Distance between the most laterial point of ext. auditory orifice and sagittal plane passing through auriculare	10.0
Height of external auditory orifice	10.0
Breadth of auditory orifice	10.0
Breadth of auditory orifice	8.6
Angle formed by axis of pyramid and sagittal plane of the skull	ca. 50°
Angle formed by axis of the pyramid and the tympanic plate	ca. 30°
Thickness at parieto-mastoid suture behind parietal notch	16
Thickness at occipito-mastoid suture medial to mastoid process	6

According to Wu (1980)

of the temporal belonging to male cannot be excluded. The distinct sutures around the temporal margins indicate that it represents a subadult. The temporal squama is 16 mm thick at the parieto-mastoid suture behind the parietal notch, and 6 mm thick at the occipito-mastoid suture medial to the mastoid process.

The length of the temporal squama is 69 mm, and its height vertical to the length of the squama is 44.5 mm. The height-length index is 64.5. The groove between the squama and the zygomatic process is very broad and deep. The attachment of the zygomatic process to the temporal squama is rather long. The superior branch of the root of the zygomatic arch runs backward to continue with the supramastoid ridge, which terminates at the parietal notch of the temporal bone. The supramastoid ridge is much less robust than that of the Dali skull, but it is more robust than in modern humans. In Xujiayao the long axis of the supramastoid process continues with the long axis of the zygomatic process of the temporal bone. Based on the less forward and downward inclination of the zygomatic process, M. Wu (1980) supposed that the Xujiayo face should be as short as in modern humans.

The mandibular fossa of Xujiayo is 27.2 mm long, 39 mm wide, and 14 mm deep, it is as deep as in "Peking Man." However, Xujiayao's mandibular fossa is longer, and its depth-length index is lower (49.2). The articular eminence is less distinct than in modern humans. The postglenoid process is similar to that of Dali.

The most lateral point of the tympanic plate is 10 mm medial to the vertical plane within which the auriculare is located. The medial and lateral margins of the tympanic plate are almost in the same coronal plane. The long axis of the tympanic plate is perpendicular to the sagittal plane. The transverse diameter is longer than the vertical diameter. This is different than the situation in "Peking Man," in which the vertical diameter of the tympanic plate is longer than the transverse diameter. The tympanic plate is similar to that of Dali. It is not convex as in "Peking Man" and modern humans.

The orifice of the external auditory meatus is elliptical with the long axis in a vertical direction, it is 11 mm long and 8.6 mm wide. At the lateral part of the tympanic plate is a notch such as found on the left side of "Peking Man" Skull XI.

The petrous portion of the temporal is thick and shorter than in modern humans. The transverse section is triangular with the base longer than each side. The upper border is rather sharp. The posterior surface is sloped instead of vertical. The transition between its anterior surface and the cerebral surface of the squama is more gradual than that which exists in modern humans. These two surfaces constitute a fossa instead of a groove. The arcuate eminence is very prominent; the sigmoid sulcus is narrow and deep.

The axis of the pyramid forms about a 50° angle with the sagittal plane. This angle is smaller than that of Hexian and intermediate between "Peking Man" and modern humans. The 30° angle formed by the pyramidal axis and the tympanic plate is larger than in modern humans.

Only a small segment of the root of the styloid process is preserved; a vaginal process is present. The distance between the styloid process and the midpoint of the external auditory meatus is 17.3 mm. This measurement is intermediate between "Peking Man" and modern humans. The internal and external orifices of the carotid canal and the jugular fossa are similar to that of modern humans.

The surface of the mastoid process is rough, and there is no developed mastoid ridge. The rounded tip of the mastoid process points downward and inward. The mastoid notch between the mastoid process and the cranial wall is narrow and deep, but the occipital groove is shallow. The mastoid notch, stylomastoid foramen, and the styloid process are in the same plane. The mastoid foramen is above the mastoid notch and close to the occipito-mastoid suture. The suprameatal spine is prominent. The thickness of the cranial bone at the parieto-mastoid suture behind the parietal notch at the occipito-mastoid medial to the mastoid process is 10 mm and 6 mm, respectively.

Maxilla (Fig. 3.6)

Only the left alveolar process and its adjacent bone have been preserved. The bone probably belonged to a young male. The newly erupted median incisor, first molar,

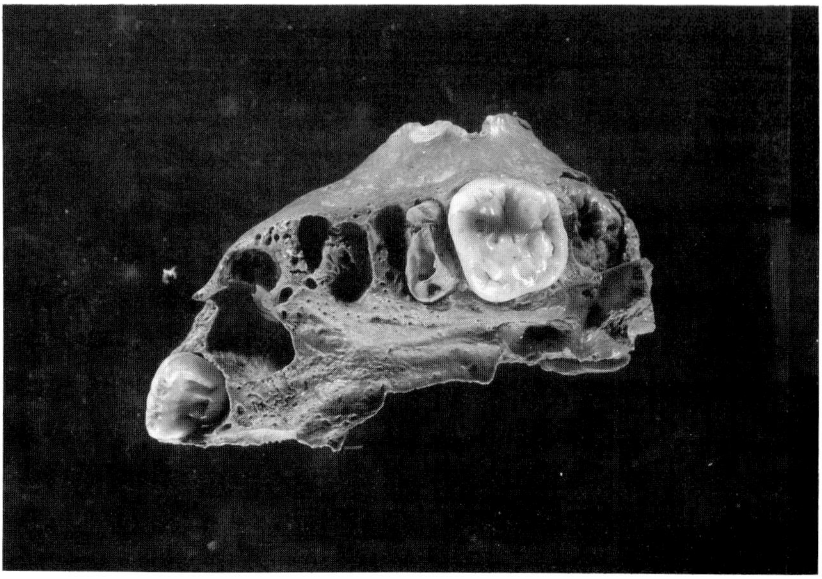

FIG. 3.6 Early *Homo sapiens* maxilla from Xujiayo (showing the permanent, just erupted first incisors, permanent first molar, and part of the root of the deciduous second molar). (Courtesy of IVPP.)

and the germ of the canine and erupting second molar, as well as the posterior root of the second deciduous molar, are attached. The germs of two left premolars are shown on an X-ray film. Jia et al. (1979) considered that the maxilla belonged to a 7- to 9-year-old child.

The distance between the supra-dentale and the coronal line passing the posterior margin of the first molar is 47 mm. The distance between the progthion and nasospinale is 13.4 mm. This is a very long palate. The alveolar process is low and the palate is very shallow. The height of the palate from the lower margin of the alveolar process is 7 mm. The alveolar arch looks like a beak due to the extremely forward protrusion of the incisor part. Although alveolar prognathism cannot be measured, progthion is probably 6 mm in front of the vertical line passing through the posterior part of the nasal spine. Although the anterior nasal spine is distinct, it is not very prominent. The incisive canal is located 10.5 mm behind the anterior surface of the alveolar process and parallel to the median sagittal line on this surface.

Mandible (no. 10A)

A mandibular ramus missing the condyloid process is preserved. The ectocondyloid ridge is not very distinct, but the endocondyloid process, triangular torus, and the pterygoid tuberosity are well developed. The mandibular foramen is rather large. Maolin Wu (1980) reconstructed the condyloid process and measured the height and width of the mandibular ramus as 63 mm and 46 mm, respectively. The index (70.7) indicates that this ramus is lower and wider than in modern humans. The mandibular angle is 104°.

The mandible was found in proximity to Parietal no. 10. This mandible and Parietal no. 10 are similar in color, so M. Wu labelled it specimen no. 10A and considered that they belonged to the same individual.

Teeth

In 1976, four teeth attached to a maxilla and an isolated upper molar were found. They were described by Jia et al., (1979). The measurements are given in Table 3.7.

Incisor (left upper). There are five small mammilary processes on the cutting edge of the incisor. The labial surface bulges vertically and horizontally. The lingual surface is shovel-shaped, with two strong marginal ridges folding toward the lingual side. The well-developed lingual tubercle has two finger-like processes extending toward the cutting edge. They become weaker at the central part of the lingual surface. Between these two fingerlike processes is a small pit that is 1.5 mm deep.

Jia et al. (1979) consider distinct yellow fovea on the left upper part of the labial surface the result of a high fluorine content in the underground water. The cervical part of the incisor is rather constricted. This incisor is just erupting and the root is not fully developed. The exposed crown is only 4.5 mm in height.

Canine (left upper). The unerupted canine was removed through the broken part of the wall of maxillary sinus. The root was not well developed. The labial surface is strongly convex in a horizontal direction and slightly ribbed with fine wrinkles. The lingual surface shows a more complicated pattern of wrinkles. The canine is wedge-shaped when viewed from the mesial or distal side. Well-developed triangular prominences on both the mesial and distal surfaces form the basal tubercle at the basal part of the lingual surface. Several rib folds are seen on the basal tubercle. An intermediate rib fold is the longest, and it extends from the basal tubercle to the tip of the tooth.

Molar (left upper). The occlusal surface is rather square; the wrinkles are rather complicated. Only part of the surface of the paracone has been worn slightly. The cusps are distinct and the cingulum is strong. The protocone is connected with the metacone by a distinct oblique ridge, but the metacone is separated from the paracone by a deep furrow.

Molar (left upper). This tooth is still embedded within the socket. The occlusal surface is almost similar to the aforementioned first molar and seems to be a little larger than the latter. All of the teeth mentioned above are attached to Maxilla no. 1. There are two additional isolated upper molars.

Table 3.7

Measurements of the Xujiayao Teeth

	Crown			Root			Author
	Length	*Width*	*Height*	*Length*	*Width*	*Height*	
Upper median incisor(l)	10.0	8.4	11.0*	–	–	–	Jia et al
Upper canine(l)	10.8	10.4	13.2	–	–	–	(1979)
Upper first molar(l)	13.4	14.0	7.4**	–	–	–	
Upper first or second molar(l)	12.0	13.7	–	9.5	13.1	16.0(m)***	Wu
						15.5(d)	(1980)

1. * Height of labial surface

2. ** Height of lingual surface

3. *** m=mesial and d=distal

Molar (left upper). The occlusal surface of this isolated tooth is so heavily worn that the number of cusps is difficult to determine. Its contour is close to ellipsoid. The root has three branches. The two buccal branches are fused along most of their height. The lingual root is 16.5 mm high; the buccal posterior root is 17.3 mm. high.

Molar (left upper, M1 or M2, no. 16). This heavily worn isolated molar was found in 1977. The hypocone has not reduced. The anterior part of the tooth is broader than the posterior part. There is no cingulum. The root is very robust and divided into three branches. The buccal branches are flattened, and its terminal end curves slightly to the distal side. The lingual root is thickest and straight.

Geology

The Xujiayo site contains lacustrine and fluvial deposits from the River Liyigou of the Datong Basin. The upper stratum is a loess-like sandy clay 4 m thick; the lower stratum is sandy or silty clay. Most of the human and nonhuman fossils and the artifacts are embedded in the yellowish-green clay of the lower stratum at a depth of 8–12 m under the local earth. The upper part of the fossil bed includes cemented sands. The geological sections are described by Jia and Wei (1976) as follows:

Holocene

8. Surface soil 0.3.–0.5 m thick
7. Yellowish-brown sandy soil, 3.5 m thick

———erosion surface———

Upper Pleistocene

6. Yellowish-brown silty clay with broken ostrich shells and wooly rhinoceros teeth, ca. 5 m thick

———erosion surface———

5. Reddish silty clay, 1 to 3 m thick

———erosion surface———

4. Yellowish-green silty clay, ca. 4 m thick
3. Reddish-brown clay, 0.3 m thick
2. Yellowish-green clay with sands and sandy concretions in the upper part; contains fossils and artifacts; ca. 5 m thick
1. Greyish-blue, grayish-green, and greyish-brown clay, about 4–6 m thick

This site is rich in fragmentary vertebrate fossils. No complete skull has been found. The mammalian fauna include 7 orders and 19 species; birds are represented by only one species. Most vertebrate fossils were found in locality 74093; the remainder were found in locality 73113. According to Jia et al. (1979), most artifacts and human fossils were unearthed from locality 74093. The vertebrate fossils follow (according to Jia et al., 1979):

Primates
Homo erectus
Lagomorpha
Ochotona sp.
Rodentia
Myospalax fontanieri
Microtus brandtioides
Carnivora
Canis lupus
Panthera cf. *tigris*
Proboscidea
Palaeoloxodon cf. *naumanni*
Perissodactyla
Coelodonta antiquitatis
Equus przewalskyi
Equus hemionus
Artiodactyla
Megaloceros ordosianus
Cervus elaphus
Cervus nippon grayi
Spirocerus hsuchiayaocus
Spirocerus peii
Procapra picticandata przewalskyii
Gazella subgutturosa
Gazella sp.
Bos primigenuis
Sus sp.
Bird
Struthio sp.

Uranium series dating of rhinocerous teeth yields a time of 104,000–125,000 years ago (Chen et al., 1984).

Archaeology

Jia et al. (1979) reported that 13,650 pieces of stone artifacts were unearthed in 1976 from this site. Large tools are very rare. According to Qiu (1989 in Wu et al., 1989), 589 pieces of stone artifacts have been studied. Scrapers constitute 38.56% of the assemblage, points 4.11%, stone balls 3.06%, burins 3.08%, borers 0.52%, and small chopper-chopping tools 0.26%, etc. There are many bone tools and some antler tools.

CHAOXIAN (CHAOHU) (117°52 E, 31°33′ N)

The human fossils were found in a limestone fissure or cave deposit (it can not be ascertained if it was a fissure or cave deposit, because no roof is now visible) at Yinshan Hill near Yinshan Village, Daishan region, Chaoxian County, Anhui Province, eastern China. The hill, which is close to the right side of the Chaoxian-Wuwei highway, is located some 6 km south of the city of Chaoxian and some 50 km southwest of the Hexian *H. erectus* site. In 1981 some fossils were found there by the Geological Survey of the Anhui Bureau of Geology. In March 1982, Xianshun Chen of the Cultural Station of Yinping District, Chaoxian County, reported the discovery of mammalian fossils from a quarry near Yinshan Village. Researchers from the IVPP in Beijing, working with archaeologists and geologists from Anhui Province, found a fragment of a human occipital in 1982 and, in the fall of 1983, a human maxillary fragment. The human fossils were studied by C. Xu et al. (1984) and Xu and Zhang (1986).

Human fossils

Occipital

Excavations in 1982 yielded an occipital fragment, which included a large part of the occipital squama. The occipital bone below the inferior nuchal line and a small part near lambda are missing. The thickness of the diploe, the inner and outer tables of the bone, are almost equal as seen at the broken line at the apical part of the bone. The lambdoidal suture is intact with serrated margins, and C. Xu et al. (1984) estimated that this individual is not more than 26 years of age. The lambdoidal suture on the right side connects with 8 mm of the occipito-mastoid margin.

The occipital torus consists of two curved eminences that are only slightly prominent at the medial point. The eminences attenuate toward the asterion, where the torus disappears; therefore, the transition between the occipital and nuchal planes is rounded instead of angled. The middle and alar parts of the torus are both 15 mm wide. No external occipital protuberance exists. The supratoral sulcus is indistinct, except for a small fovea corresponding to the suprainionic fossa above the middle of the torus. The superior nuchal line is not distinct, but the inferior nuchal line is somewhat distinct. The preserved part of the nuchal plane above the inferior nuchal line is smooth and without depressions or an eminence. C. Xu et al. (1984) provide the following measurements: ast-ast, 122.4 mm; l-i chord, 50.2 mm; distance between inion and endinion, 22.0 mm; and the thickness at the center of occipital torus is 7.0 mm

This occipital is too fragmentary to estimate the exact curvature angle, but judging from the sagittal contour, the angle is probably much larger than that of "Peking Man" and approaches that of early *H. sapiens*. According to C. Xu et al. (1984), the preserved part of the lambdoidal suture, the width of the bone, and the smoothness of the external bone surface indicate that this fragment is probably from a young female.

Maxilla

An anterior fragment of the alveolar process had the right P1-M1 and the two lateral incisors atttached. Additionally, three isolated left teeth—P2, M1, and, M2—of the same individual were unearthed.

Viewed from the lateral side, the median sagittal contour is a slightly convex line less curved than seen in *H. erectus* from Gongwangling and Zhoukoudian. Although the anterior spine is broken, it seems to be well developed. Except for one third of its left half, the lower margin of the pyriform orifice is blunt and without a prenasal fovea. The width of the pyriform orifice is probably rather large. The anterior part of the nasal floor gently slopes, with the anterior part higher than the posterior.

The exposed maxillary sinus shows that its anterior extremity is at the level of the first premolar, and it extends medially to the palatal process of maxilla. The anterior part of the hard palate and a small piece close to the first molar are preserved. Its surface is rugged. The incisive foramen is oval and very close to the alveolar margin. The incisive canal extends downward more or less vertically. The upper end of the canine jugum is distant from the nasal floor (C. Xu and Zhang, 1986; C. Xu, et al., 1984).

The roots of two lateral incisors, two right premolars, and the first right molar are preserved in the maxilla (Table 3.8). The alveoli of the two median incisors and the right canine are empty. The mesial and distal surfaces of the lateral incisors and the mesial surface of the right first premolar possess contact facets. There are three isolated teeth of the upper dentition. The isolated left second premolar and the first two left molars probably belong with the maxilla. The crowns of the left teeth are well preserved and slightly worn. The roots of the second premolar are complete, while those of the two molars are broken. The occlusal surface of the left first upper premolar can be reconstructed by its impression, which was preserved on the surrounding cemented rock. A contact facet is visible on the distal surface of the second molar. Zhang (1989) suggested that the entire dentition, including the third molar, had erupted premortem.

According to modern standards the attrition of the first two molars indicates that this individual was about 30 years old. However, because attrition in ancient humans might be heavier because of more stress on the dentition, this individual might be younger than 30 years old. Attrition on the anterior teeth is much greater than that on the molars. Y. Zhang (1989) suggests that this big difference in attrition implies that the anterior teeth functioned as tools and for chewing hard and tough food.

Geology

The Yinshan site is a small hill. Its latitude corresponds to the second terrace of the Yuxi River, which flows through this region. The second terrace is about 20 m above sea level. The cave deposits are scattered over an area of about 200 m^3. C. Xu et al. (1984) inferred that the original cave roof was destroyed by natural forces. They described the geological section of the deposits from the the surface downward as follows:

Table 3.8

The robustness of the teeth from Chaoxian

Specimen	*Robustness*
Premolar (upper first)	107.0
Premolar (upper second)	104.6-106.7
Molar (upper first)	154.1-159.6
Molar (upper second)	166.4

Cited from Xu and Zhang (1986)

1. Loose grey deposit including mainly clay and broken stones, 30 cm thick
2. Brown-red breccia including mainly clay, brecciola, and sands; human fossils found in this layer; maximum thickness, 175 cm
3. Yellow deposit, including mainly clay, breccia, and pebbles, cemented by calcium salts; a local stalactic crust of about 3.5 cm thick exists at the level corresponding to the top of the layer; 100 cm thick
4. Brownish-red deposits, mainly of cemented clay, sands, and pebbles; fossils arrayed in laminated sections of the deposit; 20 cm thick
5. Light yellow deposits mainly of clay and gravels; contains heavily mineralized fossils, maximum thickness, 23 cm

Fauna from the upper two layers include: *Homo* sp., *Hyaena brevirostris sinensis,* Felidae, Mustelidae, *Cuon* sp., *Ursus* sp, *Panthera* sp., *Stegodon* sp., *Tapirus* sp., Rhinocerotidae, *Equus* sp., *Sus xiaozhu, Sus* sp., *Megaloceros pachyosteus, Cervus* sp., Bovidae, and Caprinae. The fauna from Layers 3 to 5 are *Cuon dubius, Hyaena brevirostris licenti, Megantoreon* sp., *Panthera* sp., *Proboscidipparion* sp., *Equus* sp., *Tapirus* sp., Rhinocerotidae, *Tetralophodon* sp., *Stegodon* sp., and Cervinae.

Chaoxian has a uranium series date of 200,000–160,000 years ago (Chen et al., 1987).

Archaeology

No artifacts were recovered from the site.

MABA (113°35′ E, 24 °41′ N)

In 1958 local farmers who were digging in Shizishan (Lion Hill) cave for its fertilizer found human and nonhuman mammalian fossils. The cave is one of many located 1.5 km southwest of Maba Village, Shaoguan Municipality, Guandong Province, southern China. The human fossil, which consisted of several fragments of a skullcap, was unearthed from a cave in the middle or second of three levels in a small hill.

Human fossils

The several preserved fragments can be connected to form a skullcap with the right and middle parts of the upper face. There is a large part of the frontal and parietal bones, a large part of the right orbit, and nasal bones (Table 3.9). These have been given museum number PA 84.

The sutures are obliterated except for the anterior segment of the sagittal suture and a large part of the coronal suture, which are discernible but not clear. The external surface of the skullcap is somewhat rough, but the muscular markings are not robust. The reconstructed vault is quite large. This fossil probably belongs to a middle-aged male.

Reconstruction shows that the top of the skull is ovoid (Figs. 3.7 and 3.8). The maximum cranial breadth should be slightly above the supramastoid ridge and correspond to the posterior of the middle third of the maximum cranial length. The

Table 3.9

Measurements of the Maba Skullcap

Orbital breadth (mf-ek) rt.	44.3
Orbital height	39
Interorbital breadth (mf-mf)	20.8
Subtense to mf-mf	5.8
Minimum nasal breadth	13.3
Subtense to minimum nasal breadth	4.3
Frontal arc	134.0
Frontal chord	115.6
Parietal arc (sagittal)	114.0
Parietal chord (sagittal)	107.0
Thickness at bregma	7.0
Thickness at parietal tubercle	9.0*
Frontal sinus	
Breadth (rt)	26
(lt)	21
Length (rt)	16
(lt)	10
Height (rt)	24
(lt)	24
Bregma position index	40.6°
Bregmatic angle (b-g-op)	45°
Frontal angle (m-g-op)	70°

All measurements are cited from Wu (1959) except that 9.0* is from Woo (1980)

frontal tubercle is weakly developed. A median eminence can be seen on the frontal squama. Its upper end cannot be defined because the lateral part is missing. Its lower end almost corresponds to the level of the supraorbital groove, which is rather deep except for its median part.

Despite gnawing by rodents at its medial part, the right brow ridge is visible. The thicknesses of the middle and lateral segments of the torus are 11 mm and 12 mm, respectively. The medial segment is estimated to be about 14 mm thick. Viewed from above, the medial end of the supraorbital ridge is located more anteriorly than the lateral end. The frontal view of the left and right supraorbital tori shows that they are separated by a median downward depression. There is a trace of the fusion of the metopic suture on the anterior surface of the glabellar region. This region is slightly concave, as in the Dali skull. The frontal bone is very constricted behind the orbits. The distance between the most lateral ends of both brow ridges is 121.8 mm; the minimum width between the temporal surfaces of the fron-

FIG. 3.7 Bust of early *Homo sapiens* from Maba. (Courtesy of IVPP.)

tal bone is 96 mm according to measurements on the reconstruction. The postorbital constriction index of 80.2 is even lower than for some *H. erectus* skulls.

Gnawing by rodents also has opened the anterior wall of the rather large frontal sinus. A bony septum, which is slightly deviated to the left, makes the right part of the sinus slightly larger than the left. The upper limit of the right part exceeds the upper border of the supraorbital sulcus and protrudes into the frontal squama. Laterally the sinus reaches a point about half a centimeter lateral to the middle of the superior orbital margin. The left part of the sinus reaches the junction between the medial and middle thirds of the supraorbital torus.

A semilunar scar 14 mm long exists on the external cranial surface postero-lateral to the right frontal tubercle. On the intracranial surface, at the place roughly corresponding to this scar, there is an unexplainable bulging of about the same size.

A large part of the left temporal line is not preserved. Only the right temporal line can be seen for a considerable length. Both superior and inferior temporal lines are weak at their posterior end. The parietal tubercle is only discernible slightly supero-posterior to the temporal line. There is no parietal foramen. The parietal bone is 7.0 mm thick at the bregmatic region and parietal tubercle. Only a small part of the right temporal squama is preserved. A trace of the suture between the temporal squama and the greater wing of the sphenoid is discernible. No demarcation can be seen between the partial temporal and the parietal bone.

Most of the left orbit is lost, while most of the right orbit is preserved. The orbital index is 88.0. The superior orbital margin is evenly curved. All of its length is smooth and rounded, except for the medial part, which was destroyed by the gnawing of rodents. There are no supraorbital notch, foramen, or tubercle. The lateral orbital margin is shallowly curved. Although a large part of the inferior

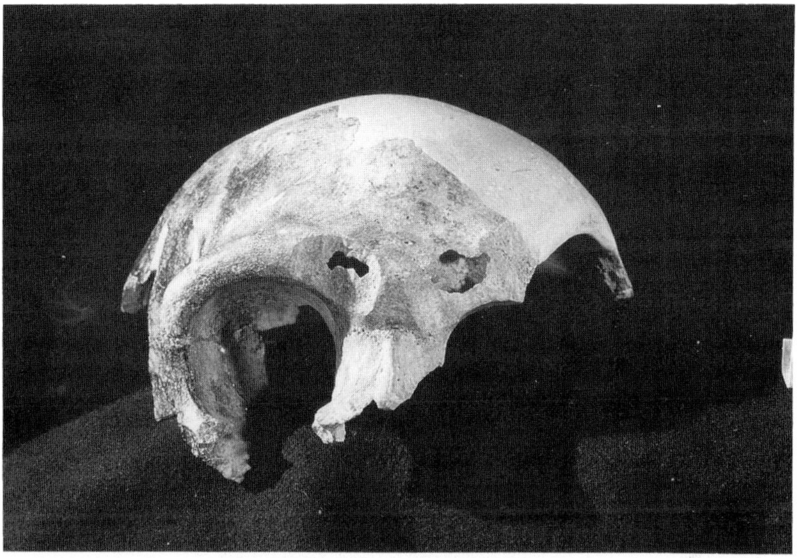

a

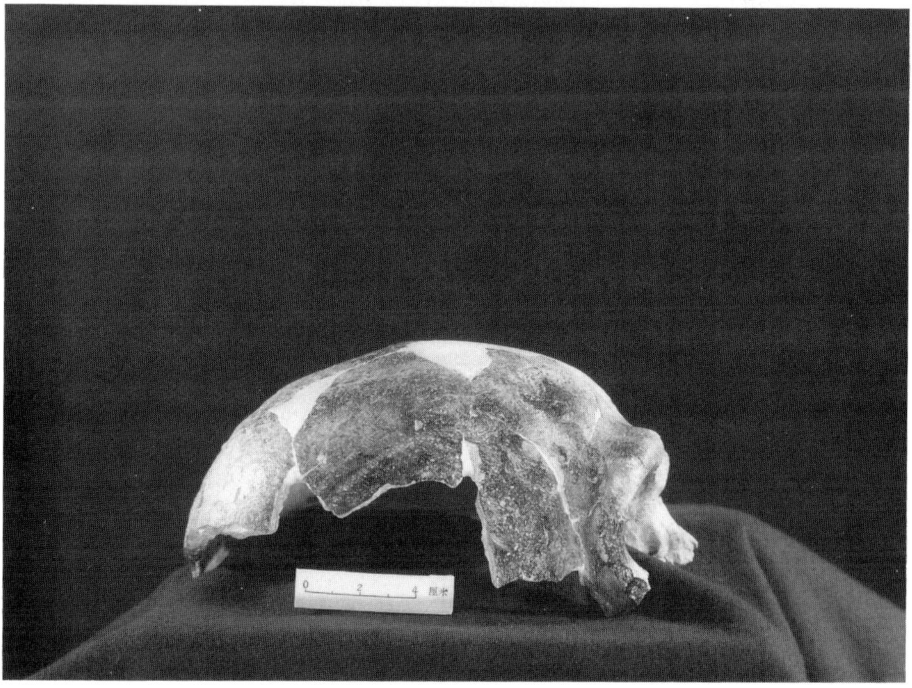

b

FIG. 3.8 Anterior view (a) and lateral view (b) of early *Homo sapiens* from Maba. (Courtesy of IVPP.)

orbital margin was broken, the orbit was probably more or less spherical. The inferior lateral orbital margin is sharp. This is different from other Pleistocene hominid specimens in China, which generally have a rounded margin.

The naso-lacrimal ridge is weak; the lacrimal fossa is shallow. The suture between the nasal, lacrimal, and frontal bones is distinct. The majority of the orbital roof was preserved, and the depression for the lacrimal gland is shallow. The zygomatico-orbital foramen is invisible on the preserved orbital wall. Perhaps this foramen is located lower than is true for "Peking Man."

Most of the nasal bone was preserved. The suture joining the frontal and nasal bones, and the maxillae forms a more or less horizontal line. Only the upper part of the suture between the nasal bone and maxilla is visible. The internasal suture is discernible only at its lower part. There is a midsagittal longitudinal narrow ridge on the nasal bones. The midsagittal contour of the anterior surface of the nasal bones is slightly concave. The profile of the nasal bone is close to a right angle, and the naso-malar angle is 146°.

The antero-lateral surface of the right fronto-sphenoidal process of the zygomatic bone faces more forward. Two horizontal lines representing the orientation of this surface of the reconstructed skull form an 85° angle. A weak protrusion on the upper part of the postero-lateral margin at the broken edge of the bone may represent the marginal tubercle.

Judging from the preserved part of the parietal bone, the arc-chord index is rather high (93.9). This figure is close to that of "Peking Man" (94.3) and far from that of 89.7 for modern humans. The arc and chord of the coronal margin of the parietal bones are 110.0 mm and 96.5 mm, respectively. An index of 87.7 is higher than both "Peking Man" and modern humans.

A very prominent frontal ridge on the endocranial surface becomes broader toward the lower end. No bifurcation of the ridge is seen on the preserved part (which is 44 mm long). The groove for the sagittal sinus is absent on the frontal bone but is present on the total length of the parietal. There is no arachnoid fossa.

The grooves of the branches of the meningeal arteries are distinct. Judging from a segment of the groove on the internal surface of the left parietal, the anterior branch of the middle meningeal artery is possibly thicker than the posterior branch. However, there can be no full confidence in this inference because of the incompleteness of the branches. The ramification of the grooves is less abundant than that in modern humans.

Fauna

Mammalian fossils associated with the human remains include the following (Han and Xu, 1989):

Primates
- *Homo sapiens*

Lagomorpha
- *Lepus* sp.

Rodentia
- *Rattus* sp.
- *Hystrix* sp.

Carnivora
Ailuropoda sp.
Ursus kokeni
Arctonyx sp.
Crocuta crocuta ultima
Panthera tigris
Panthera cf. *pardus*
Proboscidea
Stegodon orientalis
Palaeoloxodon namadicus
Perissodactyla
Tapirus sp.
Rhinoceros sinensis
Artiodactyla
Sus scrofa
Cervus sp.
Ovinae
Bovinae

Maba has a uranium series date of 135,000–129,000 years ago (Yuan et al., 1986). Han and Xu (1989) attribute the site to the Late Pleistocene.

Archaeology

No stone artifacts were found in the cave.

CHANGYANG (110°50′ E, 30°15′ N)

Human specimens from Changyang County, Hubei Province in central China, were found both in disturbed and undisturbed cave deposits. Longdon cave (Dragon cave) is located near Xiazhongjiawan village on the south slope of Guanlao Hill about 45 km southwest of Changyang city. The village belonged to the Huangjiatang Xiang (a unit representing part of a district) of Zhaojiayan Qu (district) of Changyang County. The site is about 1300 m above sea level. The fossils were probably transported by water into the cave.

The remains consist of a left maxillary fragment (V. 1681) and a left lower premolar (V. 1682). The maxilla was collected by a local farmer and a staff member of the Cultural Club of Changyan County. In September 1956, the latter sent the fossils to the IVPP (then called the Laboratory of Vertebrate Paleontology, Academia Sinica). In the spring of 1957, IVPP scientists found a left second lower human premolar and many other vertebrate fossils.

Human fossils

Jia (Chia and Wei, 1976), Han and Xu (1989), and Yuan et al. (1986) studied the human and other specimens from Changyang.

Maxilla (Fig. 3.9)

This is a fragment of the lower part of the left maxillary body, as well as the palatal and alveolar processes, with an attached first premolar and first molar. The anterior part of the nasal floor was preserved. The left part of the lower margin of the pyriform aperture is rather well preserved. Judging from the preserved basal part of the anterior nasal spine, the spine might have been very narrow and weak. The straight line connecting the nasospinale and prosthion forms a 60° angle with a plane deduced from the occlusal surface of the two preserved teeth. This plane forms an angle of 5–10° with the Frankfurt Plane. The alveolar prognathism is probably a 65–70° angle. The midsagittal contour of the anterior surface of the alveolar process is basically a straight line that is slightly convex at the extreme upper and lower parts.

The anterior wall of the maxillary sinus reaches a point above the canine. The alveoli of the incisors and the canine are rather large, and their anterior walls are broken. The tip of the canine root exceeds the level of the nasal floor.

The buccal cusp of the upper first premolar is slightly higher than its lingual cusp. The paracone is the highest cusp on the first molar. The occlusal surface of the isolated second lower premolar is almost quadrangularly shaped and has irregular wrinkles. The root is high, a vertical groove exists at the lingual and distal surfaces, and a short groove appears on the mesial surface near the tip of the root. Tooth measurements are cited in Table 3.10.

Fauna

Many mammalian fossils including remains, derive from the Changyang cave. They include the following (Han and Xu, 1989):

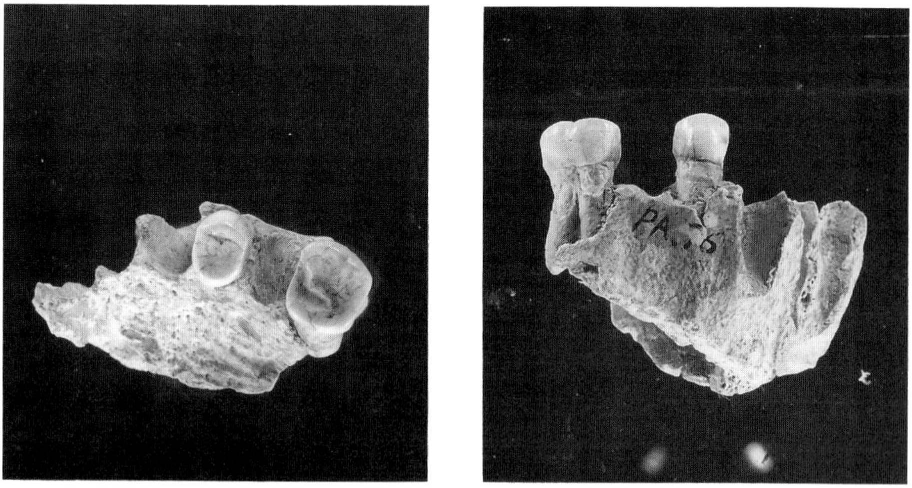

a b

FIG. 3.9 Maxilla of early *Homo sapiens* from Changyang. (a) Basal view. (b) Lateral view. (Courtesy of IVPP.)

Table 3.10
Measurements of the Teeth From Changyang

	Crown			Root		
	Length	*Breadth*	*Height*	*Length*	*Breadth*	*Height*
Upper first premolar	7.4	10.6	—	—	—	—
Upper first molar	10.8	12.8	—	—	—	—
Lower second premolar	8.3	10.6	4.8	7.2	9.8	20.5

Primates
Homo sapiens
Rodentia
Rhizomys sinensis cf. *troglodytes*
Hystrix cf. *subcristata*
Carnivora
Cuon javanicus antiquus
Cuon sp.
Ailuropoda sp.
Ursus thibetanus kokeni
Meles sp.
Hyaena breviostris sinensis
Panthera tigris,
Felidae gen. et sp. indet.
Proboscidea
Stegodon orientalis
Perissodactyla
Megatapirus augustus
Rhinoceros sinensis
Artiodactyla
Sus sp.
Cervidae gen. et sp. indet.
?*Spirocerus* sp.
Bovidae gen. et sp. indet.

Changyang has a ^{230}Th date of 194,000 + 24/ −20,000 years ago and 196,000 + 20/ −17,000 years ago (Yuan et al., 1986). Han and Xu (1989) attributed the site to the Middle Pleistocene.

Archaeology

No artifacts were found in the cave.

DINGCUN (111°25′ E, 35°50′ N)

In May of 1953 workers digging sand for construction found many mammalian fossils near Dingcun Village on the east bank of the Fenhe River, Shanxi Province (northern China). The fossils were reported to the Laboratory of Vertebrate Paleontology, Academia Sinica (now the Institute of Vertebrate Paleontology and Paleoanthropology) by the Management Committee of Cultural Relics of Shanxi Province, and from September to November an IVPP field team surveyed and excavated the site. In 1954, they found three human teeth, many stone implements, and animal fossils near Dingcun Village. In 1976, a fragment of a human infantile parietal was found in locality 54100, (Fig 3.10), the same locality that yielded human teeth in 1954. The remains have been studied by J. Woo (1958d), Pei (1958), Chen et al. (1984), and Zhang et al. (1991).

Human fossils

All three teeth are from the right side and are assumed to belong to the same individual (J. Woo, 1958d). Wu's (Woo's) assessment was based on the fact that the three teeth had similar colorization and mineralization, among other features.

FIG. 3.10 Dingcun Locality 54100. (Courtesy of IVPP.)

Teeth

Incisor (right upper PA 72). The labial surface of the crown is slightly bulging in both vertical and horizontal directions. The lingual surface is depressed at its central part and thickened at the mesial and distal margins, producing a shovel-shaped incisor. The distal marginal ridge is especially thick. Ridges on the distal and mesial margins converge and connect with the lingual protuberance (or basal tubercle). From the mesial part of the lower border of the basal tubercle, a distinct fingerlike projection is sent to the central lingual fossa. A shorter fingerlike projection is sent from the distal part of the lower border of the lingual protuberance. The lower margin of the crown is slightly worn, but the crenulationlike structure is discernible (Fig. 3.11).

The neck of the tooth is slightly constricted. The demarcation lines between the crown and the root are higher on the labial and lingual sides than on the mesial and distal sides. The lines on the latter two sides are almost at an equal level. The same is true of the lines on the labial and lingual surfaces. The labial surface of the root is slightly larger than the lingual surface. The mesial and distal surfaces of the root are flattened and without a sulcus. The diameter of the root gradually diminishes from the crown to the apex (or tip).

Incisor (right upper I2 PA 73). The right lateral maxillary incisor is smaller than the right medial incisor. The labial surface of the lateral incisor is slightly bulging. The lingual surface shows a shovel shape, with a central fossa and distinct marginal ridges meeting at the upper end. The lower margin of the crown was

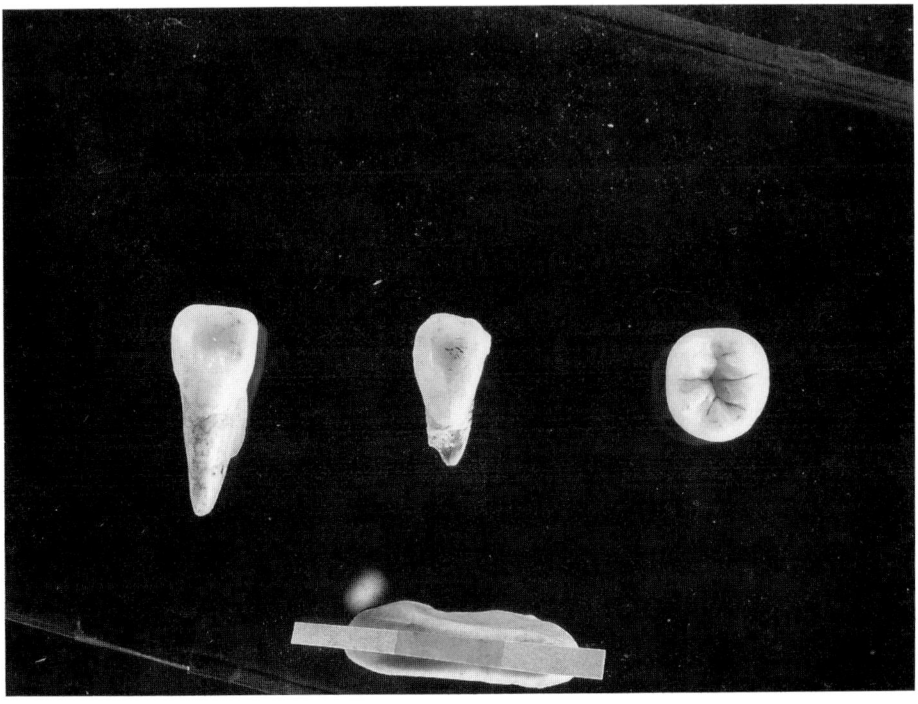

FIG. 3.11 Incisors (I1 and I2) and lower M2 of early *Homo sapiens* from Dingcun. (Courtesy of IVPP.)

slightly worn with the crenulationlike structure discernible. In newly erupted human incisors these crenulations are small mammillary tubercules on the cutting edge. The neck line is indistinct. The root is mesio-distally compressed.

Molar (right lower M2 PA 74). There are no contact facets on either the mesial or distal surfaces. The occlusal surface is unworn. This tooth may be a second newly erupted molar, assuming that all three teeth belong to same individual.

The crown height is rather high relative to its length and breadth. The buccal surface bulges in the middle. The enamel extends more downward on the buccal than on the lingual surface. The occlusal surface is elliptical and has five cusps. The longitudinal groove bifurcates at both ends. The cusp pattern is a "+", and a small accessory cusp probably represents the sixth cusp. The trigonid and talonid are of equal breadth, but the trigonid is shorter than the talonid.

Although the root has not fully developed, it obviously divided into two branches. Measurements of the Dingcun teeth appear in Table 3.11.

Parietal (Fig. 3.12)

This right parietal belongs to a child. The anterior and lateral parts were lost. The sagittal and lambdoidal sutural borders are intact and serrated. A triangular lambdoidal ossicle was inferred from the shape of the postero-medial corner of the bone.

Geology

Locality 54100, which yielded the human fossils, is one of a group of Middle Paleolithic sites. The top of the locality, at the east bank of the Fenhe River, is about 20–25 m above the water surface. The uppermost layer of the section is a loess-like deposit 5 m thick. The main deposit, about 20 m thick, is cross-bedded sands and gravels intercalated with lens of marls. Human fossils, a considerable number of stone artifacts, and vertebrae fossils were unearthed from the upper part of this layer. In the lower part of this layer, a large amount of molluscs and a few vertebrate fossils were found. Beneath this middle main layer is sandstone dating to the Early Pleistocene.

The mammalian fossils found at locality 54100 include (Pei, 1958):

Primates
- *Homo sapiens*

Insectivora
- Talpidae gen. et sp. indet.

Lagomorpha:
- *Ochotona* sp.

Table 3.11

Measurements of the Teeth from Dingcun

	Crown			*Root*		
	Length	*Breadth*	*Height*	*Length*	*Breadth*	*Height*
Upper median incisor	8.3	6.4	11.6	5.6	5.5	11.0
Upper lateral incisor	7.0	6.0	10.6	4.4	5.5	—
Lower second molar	11.2	10.1	8.0	—	—	—

Cited from Woo (1958)

Rodentia
- ?*Castor* sp.
- *Myospalax fontanieri*
- Myoidae gen. et sp. indet.

Carnivora
- *Vulpes* sp.
- *Ursus* sp.

Proboscidea
- Elephantidae gen. et sp. indet.

Perissodactyla
- *Equus przewalskyi*
- *Equus hemionus*
- *Coelodonta antiquitatis*

Artiodactyla
- *Megaloceros* cf. *ordosianus*
- *Megaloceros* sp.
- *Cervus (P.) grayi*
- *Spirocerus* sp.
- *Gazella* sp.
- *Bubalus* sp.
- *Bos primigenius*

This site was dated to the early part of the Upper Pleistocene by mammalian faunal correlation (Pei, 1958) to 160,00–210,000 years ago on the basis of uranium series dating (T. Chen et al., 1984).

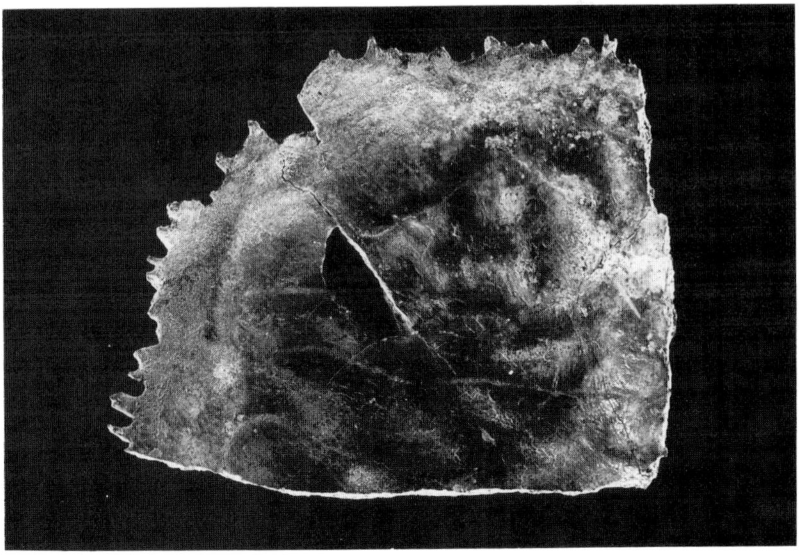

FIG. 3.12 Right infantile parietal of early *Homo sapiens* from Dingcun. (Courtesy of IVPP.)

Four samples of mollusc shells from Dingcun have been electron spin resonance (ESR) dated using the beta (β) radiation technique. The date ranges are 114,000 ± 1000, 97,000 ± 800, 84,000 ± 600, and 75,000 ± 600 years BP from 22.5 m deep to 17.5 m deep. The average sedimentary rate of the Dingcun section is 14.1 cm/1000 years (Huang and Cai, 1991). The *Lamprotula* fossils yielded a uranium series of dates of 83,000–150,000 years ago (Liang et al., 1991).

Archaeology

One hundred and seventy-one stone artifacts collected from Locality 54100 and other sites in the vicinity have been studied. The majority of the stone tools are flake tools, including points and scrapers. Points are more abundant than scrapers, and a thick prismatic point is an important tool type of this culture. Core tools include choppers, chopping tools, and bolas, although the latter are relatively rare. Most tools of the assemblage are small.

LOCALITY 4 OF ZHOUKOUDIAN (115°55′ E, 39°40′ N)

Locality 4 is on the southern slope of "Dragon Bone Hill," about 70 m south of Locality 1. In excavations before 1938, Locality 4 yielded stone artifacts and mammalian fossils dating to the early part of the late Pleistocene. In 1967 the "Red Guards" found another cave, Xindong (or New Cave), that had been obscured by deposits from Locality 4. In 1972 and 1973 the field team from the IVPP that excavated the deposits of Locality 4 and Xindong showed that the Xingdong deposits are actually part of the Locality 4 deposits.

Human fossils

Gu (1978) and T. Chen et al. (1984) described the materials from the site.

Teeth

Premolar (left, upper first PA 537). The buccal cusp of this maxillary premolar is larger than the lingual cusp. The former is more heavily worn, and the dentine is exposed as a small spot. No dentine is visible on the lingual cusp. A short groove divides the buccal cusp into mesial and distal parts. There is no cingulum on the basal part of the crown.

The root is moderately robust, and both mesial and distal surfaces have a longitudinal groove. The distal groove is the deepest. The lower part of the root is only slightly thicker than the upper part. The tip of the root is not very sharp and does not bifurcate.

The mesio-distal and bucco-lingual diameters and the height of the crown are 8.5 mm, 11 mm, and 6.1 mm, respectively. The mesio-distal and bucco-lingual diameters and the height of the root are 5.1 mm, 9.3 mm, and 18.8 mm, respectively. This tooth has been diagnosed as belonging to a male.

Geology

Deposits at the upper level of the site are looser than those of the lower level, which are cemented by calcium carbonate. The whole deposit is composed of yellow sandy clay with breccia, thin layers of travertine, and ashes. The human

reptilian fossils, there are many mammalian fossils, including the following (Gu, 1978):

Primates
- *Homo sapiens*
- *Macaca robustus*

Insectivora
- *Scaptochirus moschatus*
- *Erinaceus* sp.
- *Crocidura suaveolens*

Chiroptera
- *Rhinolophus ferrum-equinum*
- *Myotis* sp.
- *Ia io*

Rodentia
- *Sciurotamias davidianus*
- *Citellus undulatus*
- *Eutamias* cf. *sibiricus*
- *Cricetulus triton*
- *Cricetulus barabensis*
- *Myospalax wongi*
- *Microtus brandtioides*
- *Microtus epiratticeps*
- *Alticola* cf. *stracheyi*
- *Mus musculus*
- *Rattus norvegicus*
- *Apodemus sylvaticus*
- *Meriones meridianus*

Lagomorpha
- *Ochotona koslowi*
- *Ochotona daurica*
- *Lepus* sp.
- *Lepus* cf. *wongi*

Carnivora
- *Meles lucurus*
- *Mustela nivalis*
- *Felis* sp. (1)
- *Felis* sp. (2)
- *Felis lynx*
- *Cuon* sp.
- *Vulpes vulpes*
- *Ursus arctos*

Proboscidea
- Elephantidae gen. et sp. indet.

Artiodactyla
Sus cf. *lydekkeri*
Capreolus cf. *manchuricus*
Cervus elaphus
Cervus grayi
Megaloceros pachyosteus
Perissodactyla
?*Coelodonta* sp.
Equus cf. *sanmeniensis*

The faunal assemblage contains thousands of small mammals, many of which died at middle and young ages. The nature of the faunal assemblage, the size and age profile of the animals, plus burned bone, have been explained as the result of hunting and cooking. There are several uranium dates obtained from deer teeth. They are 122,000–171,000 years ago, 161,000–110,000 years ago, 171,000 + 15/ −13,000 years ago and 161,000–124,000 years ago for the third, fourth, sixth, and seventh layers, respectively (T. Chen et al., 1984).

Archaeology

Several dozen stone artifacts and two polished bone flakes were recovered. The thickest concentration within an ash layer begins from the first horizontal level, and extends northward and westward to the seventh horizontal level. It is generally 1 m wide and 0.9 m thick, but increases to 2 m wide and more than 1 m thick at the sixth and seventh horizontal levels. Within the ash layers are burnt stones and bones.

MIAOHOUSHAN (124°08′ E, 40°15′ N)

This cave site, found by limestone quarry workers in May 1978, is situated at the southern slope of Miaohoushan ("Hill Behind a Temple") near Shanchengzi village about 30 km east of Benxi Municipality and about 90 km southeast of Shenyang in Liaoning Province (northeastern China). In June a field team from the Museum of Liaoning Province found a human tooth and other mammalian fossils, as well as artifacts. Their excavations in 1979 and 1980 yielded more fossils and artifacts.

Zhenhong Zhang studied the human fossils, Yang studied the stratigraphy, Zhang et al. studied the nonhuman animal fossils, Kunshu Zhou and others did the palynological analysis, Sixun Yuan did the uranium series dating, Fang Qian did the paleomagnetic dating, and Huang et al. studied the artifacts. The results of these studies appear in the book entitled *Miaohoushan* (1986).

Human fossils

Teeth

Upper right canine (field no. B.S.M. 7802.T6-2). This heavily worn tooth preserves only 3 mm of the crown height. The occlusal surface is almost spherical, with a labio-lingual diameter of 8.7 mm. The height of the root is 15.7 mm. Its horizontal diameter gradually reduces from the neck to the tip of the root. On the basis of uranium dating of the layer yielding this canine, it seems to belong to *H. erectus*. However, the canine's shape and size suggest a closer affinity to *H. sapiens*.

Lower right molar (field no. B.S.M. 8001-T-9-1). On this moderately worn tooth, the exposure of dentine on the protoconid and hypoconid is united. The metaconid is slightly larger than the protoconid. Small areas of dentine were exposed on the metaconid and entoconid. The mesio-distal and bucco-lingual diameters are 11.56 mm and 10.80 mm, respectively. There is a Y-4 molar cusp pattern. The root was broken at the tip. Radiographic observation shows the root to have been taurodont.

Femur (field no. B.S.M. 8301-T7-1).

The 68.9 mm long femoral fragment was broken at the levels of the lesser trochanter and the nutrient foramen. The sagittal and transverse diameters measured below the lesser trochanter are 14.8 mm and 15.2 mm, respectively. The same diameters measured at the level of the nutrient foramen are 14.81 mm and 13.7 mm, respectively. The thicknesses of the bone wall at these two levels are 2.4 mm and 3.95 mm, respectively. The femoral fragment may belong to a child 8 or 9 years old, and the femoral fragment and the molar appear to belong to *H. sapiens*.

Geology

This cave is about 60 m above the water surface of the Tanghe River and corresponds to the third terrace of the river. The cave deposits were divided into eight layers. The geological strata follow:

Upper Pleistocene

8. Grayish-brown sandy clay, 0.5 m
7. Breccia with 16 mammalian species including *Megaloceros ordosianus* and *Crocuta ultima,* 5.2 m

Upper part of the Middle Pleistocene

6. Brownish-yellow sandy clay with a few breccia; rich in mammalian fossils, including a human molar, a child's femur, and 19 species of other mammals, including *Cervus canadensis, Crocuta ultima,* and *Hyaena sinensis,* 1.2 m
5. Brownish-yellow sandy clay intercalated with gray silt and a clay lens; rich with fossils, including a human canine and 20 mammalian species, including *Equus sanmeniensis, Macaca robustus, Dicerorhinus mercki,* and *Megaloceros pachyosteus,* 1.5 m
4. Brownish-yellow sandy clay with a few mammalian fossils, including *Equus sanmeniensis* and *Sinocastor anderssoni,* 1.6 m

———Erosion———

3. Breccia, 1.15 m
2. Light brownish-red gravels, 1.6 m

1. Light yellow sandy clay, 0.75 m

———Unconformity———

Ordovician limestone

Pollen and spores found in Layers 4–6 include such tree genera as *Pinus, Latrix, Abies, Picea, Quercus, Carpinus, Betula, Corylus, Alnus, Tilia, Ulmus, Zelkova, Celtis, Juglans, Rhus,* and *Salix.* Bushes and grasses are represented by *Ephedra,* Graminea, Chenopodiaceae, *Thalictrum, Humulus, Sanguisoba,* Dipsaceae, Ranunculaceae, Cruciferae, Cyperaceae, Caryophyllaceae, Liliaceae, Polygonaceae, Leguminosae, Geraniaceae, *Artemisia,* Compositae, Polypodiaceae, and *Selagenella,* among others. The flora suggests that Layer 4 was characterized by a climate slightly cooler than today. During the deposition of Layers 5 and 6, the climate corresponded to the southern part of the warm temperate zone. The faunal assemblage supports this inference.

Paleomagnetic dating suggests that the fifth through eighth layers were deposited about 0.4 mya. Partial data for the uranium series dating for this site are listed in Table 3.12.

Archaeology

The site at Miaohoushan produced 76 artifacts, among which nine are from the fourth and fifth layers and five are from the seventh layer. The remaining artifacts are from the sixth layer. Among the stone artifacts from Layers 4–6, 3 nuclei, 28 flakes, 13 scrapers made from flakes, 12 chopper-chopping tools, and 2 stone balls have been reported.

Most artifacts were made from black quartzite sandstone gravels. Andesite gravels were less frequently used, and only a few vein quartz tools were produced. The hammer percussion and stone anvil techniques were used to produce the flakes. Only one bipolar flake was found. All the stone tools are retouched by the hammer percussion technique. The scrapers are crudely made.

Some of the bone artifacts recovered from Miaohoushan were probably trimmed and used as tools.

The sixth layer of the deposits contained an ash lens, 5–10 cm thick. The ash, a deep brown powder sporadically mixed with charcoal, contained burnt nonhuman bones.

The name *Miaohoushan culture* has been proposed for the Middle Pleistocene artifacts from this site. These artifacts have both similarities and differences be-

Table 3.12

Uranium Series Dating of Miaohoushan (in ky)

No.in Lab·	*Specimen*	*Position*	*Age of ^{230}Th*	*Age of Pa method*
BKY83012	Teeth (*R. mercki*)	Upper part of 6th layer	$14.2^{+1.3}_{-1.1}$	
BKY83030	Teeth (deer)	Bottom of 6th layer	$22.8^{+5.6}_{-4.0}$	
BKY82129	Teeth (rhino)	Upper part of 5th layer	$14.8^{+5.6}_{-1.0}$	$7.4^{+5.6}_{-1.2}$
BKY83009	Teeth (pig)	Sublayer of calcified concretions of 5th layer	$24.7^{+8.3}_{-4.9}$	

According to the Museum of Liaoning Province and the Museum of Benxi City (1986)

tween this culture and that from the sites of Zhoukoudian, Kehe, and Chongokri. All of these have yielded Pleistocene tool assemblages.

TONGZI (106°45′ E, 28°15′ N)

The Tongzi fossils derive from Yanhuidong limestone cave located in Tongzi County, Guizhou Province (southern China). The cave is on the southern slope of Chaishan Hill, which is situated in Jiuba District, about 10 km northwest of the city of Tongzi. The cave is about 32 m above the local water surface.

In 1971, a field team from the IVPP were drawn to the cave because of previously discovered mammalian fossils. A further excavation in 1972 found a human tooth, stone artifacts, evidence of fire, and mammalian fossils. During this excavation several human teeth were found in the disturbed deposits. Because the teeth were very weakly mineralized and were considered as being embedded in the top stratum, they were considered recent remains and were not studied. During the cleaning and sorting in the laboratory, a human premolar was found.

After the IVPP excavations, local farmers excavated the remaining deposit. In 1983, another IVPP field team recovered four human teeth and some mammalian fossils. Based on the similarity of the 1972 and 1983 deposits, they considered these teeth to be dated to the same level as those found in 1972.

Human fossils

The following descriptions are from M. Wu et al. (1975), and M. Wu (1984).

Teeth

Incisor (right upper, I1 PA 520). This tooth was found in the 1972 excavation. The crown is nearly intact, and only a small piece of the mesial part of the cutting edge is broken. The heavily worn cutting edge forms a flat plane, with the mesial end situated at a lower level than the distal end. Dentine was exposed at the center of this flat plane. The labial surface of the crown is slightly bulging in both the vertical and horizontal dimensions. On the labial surface, the mesial part is more bulging than the distal part. The upper part of the crown is more bulging than the lower part. The mesial and distal margins are thickened and folded toward the lingual surface. A rounded basal tubercle on the lingual surface connects with the two marginal ridges. This was probably a shovel-shaped incisor. Three ridges sent from the lower margin of the basal tubercle represent the remnants of finger-like processes.

The intact root is 17.9 mm in height. Its horizontal diameter diminishes gradually toward the tip, which turns slightly to the distal side. There is no groove on either the mesial or distal surfaces. No constriction appears at the cervix. The demarcation between the crown and the root is at the same level on both the labial and lingual surfaces.

Premolar (right upper, P1 PA 521). This tooth was found during sorting in the laboratory in 1972. The small buccal part of the crown and the buccal root were not recovered. The occlusal surface is almost complete. The mesio-distal diameter is longer buccally than lingually. A deep groove divides the occlusal surface into a larger buccal and a smaller lingual part. In addition to the marginal ridge, two short transverse ridges are separated from the marginal ridge by deep transverse grooves. The lingual surface is especially bulging horizontally.

The following four teeth were collected during the 1983 IVPP excavations (Fig. 3.13).

Canine (left upper, C PA 872). M. Wu (1984) estimated that this canine with an intact and unworn crown belonged to a 10-year-old child. The buccal surface is very bulging and has a few riblike striae. A broad cingulum appears on the basal part of the tooth. Very prominent triangular eminences separated from the buccal surface by shallow grooves are found on the mesial and distal margins. The thick mesial and distal margins converge upward to form a basal tubercle, from the lower margin of which there are two thick ridges extending vertically to the middle of the lingual surface of the tooth. One of these ridges bifurcates into two and one ridge extends to the cutting edge. The lateral margins of the crown are separated from the lingual surface proper by deep grooves. The tip of the crown is rounded and tubercle-like. The cervix is not constricted.

The canine root is not fully developed. Shallow longitudinal grooves are visible on the mesial and distal surfaces of the root. The bucco-lingual diameter is obviously longer than the mesio-distal diameter.

Premolar (left upper, P1 PA 873). The premolar crown is well preserved, but a large part of the root is lost. The occlusal surface is oval-shaped. The mesio-distal diameter is much longer at the buccal than at the lingual part. A deep groove divides the occlusal surface into two parts; the buccal surface is larger and higher. The remnant of a cingulum appears as the triangular prominence basally located on the buccal surface. The lingual surface is bulging, without any ridge or fovea. Two small ridges connecting the cusps and the longitudinal groove are separated from the marginal ridge with deep grooves. The cervix is slightly constricted.

The bucco-lingual diameter of the preserved part of the root is greater than the

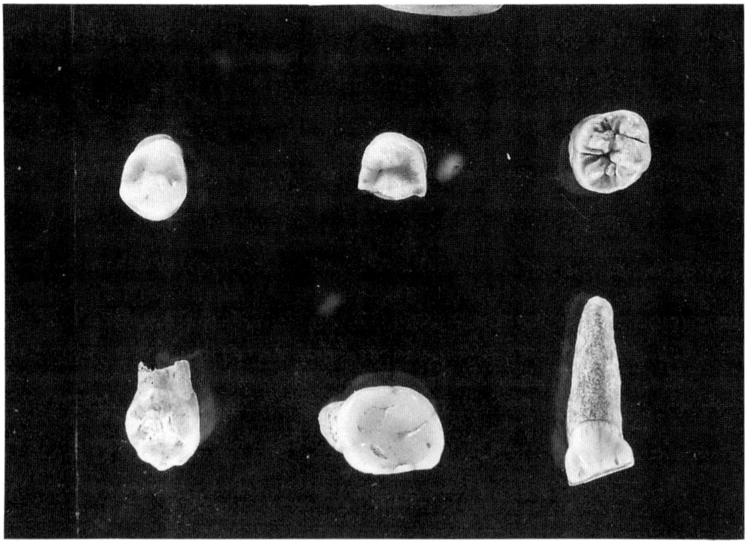

FIG. 3.13 Teeth of early *Homo sapiens* from Tongzi. (Courtesy of IVPP.)

mesio-distal diameter The buccal and lingual branches of the root are not separate at the basal portion. There is a broad shallow groove on the middle part of the mesial surface of the root.

Molar (left upper, probably M1 PA 874). The unworn crown is intact and the root is not fully developed. Wu (1984) estimated that the tooth belonged to a 6-year-old child. A vertical groove on the buccal surface extends from the cervix to the occlusal surface. The mesial and distal surfaces are slightly convex. A cingulum surrounding the basal part of the crown is especially strong at the buccal surface. The paracone is separated from the metacone by a short transverse groove that connects with the buccal vertical groove. The groove separating the protocone and hypocone continues with an oblique groove on the lingual surface. The protocone is separated from the metacone by a short oblique groove. Therefore, there is no longitudinal sulcus on the occlusal surface. On the mesial half of the occlusal surface, the tip of a V-shaped groove connects with the transverse groove. On the paracone, protocone, and the metacone, each cusp has a thick and rounded main ridge and two accessory ridges on the sloping surface.

Molar (right upper, probably M1 PA 875). Although the crown is heavily worn, four cusps are still discernible. The hypocone is well developed and almost equals the protocone in size. The protocone connects with the metacone. There is a short transverse groove at the distal part of the occlusal surface. The groove separating the protocone and paracone is almost in line with the middle line of the occlusal surface. Traces of a cingulum are visible. This tooth shares with molar PA874 the presence of a vertical groove on the buccal surface and an oblique groove on the lingual surface. The lingual root is very robust and straight. Only the basal segment of the buccal root is preserved; its two branches were not recovered.

Three of these teeth have obvious yellow pits on the crown. This coloration is especially obvious in PA872 and PA874. Wu (1984) considered this yellow pit to be the result of a high fluorine concentration in the drinking water.

The human teeth found in 1972 were considered as belonging to early *H. sapiens*. M. Wu (1984) suggested that the teeth belonged to *H. erectus,* and noted similarities between these teeth and those of "Peking Man". Futhermore, the presence of *Hystrix magna* in the cave is associated with the early Pleistocene, but not the middle and late Pleistocene. The measurements of the Tongzi teeth are provided in Table 3.13.

Table 3.13

Measurements of the teeth from Tongzi

	Crown			Root		
	Length	*Breadth*	*Height*	*Length*	*Breadth*	*Height*
Upper median incisor	(10.3)	8.3	—	—	—	17.9
Upper canine (PA872)	9.4	9.7	13.3	—	—	—
Upper first premolar (right)	9.0	(12.8)	—	—	—	—
Upper first premolar (left) (PA873)	9.0	11.1	(9.2)	—	—	—
Upper molar (PA874)	10.5	11.1	—	—	—	—
Upper molar (PA875)	11.4	14.0	—	—	—	—

Cited from Wu (1984) except those figures for the incsior and right premolar which are from Wu, et al. (1975)

Geology

Farmers had removed the deposits for fertilizer only near the entrance of the cave. From the deeper deposits M. Wu et al. (1975) recorded the geological section as follows:

1. Disturbed earth, 20–30 cm thick
2. Stalactite, 0.5 cm thick
3. Loose brown clay with a few fossils, about 21 cm thick
4. Grayish-white or grayish-yellow sandy clay, containing gravels, semicemented, with many fragments of vertebrate fossils and some stone artifacts, 60 cm thick
5. Hard yellow clay without fossils, at least 80 cm thick

The mammalian fauna include (according to Han and Xu, 1989):

Primates
- *Homo sapiens*
- *Pongo* sp.
- *Hylobates* sp.
- *Rhinopithecus* sp.

Rodentia
- *Petaurista* cf. *brachyodus*
- *Rhizomys* sp.
- *Hystrix magna*
- *Hystrix subcristata*

Carnivora
- *Canis lupus*
- *Cuon javanicus antiquus*
- *Ailuropoda melanoleuca fovealis*
- *Ursus kokeni*
- *Arctonyx* sp.
- *Crocuta ultima*
- *Panthera tigris*
- *Panthera pardus*

Proboscidea
- *Stegodon orientalis*

Perissodactyla
- *Megatapirus augustus*
- *Rhinoceros sinensis*

Artiodactyla
- *Sus scrofa*
- *Muntiacus* sp.
- *Cervus* sp. A
- *Cervus* sp. B

Capricornis cf. *sumatraensis*
Bovidae gen. et sp. indet.

Uranium dates for the nonhuman teeth perhaps derived from the fourth layer are 113,000 ± 11,000 years ago, 115,000 ± 7,000 years ago, and 181,000 ±/− 11,000 years ago (Yuan et al., 1986).

Archaeology

In 1971 and 1972, 12 stone artifacts were found in the cave. Two scrapers, one nucleus, and two pieces with obvious traces of working have been described by M. Wu et al. (1975). During the 1972 excavations charcoal and a few pieces of burnt bone were recovered. Wu et al. considered these to have been transported by water into the cave. No hearth was found.

SUMMARY

As in other parts of the world, Chinese early *H. sapiens* samples show traits of both *H. erectus* and of *H. sapiens*. The samples bridge the temporal and morphological gap between the latter two kinds of hominid. Increasingly, the Chinese sample is being used to dispute the Out-of-Africa model and is used as a strong case for regional evolution. This view has historical roots dating back to Weidenreich. It is worthy of note that some of the so-called derived characters of *H. erectus*, such as thick cranial bones, thick brow ridges, presence of angular torus, and postorbital constriction, occur also in Chinese early *H. sapiens*.

The uranium dates of the Jinniushan site raise the possibility that *H. erectus* and early *H. sapiens* might have coexisted in China. But there is a problem determining the stratigraphical position of the Jinniushan hominid in the cave deposits, hence the date is to a certain extent problematic.

It is not exactly clear how the Chinese *H. sapiens* sample relates to other Asian samples, such as the Narmada skullcap from India. However, it shares numerous traits with Narmada.

4

Anatomically Modern *Homo sapiens*

Many sites in China have yielded anatomically modern *H. sapiens* samples. A number of these sites have been found relatively recently. Interestingly, many of these fossil samples show strong morphological links with the preceding *H. erectus* samples in China. This continuity of morphology throws doubt on the so-called Out-of-Africa scenario favored by many Western paleoanthropologists. Ten of the shared traits are discussed in detail in Chapter 5.

UPPER CAVE, ZHOUKOUDIAN (115°55′ E, 39°40′ N)

The Upper Cave is situated at the top of Dragon Bone Hill, Zhoukoudian. During the 1933 excavations for "Peking Man," human fossils found in this cave included 3 nearly complete skulls (Table 4.1), 3 cranial fragments, 4 mandibles, 3 mandibular fragments, dozens of isolated teeth, vertebrae, 2 sacrum, 3 scapula fragments, 1 radius fragment, 6 fragmentary hip bones, several femoral fragments, 3 patella, 2 tibiae, 6 metatarsals, and 1 tarsal (Jia, 1951; Weidenreich, 1939). During World War II, the human fossils from the cave were lost along with the "Peking Man" remains, and only casts of some of the specimens are available for study. Xinzhi Wu (1961a) compared his measurements of the casts with figures published by Weidenreich (1939) and found that the difference between the measurements is less than 1%. Of the many researchers who studied the Upper Cave materials, we have relied here upon the work of Jia (1951), X. Wu (1961a,b), Pei (1939, 1940), X. Li et al. (1985), Pei (1985), and T. Chen et al. (1984, 1992).

Human fossils

Skulls (nos. 101, 102, 103)

Skull no. 101 (Fig. 4.1), belonging to an old male, is the most complete including a complete mandible. The superciliary arch exhibits a ridge. The mastoid process and the external occipital protuberance are robust. The coronal suture, the left wing, and a large part of the right wing of the lambdoidal suture are distinct. The sagittal suture is discernible only at the obelion region. The teeth are well worn.

Skulls no. 102 and 103 (Figs. 4.2 and 4.3) are smaller than Skull no. 101, and both have smaller mastoid processes and less distinctive muscular markings (Table 4.1). Both no. 102 and 103 belong to females, despite the fact that the superciliary arches are more robust than in modern females. The sagittal, coronal, lambdoidal, and squamosal sutures of Skull no. 102 are clearly seen. Because the sphenoid has

Table 4.1

Measurements and indices of skulls from the Upper Cave

SPECIMEN NUMBER		*101*	*102*	*103*
Linear measurements (in mm):				
Max. cranial length (g-op)		204.0	196.0	184.0
Glabella inion length (g-i)		198.0	185.0	180.0
Max. cranial breadth (eu-eu)		143.0	136.0	131.0
Min. frontal breadth (ft-ft)		107.0	102.5	101.0
Max. bizygomatic breadth (zy-zy)		143.0	131.0	137.0
Upper facial height (n-sd)		77.0	69.0	68.5
Nasal height (n-ns)		58.0	46.5	51.0
Nasal breadth		32.0	26.0	25.5
Orbital breadth (mf-ek)	lt.	48.5	40.5	45.0
	rt.	48.0	45.0	45.0
Orbital height	lt.	31.5	29.3	31.0
	rt.	33.2	31.5	32.0
Middle facial breadth (GB zm-zm)		106.2	106.4	101.0
Subtense GB from sd		35.1	36.5	33.5
IOW (fmo-fmo)		110.0	104.1	106.0
Subtense IOW from nasion		21.4	23.5	13.5
Interorbital breadth (mf-mf)		19.1	21.0	20.5
Subtense mf-mf		9.2	7.7	7.5
Breadth of nasal root (SC)		7.0	9.1	9.0
Height of nasal root (SS)		4.0	4.5	3.1
Palatal length (ol-sta)		52.0	47.0	48.0
Palatal breadth		43.0	40.0	38.0
Palatal height		13.8	10.5	8.5
Length of foramen magnum (ba-o)		39.2	44.0	40.5
Breadth of foramen magnum		35.0	32.3	32.8
Ba-pr		106.2	113.6	109.3
Ba-ss		103.0	109.6	104.0
Ba-n		111.0	116.0	108.5
Ba-b		136.0	150.0	143.0
Ba-v		137.5	154.5	144.5
Auricular height (OH v-FH)		113.0	119.0	118.0
Projectile distance from bregma to porion		148.0	120.0	118.5
Length of alveolar process		57.0	57.5	58.3
Breadth of alveolar process		69.2	72.6	66.0
Median sagittal arc (arc n⌒o)		388.5	384.5	363.0
Frontal arc (arc n⌒b)		132.0	126.0	121.0
Frontal chord (n-b)		115.5	116.2	107.0
Parietal arc (arc b⌒l)		132.0	135.5	132.5
Parietal chord (b-l)		120.8	120.4	120.0
Occipital arc (arc l⌒o)		124.5	123.0	110.0
Occipital chord (l-o)		97.6	106.0	93.0
Cranial capacity ca.*		1500	1380	1300

(continued)

Table 4.1 (continued)

Measurements and indices of skulls from the Upper Cave

SPECIMEN NUMBER		*101*	*102*	*103*
Linear measurements (in mm):				
Angles				
Angle formed by tympanic plate with				
mid-sagittal plane	lt.	80°	85°	80°
	rt.	85°	80°	75°
Total prognathism (n-pr–FH)	84°	80°	79°	
Middle prognathism (n-ns–FH)		90°	82°	80°
Alveolar profile prognathism (ns-pr–FH)		80°	75°	73°
Bregmatic angle (g-b-FH)		45°	48°	46°
Schwalbe's frontal angle (b-g-i)		52°	59°	57°
Profile angle of frontal (g-m-FH)		65.5°	65°	76.5°
Nasomalar angle (fmo-n-fmo)		135°	130°	148°
Zygomaxillary angle (zm-ss-zm)		128°	125°	131°
Indices:				
Cranial index (100 x eu-eu/g-op)		70.1	69.4	71.2
Length-height index (100 x ba-b/g-op)		66.7	76.5	77.7
Length-height index (100 x OH/g-op)		55.4	60.7	64.1
Breadth-height index (100 x ba-b/eu-eu)		95.1	110.3	109.2
Breadth-height index (100 x OH/eu-eu)		79.0	87.5	90.1
Vertical craniofacial index (100 x n-sd/OH)		56.6	46.0	47.9
Transverse craniofacial index (100 x zy-zy/eu-eu)		100.0	96.3	104.6
Zygofrontal index (100 x ft-ft/zy-zy)		74.8	78.2	73.7
Upper facial index (100 x n-sd/zy-zy)		53.8	52.7	50.0
Total facial index (100 x n-gn/zy-zy)		86.0	—	?81.4
Prognathism index (100 x ba-pr/ba-n)		95.7	97.9	100.7
Orbital index-	left	64.9	72.3	68.9
Orbital index-	right	69.1	70.0	71.1
Nasal index		55.2	55.9	50.0
Palatal index		82.7	85.1	79.2
Maxilloalveolar index		121.4	126.3	113.2
Index of foramen magnum		89.3	73.4	81.0
Frontal index (100 x Subtense IOW/IOW)		19.5	22.6	12.7
Simotic index (100 x SS/SC)		57.1	49.5	34.1
Premaxillary index (100 x Subtense GB/GB)		33.1	34.3	33.2
Maxillo-frontal index (100 x Sub.mf-mf/mf-mf)		48.2	36.7	36.6
Chord-arc index of frontal bone		87.5	92.2	88.4
Chord-arc index of parietal bone		91.5	88.9	90.6
Chord-arc index of occipital bone		78.4	86.2	84.5

Measurements by Wu (1961a, b) based on casts.

*Estimated by Weidenreich (1939).

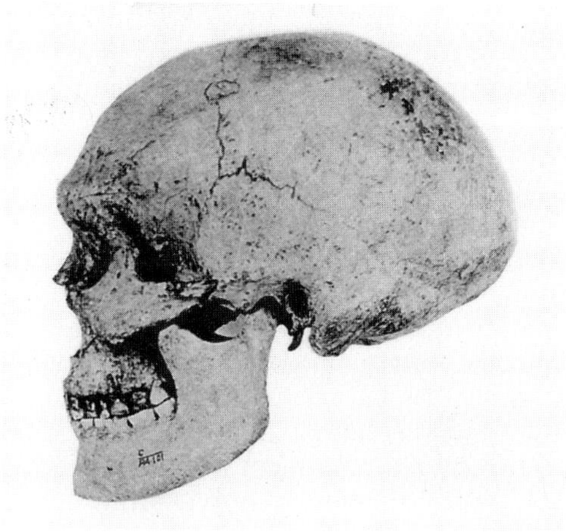

FIG. 4.1 Skull no. 101, late *Homo sapiens*, from The Upper Cave, Zhoukoudian. (Courtesy of IVPP.)

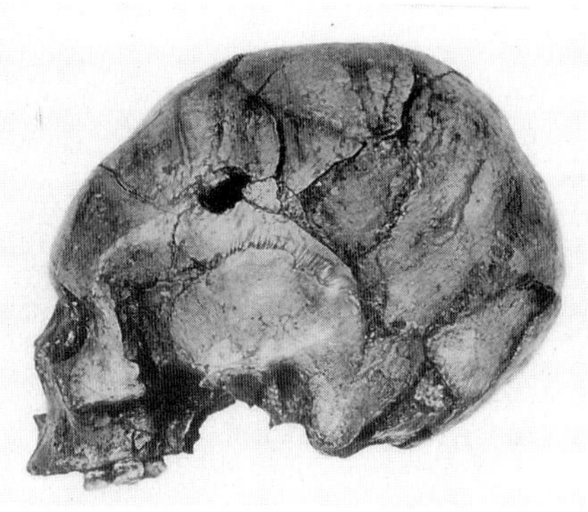

FIG. 4.2 Skull no. 102. late *Homo sapiens*, from The Upper Cave, Zhoukoudian. (Courtesy of IVPP.)

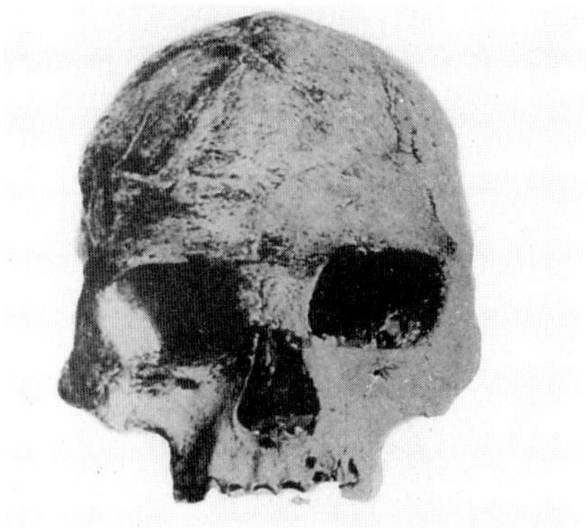

FIG. 4.3 Skull no. 103, late *Homo sapiens*, from The Upper Cave, Zhoukoudian. (Courtesy of IVPP.)

not united with the occipital bone, the third molar has not erupted, and the second molar was only slightly worn, this skull probably belonged to a young woman about 20 years old.

Skull no. 103 was cracked. The main cranial sutures are distinct. Because the sphenoid has united with the occipital bone, and because all the teeth have erupted and are moderately worn, this skull probably represents a middle-aged woman.

The three skulls have some common morphological characters. Compared with late modern humans in northern China, the Upper Cave skulls are larger, they have a lower cranial index, more robust superciliary arches, a less developed frontal tubercle, a more receding forehead, a rhomboidal depression at the region of obelion, a broader interorbital breadth, lower orbits, shallower lacrimal fossa, and broader noses, among other traits. These three skulls share a degree of prognathism, the quadrangular shape of the orbit, the zygomaxillary angle, the palatal index, a low upper face and prenasal fossa, and the absence of a thin ridge along the lower margin of the pyriform aperture. The maximum breadth in all three skulls is close to the temporal suture, and the foramen magnum opens downward and slightly backward.

Skull no. 101 has especially strong brow ridges and a strongly protruding glabella region with a shallow groove behind the ridge. The temporal squama is triangular in shape and the fronto-sphenoidal margin is perpendicular to the parietal margin. The infratemporal surface of the greater wing of the sphenoid merges with the temporal surface without a distinct infratemporal ridge in between. When held in the Frankfurt Plane, the inion is more posteriorly located than lambda, and the parietal tubercle appears above and behind the mastoid process. There is a rhombdoidal depression at the obelion region. Above the junction of the left temporal line and the coronal suture, there exists an ovoid depression (whose sagittal dimension is 1.5 cm and whose vertical dimension is 1.2 cm) with cracks along its margin and on the bottom.

The nasomalar angle of Skull no. 101 is rather low compared to that in eastern Asians (Mongoloids) in general. The nasal bones are pinched and have a longitudinal narrow ridge along the midsagittal line on the anterior surface. The midsagittal contour of the anterior surface of the alveolar process is convex. A mandibular torus is visible.

It has been suggested that Skull no. 102 represents a Melanoid type because of such features as a high vault, broad nose, and low orbits. In Weidenreich's (1939) opinion, the upper facial, total facial and maxillo-alveolar indices, total prognathism, middle prognathism, and the alveolar profile prognathism are close to that typical of New Caledonians. Wu (1961b) analyzed each of these traits and concluded that the features of Skull no. 102 are within the range of variation of modern eastern Asians. The extraordinarily high vault results from artificial deformation, indicated by a transverse shallow groove on the frontal bone above the frontal tubercle. The broad nose, low orbits, and low upper facial index are common to Late Paleolithic samples in China. Weidenreich's (1939) assessment that Skull 102 belongs to a Melanoid is incorrect because his calculations are based on the mistaken assumption that Mandible no. 104 belongs with Skull no. 102. This is contradicted by such dental differences as differential attrition and eruption of the molars, and other features suggesting that the skull and mandible belong to different individuals. Furthermore, the maxillo-alveolar index of Skull no. 102 is much higher than that of New Caledonians but is close to that of modern Chinese and Eskimos. Other features, such as the middle and alveolar profile prognathisms, are within the range of modern eastern Asians.

Because other features of Skull no. 102 approximate the condition found on other Upper Cave skulls, X. Wu (1961a,b) considers that it belongs with the other members of the Upper Cave population. He rejects Weidenreich's notion that no. 102 represents a Melanesoid type that is different from the other Upper Cave skulls.

Skull no. 102 has a perforation on the left temporal line just behind the coronal suture. The hole is about the same size and in a similar position as the depression on Skull no. 101. Weidenreich (1939) reported that all four preserved skullcaps show evidence of a violent death. The skulls have depressions, fractures, or holes, apparently caused by heavy blows with sharp, blunt implements.

Skull no. 103 has been described as representing an Eskimoid type. However, the features prompting this diagnosis are not inconsistent with a description of east Asian populations generally. A reconstructed bust of an Upper Cave inhabitant appears in Figure 4.4.

Maxilla (no. 110)

This specimen includes the right alveolar process, small parts of the maxillary body, and a hard palate. Alveolar prognathism is estimated as about 70°. The nasal floor is slightly convex. Its highest point is at the position corresponding to the premolar, from here the nasal floor slopes forward and backward. A thin ridge along the lower margin of the pyriform aperture was not seen in the three skulls, but it does exist in this maxilla. The anterior end of the maxillary sinus corresponds to the position of the second premolar. The posterior end of the sinus extends to correspond to at least the third molar. The floor of the maxillary sinus is lower than the hard palate.

A maxillary torus on the buccal surface of the alveolar process corresponds to the position of the first molar and the anterior half of the second molar. The torus is elongated. The dental arch is V-shaped, judging from its preserved half. There is an extra small tooth behind the median incisor and medial to the lateral incisor.

Mandibles (nos. 101, 104, 108, 109)

All four mandibles that were cast belonged to adults: No 101 belonged to an old male, nos. 104 and 109 belonged to middle-aged women, and no. 108 belonged to a younger adult. Mandibles nos. 101 and 104 are almost complete. Both nos. 108 and 109 include the right half of the mandibular body and a part of the left half (Table 4.2).

In four instances mental foramen are located beneath the second premolar, and in three cases they are located beneath the second premolar and first molar. Xinzhi Wu (1961a) indicated that the average height of the mental foramen in the Upper

FIG. 4.4 Bust of late *Homo sapiens* from The Upper Cave, Zhoukoudian. (Courtesy of IVPP.)

Table 4.2
Measurements and Indices of the Mandible From the Upper Cave

SPECIMEN	101		104		108	
SIDE	lt	rt	lt	rt	lt	rt
Linear measurements:						
Body height at mental foramen	35.0	36.8	29.9	30.0	—	31.3
Body height between M1 and M2	33.6	34.1	26.2	26.8	30.8	29.1
Body thickness at mental foramen	12.0	12.6	12.3	12.9	11.3	11.2
Body thickness between M1 and M2	13.6	14.8	15.0	15.0	12.4	12.0
Height of mental foramen	16.5	16.0	13.9	13.0	12.8	13.1
Height of ramus	61.0	60.8	54.5	—	—	—
Breadth of ramus	41.0	41.6	40.5	—	—	—
Length of semilunar notch	38.0	34.0	36.0	38.0	—	—
Depth of semilunar notch	13.0	15.0	16.0	16.0	—	—
Bicondylar breadth	130.0		126.0		—	
Bigonial breadth	116.0		—		—	
Id-gn	37.8		31.2		31.9	
Arc id-gn	42.7		34.2		35.9	
Bimental breadth	49.0		48.0		49.0	
Bimental arc	59.0		59.0		65.0	
Mandibular angle	117°		120°		—	
Indices:						
Index of mental height	47.1	43.5	46.5	43.3	—	41.9
Robustness at mental foramen	34.3	34.2	41.1	43.0	—	35.8
Robustness between M1 and M2	40.5	43.4	57.3	56.0	40.3	41.2
Depth of semilunar notch	34.2	44.1	44.4	42.1	—	—
Protrusion of chin	112.9		109.6		112.5	
Height-breadth index of ramus	67.2	68.4	74.3	—	—	—

Measurements according to Wu (196la, b) based on casts.

Cave material (average of 44.5) is lower than that of modern Chinese (around 50) and close to the average of European Late Paleolithic specimens.

In Mandible no. 101 a mandibular torus exists on the lingual surface just beneath the alveolar margin and corresponding to the position of the right second premolar. There seems to be a trace of a mandibular torus near the left first premolar. The mandibular angle of nos. 101 and 104 is everted. The mandibular surface is rough.

Femora

Weidenreich (1941) provided some measurements of the Upper Cave femora. These figures are listed in Table 4.3. Weidenreich (1939) estimated the statures of the Upper Cave hominids as 174 cm for males and 159 cm for females.

Patella

There is a rather robust patella, that probably belonged to a male. The maximum height, breadth, and thickness are 44.5 mm, 45.0 mm, and 21.8 mm, respectively. The vastus notch is weakly developed.

Table 4.3

Femoral measurements and indices from the Upper Cave

	Sex	No.	Range	Average
Linear measurements:				
Maximum length	F	2		438
Vertical diameter of neck	M	2	26.4-35.0	30.7
Transverse diameter of neck	M	2	22.3-26.0	24.2
Vertical distance between distal basis of lesser trochanter and upper margin of neck (parallel to axis of diaphysis)	M	2	60-73	66.5
Circumference of neck	M	2	83-107	95
Subtrochanteric level:				
Transverse diameter	M	2		32.0
	F	2		28.0
Sagittal diameter	M	2		30.2
	F	2		23.1
Circumference	M	2		94.0
	F	2		80.0
Mid-shaft level:				
Transverse diameter	M	2		28.1
	F	2		23.7
Sagittal diameter	M	2		32.0
	F	2		29.3
Circumference	M	2		95.0
	F	2		83.0
Supracondylar level: (40mm above the condyle)				
Transverse diameter	F	2		36.4
Sagittal diameter	F	2		28.0
Circumference	F	2		104.0
Transverse diameter of canal	M	1		12.3
	F	1		10.2
Transverse diameter of shaft	M	1		24.4
	F	1		23.2
Sagittal diameter of canal	M	1		16.1
	F	1		15.4
Sagittal diameter of shaft	M	1		32.1
	F	1		35.1
Indices:				
Platymeric index	M&F	6	81.3 – 104.8	86.1
Pilastric index	M&F	6	112.0 – 139.5	125.6
Popliteal index	M&F	3	74.7 – 81.4	76.9
Collum index (ellipticity)	M&F	4	72.6 – 82.3	78.3
Index of robusticity of transverse diameter	M	1		50.2
	F	1		43.6
Index of robusticity of sagittal diameter	M	1		50.2
	F	1		43.8
Angles				
Torsion angle	M&F	3°	—	8°
Collar angle	M&F	4°	123 – 137°	129°

According to Weidenreich (1941)

Geology

The following is paraphrased from Pei's (1939) original description of the stratigraphy of the Upper Cave. The Upper Cave, a limestone cave situated above the "Peking Man" cave, is comprised of four parts: the entrance, upper room, lower room, and lower recess. The total thickness of the deposits is over 10 m.

The entrance formed an archway in the limestone that measured about 4 m high and 5 m wide at the floor. The upper room is limited southward (opposite the entrance) by a steep east-west wall of limestone. The distance from the entrance to the wall is 8 m, and the east-west length of the southern wall is 14 m. The lower room is located beneath the western half of the upper room and is 8 m deep. The lower recess is the lowest part of the cave with a south to north length of about 3 m, and a width of 1 m. Most of the limestone walls are almost vertical and covered by a thick layer of stalagmitic incrustation. The bottom is formed by the hard and red colored breccia of the "Peking Man" site.

The cave deposits consist of a gray loam mixed with pieces of a roof that had gradually collapsed. A few consolidated layers exist at the entrance, just over the floor formed by the deposit of the "Peking Man" site. This is near the northern wall, within the middle of the deposits. These cemented patches are localized and never thicker than 50 cm. The sediments appear almost loose. Regular concretions (4.5 cm in diameter) were found isolated near the bottom of the lower room. In the deepest part of the cave a few patches of crystalline calcite are interbedded in the loam.

Cultural layers within the Upper Cave are distributed as follows:

A. Entrance and Upper Room

Layer 1 (uppermost layer): About 30 cm thick at 3 m above the floor near the entrance. A few human bones, a perforated tooth, and two flint implements were found here.

Layer 2: Subdivided into several thin beds located 1 m above the floor of the Upper Room. A few human bones and 28 perforated animal canines were found.

Layer 3: At the very bottom of the Upper Room. Evidence of burning on the stalagmitic floor and limestone suggests human occupation. Only a few cultural remains were found in this layer.

B. Lower Room

Layer 4: Five meters above the floor of the Lower Room.

Layer 5: Situated on the floor of the Lower Room.

These layers yielded a few isolated human teeth, numerous perforated nonhuman teeth, bone pendants and a flake chert slightly above the bone pendants, three complete human skulls, and many other human bones. The Lower Room is probably a burial place that experienced postburial disturbance.

The lower recess yielded no human nor cultural remains, but many complete nonhuman skeletons were found. The completeness of the nonhuman skeletons indicates that the lower recess might be a natural bone trap.

The mammalian fauna from the Upper Cave include the following (Qi, 1989):

Primates
- *Homo sapiens*

Insectivora
- *Erinaceus* sp.

Chiroptera
- *Scaptochirus* sp.
- *Myotis* sp. (A)
- *Myotis* sp. (B)

Rhinolophus sp.
Chiroptera fam. gen. et sp. indet.
Lagomorpha
Lepus europaeus
Ochotona daurica
Rodentia
Sciurus sp.
Petaurista sulcatus
Cricetinus varians
Cricetulus barabensis
Myospalax armandi
Alticola cf. stracheyi
Microtus epiratticeps
Meriones sp.
Apodemys sylvaticus
Rattus rattus
Hystrix sp.
Carnivora
Canis lupus
Nyctereutes procyonoides
Vulpes corsac
Vulpes vulgaris
Cuon alpinus
Ursus thibetanus
Ursus spelaeus
Mustela cf. *altaica*
Mustela eversmanni
Meles leucurus
Paguma larvata
Crocuta ultima
Felis chinensis
Lynx lynx
Lynx sp.
Panthera tigris
Panthera pardus
Acinonyx cf. *jubatus*
Proboscidea
Elephas sp.
Perissodactyla
Equus hemionus
Rhinoceros sp.

Artiodactyla
Sus sp.
Cervus (P.) hortulorum
Cervus (E.) canadensis
Capreolus manchuricus
Gazella przewalskyi
Ovis sp.
Bos sp.

This fauna, especially with the presence of cheetah *(Acinonyx)* and hyena *(Crocuta),* suggests a warm climate.

Carbon dating was done on the animal bones. The sample from the fourth cultural layer in the Lower Room was dated as 10,470 ± 3600 BP, and that from the lower recess was dated at 18,340 ± 4100 BP (Y. Li et al., 1985). Uranium series dating for the cultural layer is 19,000 ± 1000 BP, and the date for the lower recess is 21,000 ± 4000 BP (T. Chen et al., 1984). TL (thermoluminescence) dating for the lower recess is earlier than 32,500 BP, but no later than 50,000 BP (Pei, 1985). The recent accelerator radiocarbon date for the cultural layer of the Upper Cave is 29,000 to 34,000 years ago; that for the lower recess is about 34,000–33,000 years ago (T. Chen et al., 1992). These dates would better fit the composition of the fauna, including tropical animals, in this cave.

Archaeology

The following description of the archaeology of the Upper Cave relies on Pei (1939). Only five pieces of flint and chert were found in the Upper Cave. They include a flint concave side scraper, a flint end scraper, a broken flint tool, a chert flake without any secondary work, and one flint bearing no traces of human handiwork. There is a considerable quantity of river boulders and pebbles of different sizes and of various kinds of rocks. Three large rocks bear traces of human handiwork. Seventeen pieces of quartz with definite traces of handiwork were found, including a nucleus, side scraper, and a bipolar tool, among others.

Unquestionable bone tools are extremely rare. However, a bone needle was found in Layer 1, broken just at the eye. The shaft is slightly curved, 82 mm long, 3.1 mm in diameter just above the eye, and 3.3 mm at the thickest part of the shaft. The shaft is rounded and smoothed by abrading or grinding. The tip is sharply pointed. The eye was made by excavating or scratching with a pointed tool instead of by drilling.

Two polished objects have been found: a lower jaw of the large Sika deer *(Pseudaxis hortulorum)* and a basal part of a *Cervus canadensis* antler. The second brow tine was removed before polishing. The shaft, the burr, and the first brow tine were reduced by surface abrasion. The first brow tine has been flattened. The surface of the antler is highly polished and shiny, but the polishing technique is unknown. A considerable number of transverse undulations appear on the polished surfaces, and fine, straight, and frequently parallel scratches cover the antler. A few heavy but shallow lines, which are curved, zigzag, or parallel, are clearly marked on the shaft.

The polished mandible is extraordinarily large and is broken about 65 mm anterior to the first premolar. The coronoid process and the condyle were chopped off and the teeth were knocked out.

Many ornaments made by grinding and drilling were recovered at the site. Seven stone beads, found near Skull no. 102, were made of white calcareous stone and deeply colored by red hematite. They are small and of similar size, but they differ in shape. They are perforated and were probably originally strung on some kind of a string. Since they were found near the skull; they may have been part of a headband.

A perforated stone pebble was collected from Layer 4 in the lower part of the cave. It measures 39.6 mm in length, 28.3 mm in width, and 11.8 mm in thickness. The perforation, made by drilling, is 8.4 mm in maximum diameter on one side and 8.8 mm on the other.

The most abundant and characteristic ornaments, which may be the remains of a necklace, are perforated animal teeth. Of the 125 teeth recovered, 25 were colored and 100 were not painted or the coloration is no longer visible. They were fashioned from the teeth of common deer and corsac deer of different sizes, foxes, wildcats, badgers, weasels, and tigers. Most of the perforated teeth were found in groups within a small area in Layer 4, but some were recovered in Layer 2. One, a deer canine, found with the beads near Skull no. 102, was probably made by scratching or carving from both sides of the tooth root.

One bone pendant was found in Layer 2; three others sieved from material for which the exact level could not be located might have come from Layer 4. They probably belonged to the long bone of some large birds. Short transverse furrows were visible on some of the specimens.

Each of the three perforated marine shells *(Arca)* collected in Layer 4 shows a perforation near the beak. The perforation was probably made by rubbing the shell against some abrasive stone. The shells were found in the vicinity of the perforated teeth and are possibly elements of the same headband, necklace, or armlet. The nearest place where other *Arca* is found is about 200 km southeast of the Upper Cave. This distant location and the presence of large Unio shells *(Lamprotula)* and oolitic hematite in the cave suggest that the activities of the human inhabitants of the Upper Cave extended over a large area.

Three abdominal vertebrae of a very large fish with hard bones, the Teleostei *(Cyprinus carpio?)*, and six caudal vertebrae of a moderately large fish were found in Layer 4. A piece of a supraorbital bone belonging to a large fish *(Ctenopharygodon iddellus)* and showing a small marginal perforation, was also found in this cave. The perforation was probably made by drilling. Traces of red coloration observable on certain parts of the bone suggest that it was painted with hematite.

Pieces of oolitic hematite were frequently observed throughout the cave deposits. The largest piece, about 20 cm long, was found in the lower part of the eastern section. Evidence indicates that the pieces were worn by grinding to produce fine particles for painting.

The sandy clay matrix adhering to the three human skulls contains small particles of hematite. This favors the hypothesis of a burial site. An oval-shaped limestone fragment shows some dubious traces of coloration with hematite. However, the limestone is so weathered that the preservation of an artificial surface staining is rather unlikely.

Apparently, only the eastern uppermost zone of the cave was occupied by humans, possibly at two different periods. At least two definite ash layers containing implements are restricted to this area. Judging from the rather thin cultural layers, occupation was of short duration (Pei, 1939).

SALAWUSU (SJARA-OSSO-GOL) (108°26′ E, 38°30′ N)

Salawusu (Sjara-osso-gol in the original literature), is the name of a small branch of the Yellow River (Huanghe). Human fossils have been eroded from the banks of the Salawusu in the Wusheng Qi of the Inner Mongolia Autonomous Region. The area yielding human fossils is located on the Ordos Plateau; therefore, the human fossils are sometimes called *Ordos Man*. In 1922–1923 E. Licent and P. Teilhard de Chardin discovered a great number of mammalian fossils, among which a human incisor was recognized among the materials obtained in 1922. These materials come from a site where Pleistocene sands are mixed with the basal gravels of a modern terrace of the Salawusu River. Nevertheless, the grade of fossilization of the specimen and the fact that it was associated with undoubtedly fossil bones from *Rhinoceros, Elephas,* and so forth, suggests a Pleistocene date (Licent et al., 1926).

Mr. Yupin Wang of the Bureau of Cultural Affairs of Nei Mongol Autonomous Region found a human parietal and femur near Dishaogouwan from a 15-m-high terrace of the Salawusu River. He submitted them to the Institute of Vertebrate Paleontology and Paleoanthropology for study in 1956. He reported another parietal found in this region in 196l (Y. Wang, 1961). In 1963 Youheng Li reported another cranial fragment from the Salawusu Formation in Hengshan County, Shaanxi Province, adjacent to Wusheng Qi.

In 1978 and 1979 Guangrong Dong and others found and reported six pieces of human bone on the bank of the Salawusu River near Yangsigouwan and Shaojingouwan. The specimens included two complete frontals, a frontal squama fragment, the right part of a child's mandible, a right femur, and a left tibia. They were found in situ excepting the frontal and femur. Dong et al. (1981) described this material.

Human fossils

Parietal bone (PA 631)

Part of a right parietal bone was reported in 1958 by Rukang Wu. Its inferior part was lost. About half of the coronal margin and a large part of the lambdoidal margin are preserved. The sagittal arc and chord are 125 mm and 110 mm, respectively. The chord-arc index is 88.0. The thicknesses near bregma and the parietal tuberosity are 6.5 mm and 6.0 mm, respectively.

There is an obelion depression but no parietal foramen. The impressions of the middle meningeal artery show that its posterior branch is larger than the anterior one.

Two complete frontal bones and a frontal fragment

These bones found in the 1970s were described by Dong et al. (1981). The chord-arc indices of the two frontal bones are 89.77 and 89.92, respectively. The thickness near bregma is within the range of 5.5–7.0 mm.

The length of the superciliary arches equals or is shorter than the supraorbital margin. There is no groove above the superciliary arch. The superciliary arches of both sides are separate. The glabella protrudes forward, and the index of glabella protrusion is 9.1. The eye orbits are nearly circular. The fronto-nasal suture is not horizontal. The frontal sinus is large.

Anterior left wall of the neurocranium (PA 114)

This fragment includes the supero-anterior part of the temporal fossa and small parts of adjacent bones. This fragment is 7.8 mm thick at its upper part. The junctions between the frontal, zygomatic, and sphenoid bones are indistinct. They form part of the orbital wall. Y. H. Li (1963) considered this fragment to belong to a middle-aged or old individual.

Mandible of a child

The mandible that was found in the 1970s has two premolars, the erupting median incisor and an attached first molar. The mental tubercle is indistinct. The index of mental protrusion is 104.1. (This is 100 × id-gn arc/id-gn chord, where id = infradentale and gn = gnathion). The lateral eminence on the lateral surface of the mandibular body divides into two branches. The lower branch extends forward to the mental tubercle; the upper branch extends horizontally and terminates at the canine jugum instead of the mental foramen. Mandibular tori occur between the canine and lateral incisor, and behind the first premolar. A shallow depression exists above the mental spine at the median portion of the internal mandibular surface. The index of the height of a large mental foramen is 40.9. The height-breadth index of the mandibular ramus and the index of robustness of the mandibular body are 65.6 and 48.6, respectively.

Incisor (permanent, left upper lateral)

The well-formed and perfectly preserved permanent left upper lateral incisor with a partly formed root is attributed to a child about 6 years old. Its unworn crown is shovel-shaped. Five mammelon elevations are clearly seen along the cutting margin (Licent et al., 1926).

Femur (PA 632)

This lower half of the left femur (Fig. 4.5) was reported in the 1950s (Woo, 1958c) The length of the preserved part is 203 mm. It bends slightly forward. The transverse and sagittal diameters and the circumference at midshaft are 25 mm, 24.6 mm, and 84 mm, respectively. The pilastric index is 98.4. The sagittal and transverse diameters of the popliteal part are 23.6 mm and 37.0 mm, respectively. The popliteal index is 63.8. Because the specimen is gracile and not long, it is considered to belong to a female.

Femur

This bone was found in 1970s. The greatest thickness at mid shaft is 8.4 mm. The pilastric index is 112.

Tibia

There are no morphological differences between this bone found in the 1970s and that of recent humans. The tibial index is 81.

FIG. 4.5 Left femoral shaft of *Homo sapiens sapiens* from Salawusu. (Courtesy of IVPP.)

Geology

Four or five terraces have developed along the banks of the Salawusu River. The Salawusu Formation could be divided into two parts. The lower part is about 40 m thick and consists of grayish-yellow and grayish-green silty fine sands. The human fossils, cultural remains, and most of the other mammalian fossils are found in this layer. The upper part is about 20 m thick and consists of greyish-yellow to brownish-yellow fine sands intercalated with greyish-green clay-like silt and greyish-yellow silty fine sands. This part also contains a small amount of mammalian fossils. The lower terrace of the formation also contains some human fossils, stone artifacts, and other animal fossils that were eroded from the Salawusu Formation and secondarily deposited in the lower terrace (Dong et al., 1981).

According to J. Woo (1958c) the human parietal was found about 180 m north of the natural bridge of Dishaogouwan. This terrace yielding the parietal contains six layers of yellow sandy clay and concretions. The first layer is yellow sandy clay, 2.44 m thick; the second layer is sandy concretions, 0.16 m thick. The parietal was found between these two layers. The human femur was found in a higher layer solitary terrace, 60 m from the site of the parietal, near the village of Dishaogouwan. This terrace is composed of sandy clay and sandy concretions. The femur was recovered 17.4 m above the water surface of the river.

The mammalian fauna include the following (Qi, 1989):

Primates

Homo sapiens

Insectivora

Erinaceus sp.

Scaptochirus moschatus

Chiroptera

Chiroptera gen. et sp. indet.

Lagomorpha

Lepus sp.

Ochotona sp.

Rodentia

Citellus mongolicus

Cricetulus cf. *barabensis griseus*

Cricetulus sp.

Myospalax fontanieri

Alticola cf. *cricetulus*

Alticola cf. *stracheyi*

Eothenomys sp.

Microtus ratticeps

Microtus sp.

Meriones meriodianus

Allactaga cf. *annulata*

Dipus sowerbyi

Carnivora

Canis lupus

Meles meles

Crocuta ultima

Panthera tigris

Palaeoloxodon

Palaeoloxodon naumanni

Perissodactyla

Equus cf. *przewalskyi*

Equus hemionus

Coelodonta antiquitatis

Artiodactyla

Sus scrofa

Camelus knoblocki

Megaloceros ordosianus

Cervus mongolicus
Cervus elaphus
Spirocerus hsuchiayaocus
Gazella przewalskyi
Gazella subgutturosa
Ovis armmon
Bubalus wansjocki
Bos primigenius

The coexistence of hyena, elephant, and woolly rhinoceros complicate making environmental inferences.

The Salawusu Formation is of late Pleistocene age on the basis of faunal correlation. Human fossils have been reported from different levels of the lower part of this formation. Based on uranium series dating of animal bones from different levels of this part of the formation, Yuan et al. (1983) concluded that the human bone might date from 37,000 to 50,000 years BP. Xinguo Li et al. (1985) reported the radiocarbon date (using carbon particles from the deposits) of the lower part of the Salawusu Formation as 35,340 ± 1900 years BP.

C. Chen et al. (1991) analyzed the palynological data for Salawusu and divided the spectrum into five zones as follows:

1. Late period of the Holocene: *Pinus*-Chenopodiaceae, *Artemisia* zone, indicating a cool and dry forest-grassland landscape.
2. Middle and early periods of the Holocene: *Quercus-Pinus,* Gramineae zone, indicating a sparse tree-grassland landscape and a transition from warm and wet to cool and dry times.
3. Late period of the Late Pleistocene: *Pinus*-Chenopopiaceae zone with rare spores and pollens of *Quercus* and Gramineae, among others. The flora suggests a sparse grassland or a dry and cold desert-grassland landscape.
4. Middle period of the Late Pleistocene: *Pinus-Betula*-Gramineae zone with spores and pollens from the sands and cross-bedding of clay and sands, including *Pinus, Betula, Quercus, Castanea, Celtis, Typha,* Apocynaceae, Gramineae, Cyperaceae, Polypodiaceae, and others. The flora suggests a wet and warm forest-grassland landscape.
5. Early period of the late Pleistocene: *Pinus-Artemisia*-Chenopodiaceae zone with spores and pollens from the clay, sandy clay, and gravel, including *Pinus, Carya,* and a host of others. The flora suggests a cool and dry climate.

Archaeology

The late Paleolithic archaeological assemblage consists of small stone artifacts, mostly made of small pebbles 20 to 40 mm in diameter. Most are flake stone tools made by direct percussion and minimal retouching. Typical implements include borers, different kinds of scrapers, points, and burins. The side scrapers exhibit the most variety.

HUANGLONG (109°52′ E, 35°35′ N)

Human fossil fragments from Huanglong were collected by a farmer in 1975 during construction of a reservoir. The pieces are said to have been unearthed at the southern slope of Xujiafenshan Hill (the hills with graves of the Xu family), Huanglong County, Shaanxi Province. After receipt of the fossil fragments and a fragment of a deer antler, Shaohua Zheng and Yi Li of the Institute of Vertebrate Paleontology and Paleoanthropology, Academia Sinica, did a geological survey of the site in 1980.

Human fossils

Skullcap (PA 842)

The preserved fossils include a large part of the frontal squama, the lateral half of the superior part of the left orbit, and the anterior two thirds of the left and right parietal bones. The specimens probably belong to a male estimated to be over 30 years of age.

Wang and Li (1983) described these fragments. There is a sagittal ridge along the median line of the frontal bone and the sagittal suture. The coronal and sagittal sutures were fully obliterated on the interior surface, but they were discernible on some parts of the exterior surface. The frontal tubercle is less distinct and located in a higher position than in the Upper Cave and Liujiang specimens. The skull vault is rather constricted behind the orbit. The middle meningeal artery appears to have been thick. The frontal ridge is probably well developed. Overall, the bones are thick (Table 4.4).

Wang and Bräuer (1984) derived 10 measurements and did a principal component analysis on the restored frontal bone from Huanglong (Table 4.5). Based on the principal component analysis, they suggested that Huanglong is more primitive than modern Chinese and closer to late European Paleolithic skulls with respect to the frontal bone.

Geology

The geology of the site from the surface down is as follows (Wong and Li, 1983):

Table 4.4

Thickness of the frontal and parietal bones from Huanglong

Frontal tubercle	7.0
Center of frontal squama	10.3
Temporal surface of frontal	4.2-8.0
Center of temporal surface of frontal	6.0
Parietal bone near bregma	12.5
Parietal tuberosity	8.5

According to Wang and Li (1983)

Table 4.5
Measurements of the frontal bone from Huanglong

Minimum frontal breadth	96.3
Glabella-bregma arc	116.5
Supraglabellare-bregma arc	103.5
Glabella-bregma chord	109.1
Glabella subtense fraction	52.0
Frontal subtense (g-b)	19.3
Supraglabellare-bregma chord	92.5
Pars cerebralis subtense	17.0
Angle g-H-b*	141.0°

According to Wang and Bräuer (1984)

* H is the furthest point on the arc g-b line

5. Secondary deposits: pale yellow, rich in small calcified concretions; lamination of the deposits follows the slope; ceramics found in this layer; 6 m thick
4. Reddish layer, the upper part of which is composed of sands, the lower part contains a large amount of calcified concretions, 6 m thick
3. Pale green shale and sandstone, 2 m thick

———unconformity———

2. Purplish-red clay with fragmentary animal fossils, 8 m thick
1. Basal complex

The human fossil and antler of *Pseudaxis* sp. were found at the junction between Layers 5 and 4 (Wang and Li, 1983). Very fragmentary fossils of other animals were found in Layer 2, and there is no trace of fossils in Layers 4 and 3.

JINCHUAN (107°23′ E, 35°38′ N)

The human fossil from Jinchuan was reported by Yulin Liu et al. in 1984. It was found in a small gully named Niujiaogou (Cow Horn Gully) near Baijiayuan village, 35 km east of Jinchuan City, Gansu Province, in northwestern China.

Human fossils

The remains consist of a fragment of the right part of the frontal squama, a large part of the right parietal, a small part of the left parietal, a complete right temporal, and a large part of the occipital. The sagittal and lambdoidal sutures are deeply serrated. Obliteration appears on the external surface of the obelion segment of the sagittal suture. The endocranial sutures are clearly seen. There are two parietal foramina but no obelion depression. Liu and Huang (1984) diagnosed the specimen as a young adult female about 20 years old.

The anterior branch of the middle meningeal artery is larger than the posterior branch. The superior margin of the temporal squama is curved. The distance between asterion of both sides and the distance between asterion and lambda are 107 mm and 91 mm. On the right side the distance between the sagittal suture and the temporal line, and the distance between the temporal line and squamosal suture, are 64 mm and 50 mm, respectively. There are no Inca bones and no occipital ridge. The cranial capacity seems small.

Geology

According to Liu and Huang (1984), the human and other mammalian fossils, and some artifacts, were unearthed from a greyish-brown sandy clay. The strata above this layer is Malan loess, probably of Late Pleistocene derivation; that below the fossil bed is Lishi loess intercalated with a Middle Pleistocene paleosol.

Recognizable mammalian fossils associated with the human fossils are *Equus* sp., *Cervus* sp.?, *Coelodonta antiquitatis,* Bovidae, and *Myospalax fontanieri*. This site is geologically and archaeologically comparable with Salawusu or Shiyu (Liu and Huang, 1984).

Archaeology

Stone artifacts reported from Niujiaogou and from Hezhigou, another site in the same county, appear to belong to the same culture. Only the material from Niujiaogou is listed and it yielded nuclei, flakes, choppers, scrapers, and points, all made from quartzite and vein quartz. Most artifacts are flakes. The artifacts are manufactured mainly by hammering and the bipolar technique, with the former predominating.

SHIYU (112°17′ E, 39°25′ N)

Shiyu village is located 15 km northwest of Shuoxian City, Shanxi Province, in northern China. The site is on a small hill situated between the Shiyu River and one of its tributaries, the Xiaoquangou. Zeyi Wang and his colleagues excavated this site in 1963.

Human fossil

The only human fossil is a small fragment of an occipital bone that has not been studied.

Geology

The site is on the second terrace of the Shiyu River. The upper surface of this terrace is 25 to 30 m above the river's bottom. The fossils, most of which are isolated teeth, are not well preserved. Jia et al. (1972) described the geology as follows:

Upper Pleistocene

4. Pale yellow silt, 18 m thick
3. Grey or pale grey sands cross-bedded with gravel lens; a few animal fossils come from the basal part of this layer; 8.9 m thick

2. Cultural layer of grey or greyish-black loam intercalated with an ash layer up to 2.5 cm thick; most animal fossils found here; 0.9–1.5 m thick
1. Grey or brownish gravels, 0.5–1 m thick

———unconformity———

Basal complex

According to Jia et al. (1972) the fauna include:

Mammals

Primates
Homo sapiens
Insectivora
Erinaceus sp.
Rodentia
? *Myospalax* sp.
Carnivora
Crocuta sp.
Panthera tigris
Perissodactyla
Equus przewalskyi
Equus hemionus
Coelodonta antiquitatis
Artiodactyla
Cervus elaphus
Megaloceros ordosianus
Procapra picticaudata
Gazella cf. *subgutturosa*
Bovidae gen. et sp. indet. 1
Bovidae gen. et sp. indet. 2
Bubalus cf. *wansijocki*
Bos sp.

Bird

Struthio sp.

Radiocarbon dating of bovid bones from the cultural layer has been reported as 28,945 ± 1370 years BP (Laboratory of Institute of Archaeology of the Chinese Academy of Social Sciences, 1977).

Archaeology

More than 15,000 pieces of stone artifacts were recovered, including nuclei, flakes, choppers, points, scrapers, and burins. A long flint flake with a pointed tip has been considered as the pointed part of an arrow. A disc-like fragment of a possible ornament made from graphite has been polished on the margin and at least on one side. There is a hole at its center (Jia et al., 1972).

XUETIAN (127°33′ E, 44°47′ N)

In the autumn of 1986, Huili Yu excavated site H8601, collecting human and other mammalian fossils from the water-eroded bottom of an irrigation canal near Xuetian Village, Longfenshan (Hill of the Dragon and Phoenix) district, Wuchang County, Heilongjiang Province. The site is situated at the southeastern margin of the Songnen Plain.

Human fossils

Parietal bone (86 WXT 1: 357)

This fragmentary lower part of the left parietal consists of three pieces. The depressions of the middle meningeal artery are distinct. Yu (1988) considered the fragments as belonging to a 5 to 6-year-old child.

Tibia (86 WXT 1: 356)

There is a 23-cm-long fragment of a left tibia. The bone wall is thin. Yu (1988) considered that the fragment belonged to a young adult female. The upper part of the anterior ridge is sharp. The maximum diameter and transverse diameters at the midshaft and the minimum circumference are 34.2 mm, 24 mm, and 70 mm, respectively.

Geology

The geological section was recorded at an exposure on the west bank of the canal as follows: Layer 5 is black earth, 1 m thick; Layer 4 is a loesslike loam, 0.7 m thick; Layer 3 is a loesslike sandy clay, 0.9 m thick; Layer 2 is a greyish-green silt that contains human and a large amount of other mammalian bones as well as stone artifacts, 3.5 m thick. Layer 1 is a dark green silt thicker than 1 m.

Mammalian fossils from this site include: *Mammuthus sungari, Mammuthus primigenius, Coelodonta antiquitatis, Bison* sp., *Equus przewalskyi, Cervus* sp., Carnivora, and *Myospalax* sp.

Faunal analysis suggests that this site is of Late Pleistocene age and indicates a cold climate. The flora includes: *Picea, Abica, Pinus, Ulmus, Betula, Alnus, Juglans, Quercus, Castanea, Salix,* Rosaceae, *Lonicera,* Cyperaceae, Umbelliferae, Gramineae, *Polygonum,* Coryophyllaceae, *Artemisia,* Chenopodiaceae, *Humulus,* and Polypodiaceae, among others. The flora suggest a forest-grassland landscape (Yu, 1988).

Archaeology

In addition to 3 stone artifacts, 45 pieces of broken animal bone with possible chipping were recovered.

YANJIAGANG (45°36′ E, 126°18′ N)

In 1982 a field team organized by the Heilongjiang Provincial Museum and other local organizations excavated this site located on Yanjiagang farm, 25 km southwest of Harbin, the capital of Heilongjiang Province. The team found stone artifacts and mammalian fossils. A human cranial fragment was exposed on the surface of Zone A in the eastern area of the site, a locale where the stratum was heavily eroded by digging for sands. The site and the remains were studied by Yuzhuz You and others (Management Committee of Cultural Relics of Heilongjiang Province et al., 1987).

Human fossil

The cranial fragment (HY 82 C 1:1) measures 76 mm × 43 mm × 8 mm from the posterior part of the right parietal. The bone's surface was so severely eroded that no important features could be observed.

Geology

Between 1982 and 1985 archaeologists dug 44 test pits (a total dimension of 1150 m^2). The representative section in the eastern area, Zone A, consists of four layers: black earth, yellow earth, sandy clay, and fine sands, from the top downward. Mammalian fossils mainly derive from the lower part of the yellow earth and the upper part of the fine sands. Fauna from the upper bed is later in time than that from the lower bed. The upper fossil bed contained *Mammuthus primigenius, Coelodonta antiquitatis, Equus przewalskyi, Megalocervus ordosianus, Bison exiguus, Gazella przewalskyi, Crocuta crocuta ultima, Canis lupus, Marmota bobac, Cuon* sp., and *Mustela sibirica.* The flora and fauna collected from Test pit Hy84T4 suggest a dry and cool forest and a grassland habitat. Radiocarbon dating of the site was reported as 22,370 ± 300 BP (Management Committee of Cultural Relics of Heilongjiang Province et al., 1987).

Archaeology

The nine stone artifacts include one chopper, one scraper, and seven flakes.

JIANPIN (119°41′ E, 41°25′ N)

In 1957, Shuodao Sun of the Liaoning Provincial Museum found a human humerus among bones kept in storage at the Jianpin Cooperative. The humerus was sent to the Institute of Vertebrate Paleontology and Paleoanthropology, Academia Sinica, Beijing, in the early 1960s.

This right humeral shaft, which lacks both ends, is 235 mm long. The upper end was broken at the surgical neck; the lower end was broken slightly above the epicondyles. J. Woo (1961) inferred that it belonged to a male. The shaft bends backward, and although the torsion angle could not be measured because of the missing ends, the shaft is probably rather twisted. The spiral groove is very shallow. The deltoid tuberosity is rough and prominent. Both lips of the bicipital groove are slightly more prominent than that in modern humans.

The sagittal and transverse diameters at midshaft are 21.6 mm and 16.7 mm, respectively. The transverse section of the lower part of the shaft is bilaterally triangular, with the medial angle somewhat rounded and the lateral angle sharp.

In the spring of 1957, before the discovery of the humerus, the Liaoning Provincial Museum collected mammalian fossils considered to belong to the Late Pleistocene. They included *Coelodonta antiquitatis, Spirocerrus* sp. *kiakhtensis, Bison* sp., *Equus hemionus,* and *Equus przewalskyi* from the same store.

MIAOHOUSHAN DONGDONG (124°08′ E, 40°15′ N)

Dongdong (East Cave) is located at the southern slope of Miaohoushan Hill, about 90 km southeast of Shenyang, the capital of Liaoning Province. East Cave is about 100 m east of and 25 m higher than the Miaohoushan site, which predates it. Dongdong was excavated during the excavation of the Miaohoushan Hill in 1978 to 1980. The following description is based on the book entitled *Miaohoushan* (1986).

Human fossils

Parietal bone 1 (B.S.M. 7901 B-2.6-1)

This posterior part of a right parietal is 1.8 mm thick and 38 cm^2 in dimension. The preserved part of the lambdoidal suture is serrated. The parietal tuberosity is not distinct, and there are no traces of temporal lines. The middle meningeal artery is not extensively ramified.

Parietal bone 2 (B.S.M. 7901 B-2.6–2)

This may be the postero-inferior part of a right parietal. However, the possibility of it belonging to a temporal bone cannot be excluded. This specimen is 15 cm^2 in dimension.

Radius (B.S.M. 7901 B-1.8-3)

This almost complete child's radius lacks the epiphyses. The length, transverse, and sagittal diameters at mid shaft, and minimum circumference are 170.8 mm, 12.2 mm, 9.1 mm, and 31.9 mm, respectively. The tuberosity is low and flat.

Geology

The geology of the site from the surface down is as follows:

Holocene

5. Recent artifical deposits, 0.2–0.3 m thick
4. Greyish-black sandy clay with bronze and polished stone artifacts, 0.5–1.6 m thick

Late Pleistocene

3. Breccia and brownish-yellow sandy clay containing fossils such as *Cervus* sp., *Crocuta ultima, Myospalax fontanieri,* and *Sus scrofa,* among others, in its basal part, 0.8–2 m thick
2. Brownish-yellow earth with breccia containing human parietals in the basal part of this layer and fossils of *Crocuta ultimata, Ursus arctos, Hydroptes pleistocenica, Cervus* sp., and *Lepus europoeus.* A stone artifact was found in its basal part. The human radius was in the top part of this layer associated with *Cervus* sp. and *Sus scrofa.* Because the radius was more lightly mineralized than the parietal, the former is probably later than the parietal. The layer is 0.4–0.6 m thick.

———unconformity———

Pliocene

1. Red clay with an exposed thickness of 3 m.

The radiocarbon date for the animal remains is 28,040 ± 680 years BP (Museum of Liaoning Province and Museum of Benxi City, 1986).

Archaeology

The 3 m^3 of the test pit yielded only a quartzite flake.

WUSHAN (YUANYANG) (104°45′ E, 34°47′ N)

Parts of a skull were collected by Fuxin Yang during a geological survey in Gutuogou, 2 km southwest of Yuanyang Town, Wushan County, Gansu Province (northwestern China) in the summer of 1984. Xie et al. (1987) described the fossil and the Yuanyang site.

Human fossil

The specimen consists of part of the skull vault, including a complete frontal bone and two parietals and parts of the nasal and sphenoid bones. Xie et al. (1987) suggested that it represented a male around 20 years of age. The skull vault is between pentagonoid and ovoid in superior view. The frontal tubercles are not very prominent, whereas the parietal tuberosities are prominent. The segments of the cranial contour in front and behind the tuberosities are more or less straight. The sutures near the bregma are simple. The posterior segment of the sagittal suture and the preserved part of the lambdoidal suture are more complex than the anterior segment of the sagittal and coronary sutures. Wormian bones exist along the lambdoidal sutures of both sides. The parietal foramina are small, and the distance between them is only 7 mm. There is an obelion depression, but no sagittal ridge (Tables 4.6 and 4.7).

The superciliary arches are prominent with a shallow groove behind. An angular torus exists at the postero-lateral angle of the parietal where the bone is thickened

Table 4.6

Measurements and indices of human calva from Wushan (Yuanyang), Gansu

Frontal arc	126.0
Parietal arc	120.0
Interorbital breadth	27.5
Nasomalar angle	147°
Bregma position index*	39.0
Calvarial height index**	46.3

According to Xie, et al. (1987)

*Bregma position index = $100 \text{ X } \dfrac{gy}{g-op}$. Where g is glabella; y is the juncture between the line g-op with a vertical line from bregma to line g-op.

** Calvarial height index = $100 \text{ X } \dfrac{vx}{g-op}$ Where v is the vertex of the skull; x is the juncture between the line g-op with a vertical line from point v to line g-op.

Table 4.7

Thicknesses of the human frontal and parietal from Wushan (Yuanyang), Gansu

Center of frontal squama	10.0
Temporal surface of frontal bone	4.0
Frontal tubercle	7.0
Parietal tuberosity	7.0
Parietal bone near bregma	12.0
Parietal bone at mastoid angle	12.0

According to Xie, et al. (1987)

on both the external and internal surfaces. The anterior branch of the middle meningeal artery is thicker than the posterior branch.

Geology

The human fossil derives from the middle part of a layer of rather hard greyish-brown calcified clay intercalated with small lens of gravels. This layer is 15 m thick. It is covered by greyish-white sandy clay (3 to 5 m thick) and greyish-brown clay (about 20 m thick, intercalated with purple clay and sands as well as small stones). A clay sample collected from the fossil layer was radiocarbon dated to 38,400 ± 500 years BP by the Laboratory of Geography of Lanzhou University.

QUWO (111°27′ E, 35°44′ N)

The site of Chaoyang Xigou is located about 11 km northwest of the city of Quwo in southwestern Shanxi Province, northern China. The site is on a terrace of the north bank of the Fuhe River, a branch of the Fenhe River, itself a tributary of the Yellow River. Chaoyang Xigou is 4.5 km from the Fenhe River and 12 km southeast of Dingcun, a site yielding early *H. sapiens* fossils. The following description relies on the work of Liu (1986).

Human fossil

Only a right upper deciduous canine was found.

Geology

The site is located at the third terrace of the Fuhe River. The top of this terrace is 20–40 m above the local water surface. The geological sections of the terrace from the top downward follow: Layer 12 is a brownish-yellow soil 0.3 m thick; Layer 11 is a brownish-yellow silt containing gastropods and is 3 m thick; Layer 10 is a light brownish-red silt 2 m thick; Layer 9 is a brownish-red loam 1 m thick; Layer 8 is a greyish-green, greyish-blue clay containing polished stone artifacts and is 4 m thick; Layer 7 is a brown clay containing fragments of vertebrate fossils and is 3 m thick; Layer 6 is a greyish-blue loam with gastropod fossils and is 0.8 m thick; Layer 5 is a yellow loam 1 m thick; Layer 4 is a greyish-blue loam 0.3 m thick; Layer 3 is a light brown silt 1.4 m thick; Layer 2 has pale brown coarse sands intercalated with yellowish-brown silt with concretions. Within this layer a human canine and fossils of *Struthio* sp., *Equus hemionus, Equus przewalskyi, Coelondonta antiquitatis,* Bovinae, and Antilopinae were found. The layer is 0.9 m thick. Layer 1 is yellow sand with an exposed thickness of 0.5 m.

Gramineae, Chenopodiaceae, and *Artemisia* are the predominant flora. Other plants include Compositae, Cyperaceae, Umbelliferae, and *Pinus.*

Archaeology

The associated cultural remains are 514 pieces of stone artifacts including 69 scrapers, 10 points, 2 choppers, 2 bifaces, 1 ball, 8 burins, flakes, and nuclei. These are made predominantly from flint pebbles and quartz. Based on faunal and archaeological analyses, Yuan Liu (1986) concluded that this site belonged to the Late Pleistocene and the Late Paleolithic.

CHANGWU (107°50′ E, 35°15′ N)

The human fossil was collected among mammalian fossils sold by local farmers. The following description is from Huang and Zheng (1982). The human fossil is probably an erupting left upper second molar. Enamel on the mesial and lingual surfaces is destroyed. The protocone is large and the hypocone is small. The paracone is slightly larger than the metacone. The mesio-distal and bucco-lingual diameters are 10.0 mm and 9.9 mm respectively.

In the article describing this tooth, Huang and Zheng (1982) mention two Late Pleistocene sites located near Changwu City, Shaanxi Province, northwestern China. They considered one of these, Yaergou (Duck Gully), the possible derivation for this tooth. From the top down, the geology of Yaergou includes: Layer 8, which is a greyish-yellow loess-like sandy clay 9 m thick; Layer 7 is a dark red paleosol 1 m thick; Layer 6 is a light yellow sandy clay with snail fossils and is 7 m thick; Layer 5 is a grey sandy clay with mammalian fossils and stone artifacts and is 2 m thick; Layer 4 is a greyish-yellow sandy clay with mammalian fossils and is 10 m thick; Layer 3 is a light grey sandy clay and has fine sands and a thin lamination of mammalian fossils and is 2 m thick; Layer 2 is gravel and concretions 5 m thick; and Layer 1 is a reddish formation dating to the Middle Pleistocene.

XINTAI (117°45′ E, 35°55′ N)

In 1966 farmers from Wuzhutai village, Xintai County, Shandong Province in eastern China found a human molar while searching for water sources. A field team led by Xinzhi Wu from the Institute of Vertebrate Paleontology and Paleoanthropology, Academia Sinica, and the Shandong Provincial Museum found more mammalian fossils in a limestone fissure where the human molar was reportedly found.

The fossil, a left lower molar, probably belonging to a female has five slightly worn cusps. Its length, trigonid and talonid breadths, and crown height are 11.6 mm, 10.2 mm, 10.4 mm, and 7.3 mm, respectively. There is no cingulum. The buccal surface is not bulging at its basal part. A large part of the root is missing (X. Wu and Zong, 1973). Mammalian fossils from this fissure are *Felis cf. tigris, Equus sp., Coelodonta antiquitatis, Sus* sp., *Cervus sp.,* and Bovinae gen. et sp. indet.

ANTU (128°56′ N, 43°05′ E)

In May 1964 a human tooth was discovered in a limestone cave in Jilin Province by a joint team from the Bureau of Geology of Jilin Province, Northeastern Institute of Geology, and the Jilin Provincial Museum. The cave is in the Shimenshan quarry, located 2.5 km southeast of Mingyue Town, Antu County, northeastern China. The materials from the site were described by Jiang (1982). The geology of the site is based on strata in a nearby smaller cave.

Human fossil

Premolar (no. 501)

The following description relies on Jiang (1982). With regard to general morphology, this heavily worn right lower premolar is similar to that of modern humans. A shallow fovea of a deep grey color located on the lingual side of the neck indicates caries. The rough surfaces on the lingual, mesial, and distal sides of the neck indicate atrophy of the alveoli and the gingiva. The length, breadth, and height of the crown are 6.9 mm, 7.1 mm and 4.0 mm, respectively. The tip of the root was slightly absorbed. The height of the root is 16.0 mm.

Geology

The geology was reported by Jiang (1982). The cave is located 365 m above sea level and 25 m above the local river bed. Because the cave had been destroyed by quarrying, its stratigraphy could not be recorded. In 1973 a second smaller cave was found 9 m away at about the same level. The strata of the small cave is discernible, from the surface downward they are as follows:

4. Brownish-grey sandy clay with gravels. It is 11 cm thick and contains *Equus przewalskyi, Elephus canadensis,* etc.
3. Greyish loam with small gravels. The clay is hardened by calcification and is 61 cm thick. This layer is rich with vertebrate fossils, including *Coelodonta antiquitatis, Equus przewalskyi, Bison exiguus,* and *Elephus canadensis.*
2. Brownish-yellow loam with gravels and also calcified. The layer is 52 cm thick and contains such vertebrate fossils as *Equus przewalskyi, Bison exiguus,* and *Capreolus* sp.
1. Brownish-green sandy clay with gravels. This layer is 9 cm thick. Fossils such as *Equus przewalskyi* are occasionally found.

The human tooth was found in the fossil assemblage collected from the large cave, but its provenance is uncertain. Jiang (1982) infers that it was probably unearthed from a layer resembling the greyish-yellow loam in the small cave.

LIUJIANG (109°25′ E, 24°09′ N)

The site is in Tongtianyan limestone cave located in Liujiang (formerly Liukiang) County, Guangxi Zhuang Autonomous Region in southern China. The cave is in a hill located south of Liuzhou, and along the west side of the Liuzhou-Shilung Highway. In mid-September, 1958, while digging the cave deposits for fertilizer, workers found a fossilized human skull (Fig. 4.6) and postcranial bones. The fossils were collected by Dian Li, the head of Xinxin Farm where the cave is located. The human fossils were studied by Jukang Woo (1959a, b)

Human fossils

Skull (PA 89)

The specimen is almost complete, except for parts of the zygomatic arch on both sides. The mandible is also missing. The moderately sized skull has well-developed superciliary arches and a slightly receding forehead. Neither the frontal nor parietal eminences are well marked. The mastoid process is small. Muscular markings on the skull are weak. It was diagnosed as belonging to a male by J. Woo (1959a), but others have questioned this status. All the main sutures are moderately closed and the teeth are well worn. It has been given an age of about 40 years old.

The skull is ovoid in top view and mesocranic with a cranial index of 75.1. The maximum breadth occurs close to the junction between the middle and posterior one third of the skull. The skull vault is rounded. A depressed area appears at the posterior part of the sagittal suture just anterior to the lambda. The lateral borders of this depression are indistinct. There is weak occipital bunning. There is a well-defined supramastoid groove and a definite supramastoid crest above the diminutive mastoid process.

The frontal arc is much longer than one third of the total median sagittal arc

a

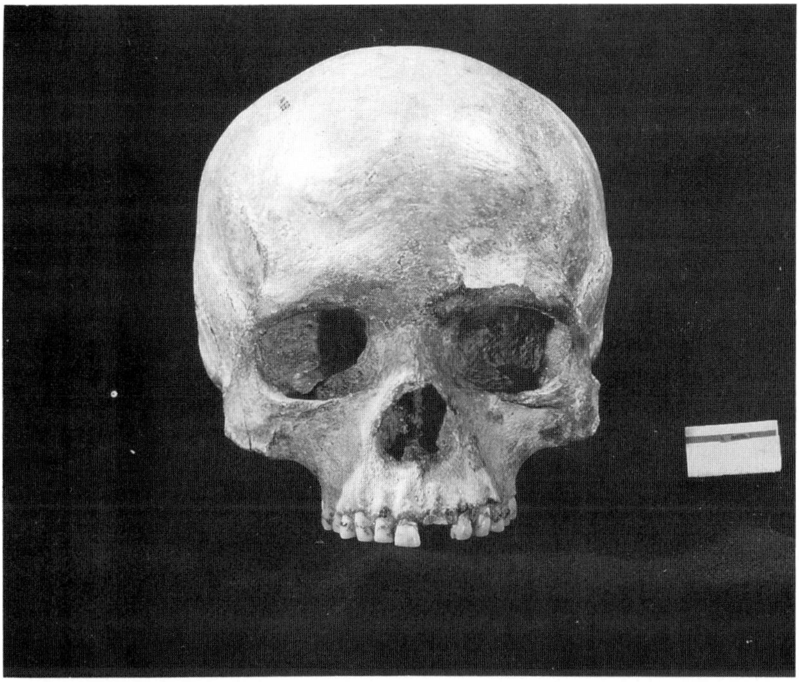

b

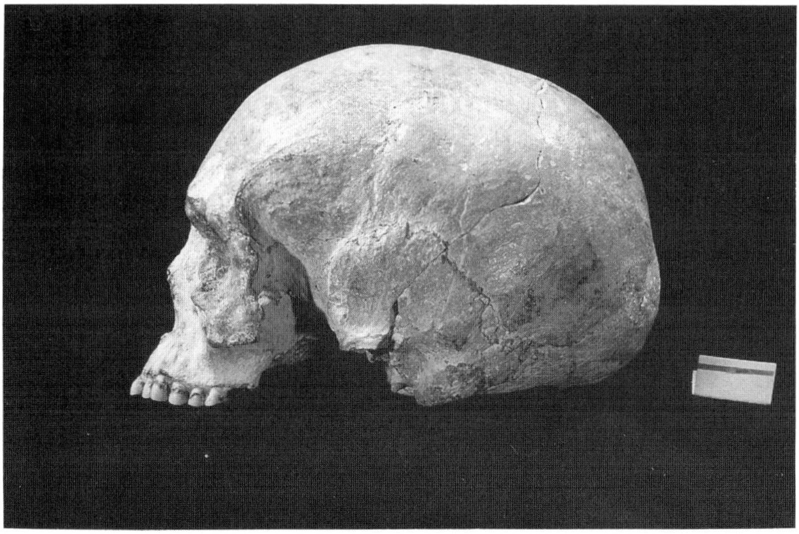

FIG. 4.6 Face and skull of *Homo sapiens sapiens* from Liujiang. (a) Anterior view; (b) lateral view (Courtesy of IVPP.)

from nasion to opisthion. The bregma is located more posteriorly than in late modern humans. A midsagittal ridge is discernible above the glabella.

The face and the orbits are broad and low. The orbits belong to the chamaeconch type and are obtuse-angled. The infero-lateral orbital margin is rounded. The nose is broad and low, and belongs to the hyperchamaerrhine type. The nasal bones are large and broad, and the nasal saddle is slightly concave. There is no depression or notch at nasion. The lower borders of the pyriforme aperture are not sharp, but they have a pair of shallow and broad prenasal fossa. The anterior nasal spine is rather small.

The anterolateral surface of the frontal process of the zygomatic faces rather forward. The angle formed by the lines representing the direction of the middle part of these surfaces could be roughly estimated as 80°. In frontal view, there is a moderately developed malar tuberosity on the zygomatic. The origin of the lower border of the zygomatic process of the maxilla is located rather high off of the alveolar margin. The lower border runs upward and laterally to join the lower border of the maxillary process of the zygomatic and forms a curve instead of a straight oblique line. There are traces of a canine fossa (Table 4.8).

Alveolar prognathism is moderate. The palate is short, broad, shallow, and horseshoe-shaped. The teeth are the size of those of modern humans. The third molars have not erupted. Because of heavy wear it is difficult to judge whether the upper right median incisor is shovel-shaped. The crown of the left median incisor was broken. The upper lateral incisors clearly show a shovel-shape.

Axial skeleton (PA 90)

The axial specimens include four lower thoracic and five lumbar vertebrae, the sacrum, and four rib fragments (Fig. 4.7). The four lower thoracic vertebrae and the first lumbar vertebra were articulated. The upper two thirds of the anterior surface of the sacrum is fairly flat, but its lower part is markedly curved forward. The sacrum is dolichohieric (sacral index: 93.8). The foramina on the anterior surface are large, in contrast to their appearance on the posterior surface. The latter is strongly convex and rough. The sacral hiatus at the termination of the sacral canal extends upward to the middle of the third sacral vertebra.

The sacro-iliac auricular surface corresponds to the lateral side of the first to third sacral vertebrae. The highest point of the sacral wing rises considerably above the promontory.

The rib fragment adhering to the right side of the last thoracic vertebra is the posterior half of the 12th rib. The last rib was likely of considerable length.

Right innominate bone (PA 93)

Except for the superior and inferior rami of the pubis and the inferior ramus of the ischium, this bone is well preserved. The articular surface is large. Inasmuch as this articular surface fits well with the sacrum from the same cave, it is assumed that they belonged to the same individual.

Femur (PA 91, PA 92)

Two femoral shaft fragments are preserved: one (PA 9l) is from the left side and one from the right (PA 92). The upper end of the right fragment is broken below the lesser trochanter. The left fragment corresponds to the upper half of the shaft. The exposed medullary canal is relatively narrower than that of late modern humans.

Table 4.8

Measurements and indices of the Liujiang skull

Linear measurements (in mm):		
Max. cranial length (o-op)		189.3
Glabella inion length (g-i)		172.0
Max. cranial breadth (eu-eu)		142.2
Auricular height		114.5
Basion bregma height (ba-b)		134.8
Basion vertex height (ba-v)		134.0
Porion bregma height (po-b)		114.8
Sagittal arc (n⌒o)		374.0
Frontal arc (n⌒b)		136.5
Parietal arc (b⌒l)		132.0
Occipital arc (l⌒o)		105.5
Frontal chord (n-b)		117.2
Parietal chord (b-l)		119.2
Occipital chord (l-o)		91.5
Min. frontal breadth (ft-ft)		95.2
Foramen magnum length (ba-o)		36.9
Foramen magnum breadth		30.5
Basis length (ba-n)		103.5
Profile length (ba-pr)		100.0
Mid-profile length (ba-ss)		95.0
Max. bizygomatic breadth (zy-zy)		136 (?)
Upper facial height (n-pr)		65.9
Orbital breadth (mf-ek)	rt.	43.1
	lt.	42.0
Orbital height	rt.	29.0
	lt.	28.7
Nasal height (n-ns)		45.8
Nasal breadth		26.8
Palatal length (ol-sta)		45.0 (?)
Palatal breadth		36.0
Palatal height		9.5
Internal biorbital breadth (IOW fmo-fmo)		98.8
Subtense IOW		15.5
Simotic chord (SC)		10.6
Simotic subtense (SS)		3.0
Bimalar breadth (GB) (zm-zm)		97.1
Subtense GB		31.5
Interorbital breadth (mf-mf)		21.2
Subtense mf-mf		6.2
Mid-orbital breadth		53.8
Cranial capacity		ca. 1480

(continued)

Table 4.8 (continued)

Measurements and indices of the Liujiang skull

Angles:		
Facial profile angle (n-pr-FH)		86.0°
Nasal profile angle (n-ns-FH)		89.0°
Alveolar profile angle (ns-pr-FH)		75.0°
Frontal profile angle (g-m-FH)		74.0°
Glabella-bregma angle (g-b-FH)		46.0°
Naso-malar angle (fmo-n-fmo)		143.5°
Zygomaxillary angle (zm-ss-zm)		138.0°
Indices:		
Cranial index		75.1
Length-height index		71.2
Length-auricular height index		60.5
Breadth-height index I		94.8
Breadth-height index II		80.5
Calvarial height index*		42.9
Bregma position index**		44.2
Cranio-facial height index		48.9
Cranio-facial breadth index		95.6
Jugo-frontal index		70.0
Superior facial index		48.5
Facial prognathic index		96.6
Orbital index	rt.	67.3
	lt.	68.3
Nasal index		58.5
Palatal index		80.0
Foraminal index		82.7
Frontal index		15.7
Simotic index		28.3
Premaxillary index		32.4
Maxillo-frontal index		29.2
Frontal chord-arc index		85.9
Parietal chord-arc index		90.3
Occipital chord-arc index		86.7

According to Rukang Woo (1959a)

*Calvarial height index = 100 x vx/g-op. Where v is the vertex of the skull; x is the juncture between the line g-op and a vertical line from point v to line g-op.

**Bregma position index = 100 x gy/g-op. Where g is the glabella; y is the juncture between the line g-op and a vertical line from bregma to line g-op.

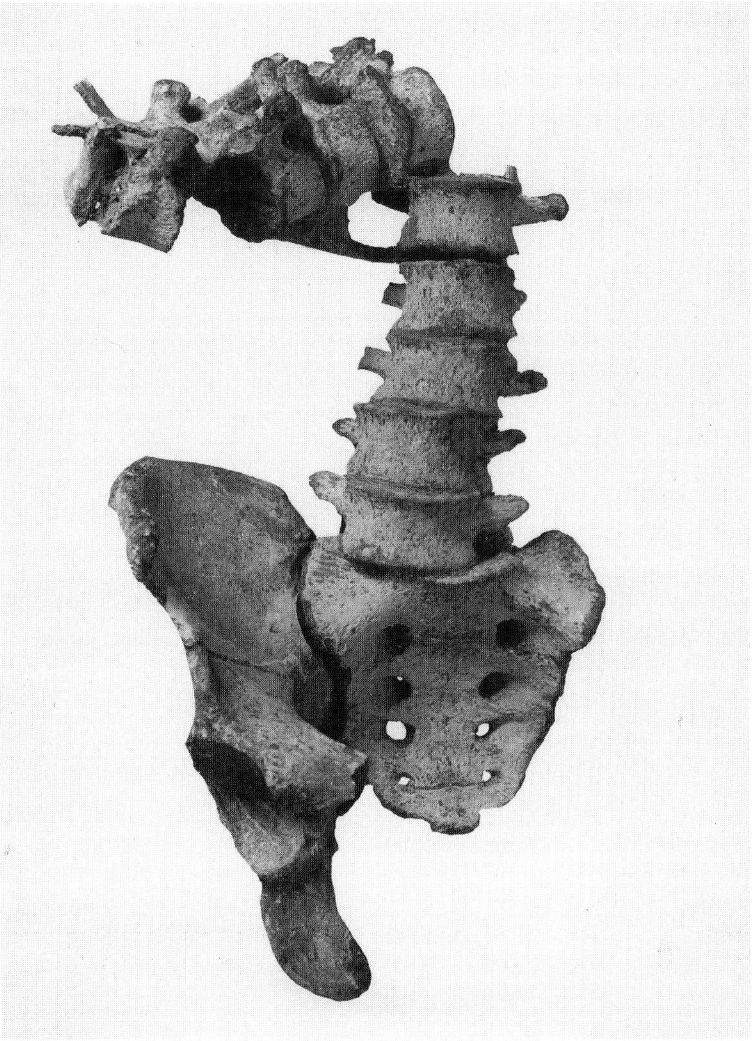

FIG. 4.7 Vertebral column and pelvis of *Homo sapiens sapiens* from Liujiang. (Courtesy of IVPP.)

The linea aspera on both fragments is distinct. The right femur bends forward with the vertex of its curvature at its middle part. The right femur is hyperplatymeric. X. Wu et al. (1984) estimate the stature of Liujiang on the basis of the reconstructed right femur as 157 ± 3.59cm or 156.69 ± 3.74cm (Table 4.9).

Geology

According to Youheng Li (pers. comm.), the geologist having initial access to the fossils, the human specimens were found near the cave entrance close to where a

Table 4.9

Measurements and indices of the Liujiang postcranial bones

Vertebrae:	
Anterior height:	
T 9	20.0
T 10	20.5
T 11	21.0
T 12	22.5
L 1	23.6
L 2	24.5
L 3	21.0
L 4	25.0
L 5	25.0
Sacrum:	
Anterior length	92.2
Anterior breadth	86.5
Index	93.8
Sacro-iliac articular surface:	
length (left)	56.0
(right)	54.0
breadth (left)	32.0
(right)	31.0
Innominate:	
Height	187.0
Height of anterior iliac spine from acetabular border	59.0
Maximum breadth of greater sciatic notch	48.0
Femur:	
Right fragment:	
Length	214.0
Transverse diameter at subtrochanteric level	28.5
Sagittal diameter at subtrochanteric level	21.0
Circumference at subtrochanteric level	83.0
Platymeric index	73.7
Transverse diameter at mid-shaft level	22.5
Sagittal diameter at mid-shaft level	27.0
Pilastric index	83.3
Circumference at mid-shaft level	82.0
Left fragment:	
Length	122.5
Transverse diameter at mid-shaft level	22.0
Sagittal diameter at mid-shaft level	26.2
Pilastric index	84.0
Circumference at mid-shaft level	80.0
Medullary canal:	
Transverse Diameter	8.0
Sagittal diameter	10.0
Index of robusticity:	
Transverse diameter	36.4
Sagittal diameter	38.2

According to Rukang Woo (1959a)

complete skull of a giant panda was discovered. When Li entered the cave, only a small amount of the fossiliferous deposits remained. The greyish-brown deposits are mainly composed of loosely cemented limestone breccia intercalated with sands and clay. The human skull was reportedly embedded in deposits of unconsolidated breccia distinctly different from the hard yellowish deposits of Middle Pleistocene age usually found in caves in Guangxi.

The mammalian fauna associated with the human remains include (Han and Xu, 1989): *Hystrix, Ailuropoda melanoleuca fovealis, Ursus* sp., *Stegodon orientalis, Megatapirus augustus, Rhinoceros sinensis, Sus* sp., *Pseudaxis* sp., and Bovidae gen. et sp. indet.

Uranium series dating yielded the following: stalagmitic crust layer II, earlier than 67,000 plus 6000 minus 5000 years BP; five specimens of animal teeth beneath (?) the above-mentioned stalagmitic crust layer, 101,000–227,000 years BP (Yuan et al., 1986).

LAIBIN (109°05′ E, 23°40′ N)

A human skull base (Fig. 4.8) was discovered in 1956 by a team from the Laboratory of Vertebrate Paleontology (now the Institute of Vertebrate Paleontology and Paleoanthropology), Academia Sinica. The specimen comes from a cave on Qilinshan (then Chilinshan) Hill, Laibin (then Leipin) County, Guangxi Zhuang Autonomous Region. The following description is from Chia and Woo (1959).

Human fossils

The human fossils include a nearly complete hard palate with several teeth and the adjoining lower part of the body of the maxilla, a large part of the right zygomatic, and an occipital fragment. The three pieces were disconnected. They are greyish-white in color. The heavily worn teeth and the rough surface of the skull indicate that the cranial fragments probably represent an old male.

The palate is high. The floor of the maxillary sinus is at a lower level than the nasal floor, which is even and separated from the clivus naso-alveolaris by a distinct ridge, forming the lower margin of the pyriforme aperture. The lower part of the pyriforme aperture seems to be rather broad. The canine fossa is shallow. Although the canine jugum is rather prominent, it does not extend upward to the level of nasal floor. Alveolar prognathism is moderate. The palate is U-shaped and its surface is rugged. The occlusal surfaces of the molars are rectangular.

FIG. 4.8 Skull base of *Homo sapiens sapiens* from Laibin. (Courtesy of IVPP.)

The right zygomatic fragment includes the main part of its body and part of its orbital process. The orbital process is nearly horizontal and forms almost a right angle with the anterior surface of the bone. Its anterior surface is smooth and slightly convex. There is a marked bending near its maxillary border.

The squama, left basilar part of the occipital, and condyle are mostly preserved. The lambda-opisthion arc is about 121 mm in length. The l-o chord is 101 mm. The length of the midsagittal arc of the upper and lower scales of the occipital are 70 mm and 51 mm, respectively. The external occipital protuberance is moderately sized, and the internal and external occipital protuberances are situated at the same level. The external occipital crest is weak. The superior nuchal line is markedly distinct, and the inferior line is less distinct.

The deeply concave interior surface of the occipital is divided into four fossae. The superior fossae are deeper and much smaller than the inferior ones. The right arm of the cruciate eminence is higher than the left.

Geology

This cave is about 7 m above the present ground surface. The cave deposits are divided into two layers: a lower layer of red loam, and an upper layer of yellowish-grey breccia subdivided by stalagmitic crusts into three layers. Human fossils and cultural remains were found in the first layer of the upper layer. Associated with the human fossils are tooth fragments of *Cervus* and *Sus*, as well as a considerable number of molluscan shells of modern species.

Archaeology

Three stone artifacts (two flakes and a quartzite tool), a thin layer of ash with burnt bone, and charcoal occur in the yellowish-grey breccia (upper layer) of the cave.

ZIYANG (104°39′ E, 30°07′ N)

The Ziyang human fossils were collected in 1951 by workers digging a pit for building a railway bridge across the Jiuqu (Nine Winding) River, Ziyang County, Sichuan Province. The fossil-bearing locality is situated about 0.5 km west of Ziyang City on the right bank of the Jiuqu River. The river is about 20 m wide in this area, a small alluvial plain formed by the Jiuqu River. In addition to human fossils, other animal fossils and fossil trees were collected. The following description relies on the work of J. Woo (1957), Pei and Woo (1957), An (1972), and X. Li and Zhang (1984).

Human fossils

The Ziyang human fossils are dark brown in color and include a large part of the neurocranium and a hard palate. The vault (Fig. 4.9) is nearly complete, except for a small part of the right half of the frontal, the right half of the cranial base, and the right temporal. The medial part of the sphenoid, tip of the left temporal pyramid, and the bone surrounding the foramen magnum are also missing.

The small vault has a very smooth surface. The forehead looks fairly full. Most parts of the exocranial sutures are distinct, but the left part of the coronal suture

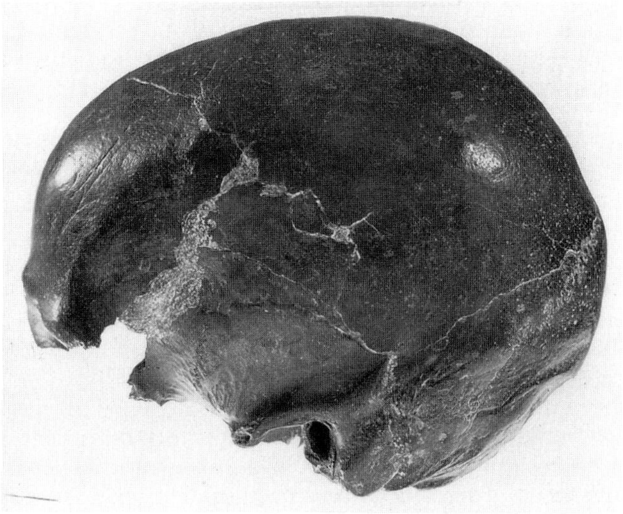

FIG. 4.9 Skullcap of *Homo sapiens sapiens* from Ziyang. (Courtesy of IVPP.)

had begun to close. Almost all the endocranial sutures were obliterated, as was the suture between the two palatal bones. The suture between the two maxillary bones is distinct. X-rays show that the posterior part of the left alveolar process suffered from chronic inflammation that had spread into the left maxillary sinus. This sinus is rather large and its walls are thin. X-rays also show that the canal of the remaining part of the root of the second premolar is very slender. Judging from the pathological change of the alveolar process and the obliteration of the endocranial sutures, Jukang Woo (1958e) suggested that these fossils represent a female past middle age.

The skull belonged to the mesocrany type, is ovoid in top view, and is slightly asymmetrical. Although the frontal tubercles and the parietal tuberosities are rather prominent, the right parietal tuberosity is more laterally protruding and located more anteriorly than the left. The left part of the occipital protrudes backward more than the right side. The occipital squama protrudes backward more on its left than its right side.

The vault is broadest close to the parietal tuberosities. The superciliary arches are much more robust than is true in late modern adult females. The lateral part of the superciliary arch gradually diminishes and disappears above the middle part of the superior margin of the orbit. The medial parts of the arches of both sides are rather prominent and connect with each other. A shallow groove is observed behind and above the arches.

A midsagittal ridge extending from the level of the frontal tubercles to the middle part of the sagittal suture widens to form a bregmatic eminence. The superior surface of the parietal bones just lateral to the midsagittal ridge is somewhat flattened, giving the vault a gable shape. The sagittal and coronal sutures are simple. The parietal foramina are distinct and located close to the sagittal suture. The right foramen is very small. There is a depressed rhombdoidal area at the obelion region. In front of the lateral end of the lamboidal suture is a distinct prominence, the angular torus, which is visible in "Peking Man" but seldom seen in modern humans.

When the skull is in the Frankfurt Plane, the parietal tuberosity is directly above the mastoid process instead of behind it. The robust and short mastoid process turns inward at its tip. The supramastoid ridge is prominent. The temporal line is rather distinct; its squamous surface is rough. Most of the zygomatic arch is missing. The root of the zygomatic arch slopes upward and backward, and forms a 20° angle with the Frankfurt Plane. The zygomatic arch is more inclined than in late modern humans. The root of the zygomatic arch continues with the supramastoid ridge, which terminates suddenly at the parieto-temporal sulcus to form a triangular sharp ridge. The inferior temporal line begins at the tip of this triangle. The external auditory orifice is ovoid, its long axis is not vertical, and its superior end inclines slightly forward.

The supraorbital margin is more or less straight. The well-preserved fronto-nasal suture is arched. The remaining small part of the nasal bone shows a high and narrow nasal root.

In posterior view, the skull's contour has five angles. The apex of the middle and uppermost angles is at the sagittal suture, the apex of the two upper angles is at both parietal tuberosities, and the apex of the two lower angles is at the mastoid portion of both sides. There is a slight occipital bun. The superior and inferior nuchal lines are well developed, and the nuchal plane is rough.

A basal view shows that the frontal sinus is large, and the right part is larger than the left. The lacrimal fossae are shallower than those in late modern humans. The mastoid notch is deep and narrow; the sulcus of the occipital artery is shallow. The mandibular fossa is deep. The articular tubercle is more developed than in recent humans. There is no postglenoid process. The tympanic plate leans slightly backwards and is thicker than in modern humans. Its horizontal axis forms an 80° angle with the sagittal plane. The lateral part of the tympanic plate protrudes forward, and the medial part is slightly concave. There is a thin infratemporal ridge between the temporal and infratemporal surfaces of the greater wing of the sphenoid.

The low and thick crista frontalis originates at the same level as the nasion and begins to disappear slightly above the level of the middle of the frontal squama, where some fovealae granulares pachioni exist. The anterior branch of the middle meningeal artery is thicker and has fewer branches than the posterior branch. The angular torus is less prominent on the endocranial than on the external surface. The cerebral fossae are much larger and deeper than the cerebellar fossae. The internal and external occipital protuberances are at the same level. The internal occipital crest (the lower arm of the eminentia cruciata) divides into two near the posterior margin of the foramen magnum. The left and right arms of the eminentia cruciata are at the same level. The sagittal sulcus lies along the median line and is connected to the right horizontal sulcus, which is wider than the left one.

The bone of the occipital squama is rather thick, and the fossa on the endocranial surface have no influence on the shape of the external surface of the occipital squama. The sigmoid sulcus is shallower and narrower than in late modern humans (Table 4.10).

The thickness of the mastoid portion of the temporal bone anterior to the parieto-mastoid suture is 8 mm. The thickness anterior to the occipito-mastoid suture is 7 mm.

Besides a few pieces of nasal bone, the facial bone includes the lower part of the maxillae and the horizontal plate of the palatines. The maxillary sinus is rather large, extending from anterior to the premolar to posterior to the third molar. The

Table 4.10

Measurements and indices of the Ziyang skull

Linear measurements(in mm):	
Maximum cranial length (g-op)	169.3
Maximum cranial breadth (eu-eu)	131.1
Auricular height	110.0
Horizontal circumference at level above superciliary arch	473.0
Median sagittal arc	354.0
Frontal arc	126.0
Frontal chord	109.0
Subtense to frontal chord (st)	22.0
Parietal arc	121.0
Parietal chord	110
Occipital arc	107.0
Interasterionic arc	116.0
Length of temporal squama	54.0
Height of temporal squama	35.0
Median arc of occipital plane	57.0
Median arc of nuchal plane	50.0
Length of hard palate	42.6
Breadth of hard palate	39.0
Angles:	
Bregma angle	47.5°
Frontal angle	81.0°
Indices	
Cranial index	77.4
Height-breadth index	84.0
Calvarial height index *	45.3
Frontal chord-arc index	86.5
Parietal chord-arc index	90.9
Index of flatness of frontal squama**	20.2
Bregma position index***	41.8
Index of height of temporal squama	64.8
Index of occipital/nuchal plane	87.7
Index of hard palate	91.3

According to Woo (1957)

*Calvarial height index = $100x \frac{vx}{g-op}$. Where v is the vertex of the skull; x is the juncture between the line g-op with a vertical line from point v to line g-op.

**Index of flatness of frontal squama = 100 x greatest height of the point on the g-b arc/g-b chord.

***Bregma position index = $100x \frac{gy}{g-op}$. Where g is glabella; y is the juncture between the line g-op and a vertical line from bregma to line g-op.

walls are thin. The anterior part of the maxillary floor is lower than the nasal floor, while the posterior part is higher than the nasal floor. The preserved broad base suggests a well-developed anterior nasal spine. There is only a trace of the boundary between the nasal floor and the anterior surface of the alveolar process of maxillae at the lower margin of the piriform orifice. A large and deep prenasal fossa extends laterally to the root of the canine prominence on each side of the skull.

The lateral margin of the zygomatic process of the right maxilla extends upward and laterally in the shape of an arch. The whole extent of the lateral margin is rounded and blunt. The lower end of the lateral margin of the zygomatic process is above the roots of the first and second molars on the right side, while it is above the root of the second premolar on the left side. The maxillary tuberosity is small.

The upper dental arch is U-shaped (Fig. 4.10). The oral surface of the hard palate is rough. There is a slight torus palatinus. The incisive foramen is very large and the palatine foramina are larger than in modern humans. There is no trace of the incisive suture in the Ziyang specimen.

The only remaining dentition is a broken root of the left second premolar remaining in its socket. Two incisors, a canine, the first premolar, and the second and third molars of the right side, and the medial incisor, a canine, and the first premolar of the left side, were lost after death. The right second premolar and the left three molars were lost before death because distinct scars of bony regeneration are shown in their sockets. It is not certain whether the right first molar and left lateral incisor were lost before or after death. Judging from the condition of the socket, the right third molar seems to have only a single root and the right second molar has its two buccal roots mostly fused.

Geology

In September 1951, Wenzhong Pei and his field team dug two test pits, one west and another east of the Jiugu bridge base no. 1, from which the skull derives. The human skull was said to be dug from midway between bridge base no. 1 and the east test pit.

Pei and Woo (1957) described the geology as follows: Layer 1 is a yellowish-red clay 6 m thick; Layer 2 is a dark grey clay with thin sandy layers and is 1 m thick; Layer 3 is yellow sands and small pebbles and has a few animal fossils, but is rich in fossil roots, branches, and leaves of big trees and is 1–1.5 m thick; and Layer 4 contains gravels in which the pebbles become larger in the lower part.

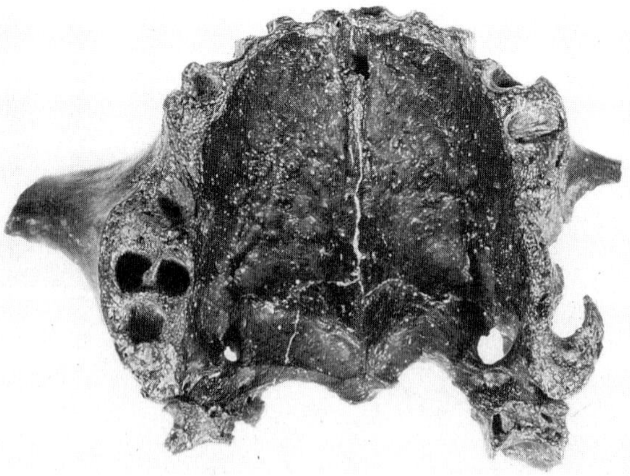

FIG. 4.10 Hard palate of *Homo sapiens sapiens* from Ziyang. (Courtesy of IVPP.)

There are no fossils. There is no sharp boundary between the adjacent layers. The human fossils were probably collected from the third layer.

Fluorine analysis and the comparison of specific weight (gravity) of various animal fossils indicate that the mammalian fossils are a mixture of two groups. One group, including *Homo sapiens, Equus* sp., *Muntiacus* cf. *reevesi,* ? *Moschus,* and *Mammuthus primigenius* is later in age, contains a lower flourine content, and the fossils are lighter in specific weight and less worn. The second group consists of *Rhinoceros* cf. *sinensis, Cervus (Rusa) unicolor,* and *Stegodon orientalis.* Other mammalian fossils that are difficult to assign to either group include: *Hyaena* sp., *Felis tigris, Hystrix* sp., *Rhizomys* sp., *Sus* sp., and *Bibos gaurus.*

Based on faunal analysis, the Ziyang hominid is probably of Late Pleistocene age. Since the application of carbon dating in Chinese archaeology, the dating of the Ziyang hominid has been debated. Zhimin An (1972) suggested that it should be of a later date than that estimated by Wenzhong Pei because the mineralized trees collected from a test pit close to the bridge base provided radiocarbon dates of 7485 ± 130 years BP and 6740 ± 120 years BP. In 1984 Xuanmin Li and Senshui Zhang presented other radiocarbon dates based on mineralized trees collected from a pit about 100 m northeast of the Ziyang site. The strata of this pit are supposedly comparable with those of the west test dug by Pei in 1951. Li and Zhang (1984) called this site Ziyang Man Site B.

Mammalian fossils unearthed from Ziyang Man Site B are similar to those associated with the human fossils in color and mineralization, among other features. Specimens from both localities probably represent parts of the same fauna. Li and Zhang (1984) described seven geological layers at the Ziyang Man B site. Fauna and stone artifacts come from the sixth layer that is radiocarbon dated to 37,400 ± 3000 years BP or 39,300 ± 2500 years BP.

Archaeology

The only artifact is a 108.2-mm-long bone awl from the basal part of the third layer of the west test pit. This awl was made of a triangular bone flint, and its dark brown color is similar to that of the human skull. Its lower end seems to have broken off before mineralization. The surface seems to be scraped longitudinally by a rough stone instrument. Although well polished on its edges and projecting parts, the blunt tip is not polished.

LONGLIN (105°20′ E, 24°47′ N)

A human skull from Longlin was found in 1979 by Changqing Li in a cave near the town of De'e, Longlin county, northwestern Guangxi, in southern China. The skull and deposits were described by Xingyun Zhang and Changqing Li (Zhang and Li, 1982). The skull has not been cleaned of the deposits. Zhang and Li described the mandible, which lacks the mandibular ramus and a small part of the anterior region of the mandible. The first lower molar from both sides was lost before death, and the right lower third molar was damaged during excavation. The teeth were heavily worn. The indices of robustness of the mandibular body were 43.3 and 40.0 at the level of the mental foramen for the left and right sides, respectively. The indices are 45.2 and 43.7 at the level between the first and second molars for the left and right sides, respectively. The single mental foramen is at the level of the middle of the height of the mandibular body between the second premolar and the first molar on both sides. Zhang and Li estimated that this individual was between 36 and 55 years of age.

The cave entrance is about 20 m above the ground surface. Much of the cave deposits had been dug for fertilizer. According to Zhang and Li (1982), the human skull, rib, vertebrae, and limb bones were covered by a deep-brown hard deposit, remains of which can be seen on the cave wall. This deposit contains carbon particles, burnt clay, ashes, breccia, and animal bone fragments, the taxonomic status of which cannot be determined.

CHUANDONG (105°45′ E, 26°18′ N)

Chuandong (Perforated Cave) is situated about 5 km southwest of Puding City and about 140 km southwest of Guiyang, the capital of Guizhou Province, southwestern China. From 1978 to 1982 teams from the Guizhou Provincial Museum and the Insititute of Vertebrate Paleontology and Paleoanthropology, Academia Sinica, excavated this cave. In 1979, Jinbiao Yu of Nanjing University found a broken calvaria, a mandible, and an isolated upper right incisor, and in 1982 the teams found a human skull.

Human fossils

Skull I (Fig. 4.11)

This skull was studied by Jinbiao Yu (1984) and Xianghong Huang (1989), who determined that it belonged to a middle-aged (45–55 years old) male (Table 4.11). The neurocranium is rather complete and is ovoid in top view. The sutures are deep and either wavy or serrated. The clear coronal and sagittal sutures are slightly obliterated on the external surface and completely obliterated on the endocranial surface. The bone walls are not thick. The superciliary arches are distinct, and the lateral end does not extend to the middle point of the supraorbital margin, which is rounded. Above the glabella is a trace of a sagittal ridge. The frontal tubercle and parietal tuberosities are weak. The occiput is rounded and has a chignonlike structure. The fronto-nasal suture is nearly horizontal. A right maxillary fragment with a median and lateral incisor and canine was considered by Yu (1984) as belonging to the same individual as the neurocranium. The upper median and lateral incisors are shovel-shaped and bear a trace of the basal tubercle. The canine fossa is shallow. Alveolar prognathism is moderate. The alveolar process of the maxilla is atrophied with part of the tooth root exposed.

Another right median incisor (no. 2) was unearthed from a disturbed clay deposit 1.2 m below the surface at the cave's entrance. This incisor has a basal tubercle and is shovel-shaped. A mandibular fragment, with three attached molars, and an isolated lower molar were also found (Table 4.12).

Skull II

The skull was studied by Maolin Wu (1989). The external surface of the skull is rather smooth, ovoid-shaped, and asymmetrical in top view. The bone wall is not thick. The right parietal tuberosity is more distinct than the left one. There is an

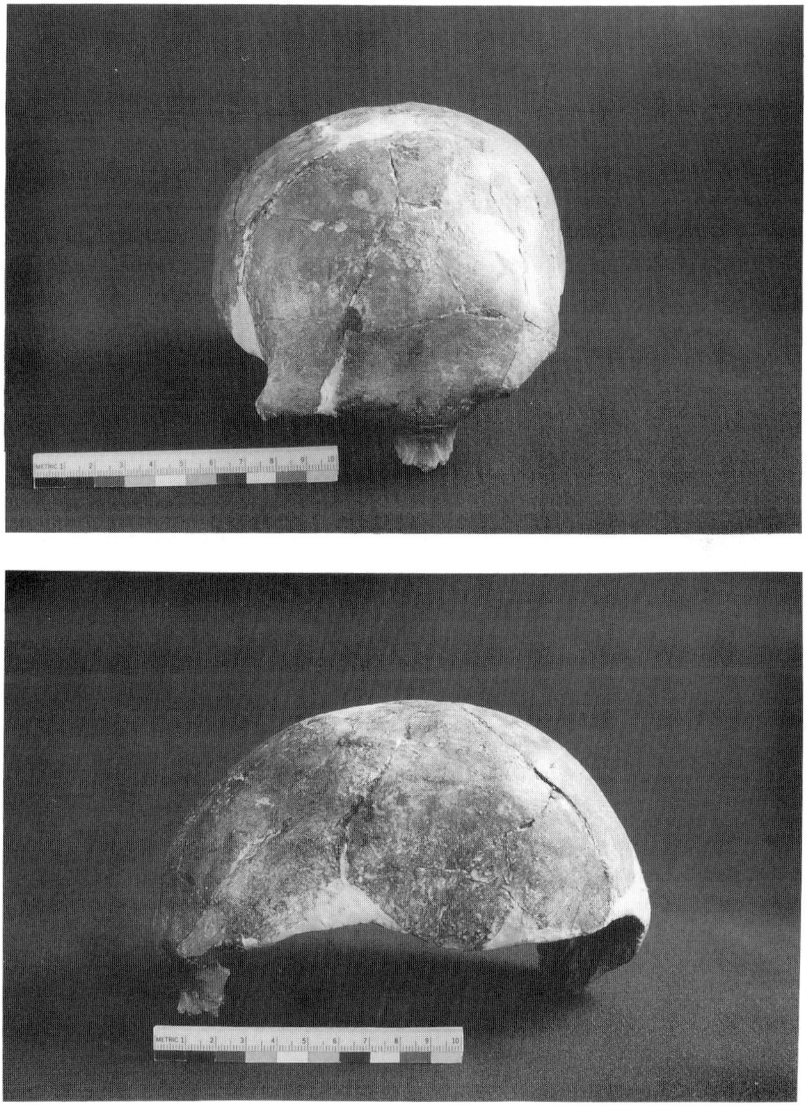

a

b

FIG. 4.11 Skullcap of *Homo sapiens sapiens* from Chuangdong. (Courtesy of Xianghong Huang.) (A) Anterior view; (B) lateral view.

interparietal depression with no midsagittal ridge. The right parietal foramen is distinct; the left one is very small. A weak angular torus is visible on the left side. The lower margin of the parietal bone indicates a curved upper border of the temporal squama. The root of the zygomatic arch inclines slightly upward. With the supramastoid ridge, they form an angle of about 20° with the Frankfurt Plane. The auditory orifice is elliptical with its long axis vertically oriented. The mastoid process is moderate sized, and the mastoid notch is deep and narrow.

The superciliary arch is rather robust for a female but is not as robust as the

Table 4.11

Measurements and indices of skull I from Chuandong

Sagittal arc (n-o)	343.0
Frontal arc (n-b)	128.0
Parietal arc (b-l)	120.0
Occipital arc (l-o)	95.0
Parietal chord (b-l)*	111.7
Frontal chord (n-b)*	107.0
Occipital chord (l-o)*	71.0
Max. cranial length (g-op)	174.5
Max. cranial breadth (eu-eu)	136.0
Simotic breadth	7.5
Simotic subtense	2.0
Simotic index	26.7
Calvarial height index	50.0**
Bregma position index	40.1***
Bregma angle (b-g-op)	51°
Frontal angle (m-g-op)	78°
Thickness:	
lambda*	10.1
occipital torus*	11.2
internal occipital protuberance*	14.1
cerebral fossa*	7.0
cerebellar fossa*	6.0
mastoid angle*	5.1°

According to Huang (1989)

* According to Yu (1984)

** Calvarial height index = $100x\dfrac{vx}{g-op}$ Where v is the vertex of the skull; x is the juncture between the line g-op with a vertical line from point v to line g-op.

***Bregma position index = $100x\dfrac{gy}{g-op}$ Where g is glabella; y is the juncture between the line g-op with a vertical line from bregma to line g-op.

Ziyang female specimen. A shallow groove sits above the arch. The superior orbital margin is straight. There is a broad supraorbital notch instead of a supraorbital foramen. Postorbital constriction is distinct.

The occipital plane is rounded and without a chignon. What remains of the sutures indicates the probable presence of an Inca bone. A broad transverse eminence in the occipital region is rather thin near the midsagittal plane and gradually attenuates toward its lateral ends. The external occipital ridge is prominent and bone surfaces on both its sides appear as deep depressions.

Behind the deep and narrow mandibular fossa is a postglenoid process. The

Table 4.12
Measurements of human teeth from Chuandong

	Length	*Breadth*	*Height*
Upper median incisor	8.2	6.6	7.3 (worn)
Upper lateral incisor	7.4	6.1	—
Canine	8.0	8.3	7.3 (worn)
Lower first molar	11.0	11.0	6.0
Lower second molar	10.5	10.0	6.0
Lower second molar	11.5	10.0	6.0
Lower third molar	10.5	9.5	7.5

Upper teeth according to Yu (1984). Lower teeth according to Huang (1989).

articular eminence is well developed. The well-preserved right tympanic plate forms a 65° angle with the sagittal plane.

The frontal sinus is large. Impressions caused by the middle meningeal artery are very clear. Its anterior branch is thicker and more ramified than the posterior branch. The preserved left part of the occipital squama shows that the cerebral and cerebellar fossae are more or less equal in size. However, the former is shallower than the latter. The internal occipital ridge bifurcates near the posterior margin of the foramen magnum.

A nearly triangular perforation is anterior and inferior to the left parietal tuberosity. The length of its base and height are about 25 mm (Table 4.13). The perforation might have been caused by a blow just before or after death; Wu (1989) prefers the latter possibility.

Geology

The cave is about 26 m above the water surface of the local river. Jinbiao Yu (1984) reported that the deposits were disturbed in recent years. The human skull and the cultural remains were located 150 cm below the surface, while the cultural layer is 30 cm below the surface. It contains small broken stones, ashes, burnt bones, and stone artifacts. Three meters below the surface are gravels cemented with calcium salts. The thickness of this layer is not known.

Archaeology

The majority of the flakes recovered are bipolar, and a minority were made by hammering. Most of the stone tools are scrapers; points and choppers are rarer occurrences. There are three kinds of bone tools: bone spades, bone awls with blunt points, and bone awls with a long apex (S. Zhang, 1980).

LIJIANG (100°17′ E, 26°47′ N)

In 1964 a letter from the Cultural Club of Lijiang County informed the Yunnan Provincial Museum that a fossilized human skull (Fig. 4.12) was found by a local farmer while he and others were widening a river bed in constructing an irrigation canal near Mujiajiao. The site is 11 km southeast of Lijiang city, Yunnan Province, southern China. The skull was later given to the Yunnan Provincial Museum

Table 4.13

Measurements and indices of skull II from Chuandong

Maximum cranial length(g-op)	170
Glabella-inion length(g-i)	163
Maximum cranial breadth(eu-eu)	130.5
Auricular height	108
Auricular bregma height	97
Median sagittal arc(n-o)	360(?)
Frontal arc (n-b)	120(?)
Cranial index	76.8
Length-height index	63.5
Breadth-height index	82.6
Calvarial height index*	45.7
Bregma position index**	33.7
Frontal angle	76.5°
Angle b-g-op	51°
Angle b-g-i	58.5°
Angle m-g-i	79°

According to Maolin Wu as cited in R. Wu, et al. (1989).

*Calvarial height index = $100x\frac{vx}{g-op}$. Where v is the vertex of the skull; x is the juncture between the line g-op with a vertical line from point v to line g-op.

*Bregma position index = $100x\frac{gy}{g-op}$. Where g is glabella; y is the juncture between the line g-op and a vertical line from bregma to line g-op.

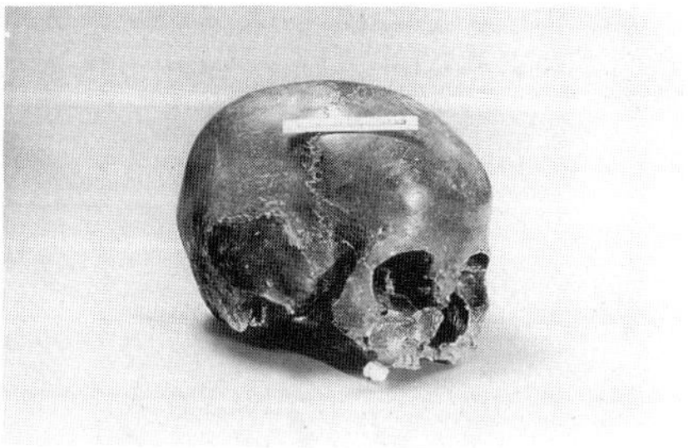

FIG. 4.12 *Homo sapiens sapiens* skull from Lijiang. (Courtesy of Xingyun Zhang.)

Human fossil

The skull, identified as Yunnan Provincial Museum cat. no. Limu 01, was described by Yipu Lin and Guoxin Zhou in 1977 (Yunnan Provincial Museum, 1977). The neurocranium is preserved except for the mastoid and tympanic portion of the left temporal and the basal and lateral parts of the occipital. Much of the face is also preserved.

Some of the sutures are visible. The rough posterior surface of the sphenoid body indicates the presence of the synchondrosis between the sphenoid and occipital before death. The mastoid, tympanic, and squamosal parts of the temporal have not united. The skull is small and its surface is smooth and gracile. The following features suggest that the skull is of a young female: the superciliary arches are weak and short, the frontal and parietal tubercles are very distinct, and the external occipital protuberance is not well developed.

The skull is pentagonal in top view, and its broadest part is slightly below the supero-posterior margin of the temporal squamae. The coronal and sagittal sutures are markedly wavey in the vicinity of bregma. The lateral third of the coronal suture and the vertex and obelion parts of the sagittal suture are serrated. Two Wormian bones are intercalated at the junction between the sagittal and lambdoidal sutures. One of the Wormian bones is 13.2×2.6 cm^2 and the other is smaller. The fronto-nasal sutures are at a higher level than the fronto-maxillary sutures. A sagittal ridge is only discernible between the left and right frontal tubercles. The parietal tuberosities are above the mastoid processes; there is no angular torus. The superior margin of the temporal squama is curved. The mastoid process is small and the supramastoid ridge is weak. A pteryon bone is located between the frontal, parietal, and temporal bones, and the greater wing of the sphenoid of the left side. On the right side two pteryon bones are between the parietal, temporal, and sphenoid bones. The inion is located somewhat more anteriorly than lambda. A distinct chignon exists on the occipital (Table 4.14).

The mandibular fossae are shallow and the articular tubercles are weak. The postglenoid processes are distinct. The preserved right tympanic plate is concave. The orbit is quadrangular. The palatal torus is indistinct. A Carabelli cusp exists on the slightly worn, erupted right upper second molar.

Geology

As reported by the Yunnan Provincial Museum (1977), the geology from the surface downward is as follows: Layer 1 is human disturbed deposits 0.5 m; Layer 2 is blackish-brown soil 0.5 m; Layer 3 is black sandy clay 2 m; Layer 4 is greyish-yellow gravel 1 m; this is followed by a pseudoconformity; Layer 5 is greyish-white sandy clay and the exposed thickness is 0.2 m.

Yipu Lin and Xingyun Zhang excavated along both banks of the irrigation canal near the village of Mujiajiao. They found a fauna including *Stegodon* sp, *Rhinoceros* sp., *Bibos gaurus, Bubalus bubalis, Axis yuannanensis,* and Elephantidae from the lower yellowish-brown gravel.

Judging from the coarse sand adhering to the human skull, Lin and Zhang inferred that the skull might come from Layer 4 and date to the Late Pleistocene (Yunnan Provincial Museum, 1977).

Archaeology

Lin and Zhang (1978, reprinted in Yunnan Provincial Museum, ed., 1991 pps. 165–173) reported that six flint flakes and nuclei were associated with the fauna. Sixteen additional stone artifacts were found near Mujiajiao in 1984 including two nuclei, four flakes, two scrapers, one chopper, and five bolas, and 2 unidentified artifacts.

Table 4.14

Measurements and indices of human skull from Lijiang

Linear measurements			
	Maximum cranial length (g-op)		167.0
	Glabella-inion (g-i)		150.0
	Maximum breadth (eu-eu)		141.0
	Auricular height		112.0
	Projectile height of bregma above porion (po-b)		105.0
	Median sagittal arc (n-o)		335.0
	Frontal arc (n-b)		115.0
	Frontal chord (n-b)		99.0
	Parietal arc (b-l)		123.0
	Parietal chord (b-l)		113.0
	Occipital arc (l-o)		119.0
	Occipital chord (l-o)		96.0
	Transverse cranial arc (po-po)		308.0
	Cranial circumference		490.0
	Minimum frontal breadth (ft-ft)		89.0
	Orbital breadth	lt.	36.0
		rt.	37.5
	Orbital height	lt.	28.0
		rt.	29.0
	Nasal height		? 40.0
	Nasal breadth		? 23.0
	Breadth of upper part of nasal bones		10.5
	Breadth of lower part of nasal bones		15.5
	Palatal breadth		39.5
	Palatal height		6.0
	Fmo-fmo		86.0
	Subtense from nasion to fmo-fmo		13.5
	Simotic breadth (SC)		8.5
	Simotic height (SS)		3.2
	Interorbital breadth (mf-mf)		18.5
	Subtense of nasal saddle to interorbital breadth		4.0
	Height of maximum cranial breadth above Frankfurt Plane		35.0
Volume			
	Cranial capacity		about 1300 cc
Angles			
	Angle g-b-FH		59°
	Frontal angle (m-n-FH)		83°
	Nasomalar angle (fmo-n-fmo)		146.5°
	Schwalbe's angle (b-g-i)		70°
Indices			
	Cranial index		84.4
	Breadth-height index		79.4
	Calvarial height index*		49.1
	Index of flatness of frontal bone**		24.0
	Index of protrusion of occipital bone***		31.3
	Frontal chord-arc index		87.6
	Parietal chord-arc index		91.9
	Occipital chord-arc index		79.1
	Simotic index		37.6
	Nasal index		? 59.0
	Orbital index	lt.	77.8
		rt.	79.2

*As previously defined.

**As previously defined.

*** Index of protrusion of occipital bone = 100 x furthest distance from l-o arc on the occipital squama to l-o chord/l-o chord.

According to Yunnan Provincial Museum (1977)

MENGZI (103°24′ E, 23°20′ N)

The cave site of Maludong is situated about 7 km southwest of the city of Mengzi in southern Yunnan Province. A field team organized by the Yunnan Provincial Museum excavated this site in the fall of 1989. The site and its contents were described by Xingyun Zhang and others (Zhang et al., 1991 in Yunnan Provincial Museum, 1991)

Human fossils

Skull I (Fig. 4.13)

The frontal and two parietals are preserved. The dolichocephalic skullcap is ovoid in top view, the maximum breadth of the skull is at the level of the parietal tubercle, and the highest point of the skull is close to bregma. The brow ridges connect at the glabella and are separated from the frontal squama by a shallow and broad supraorbital groove. The orbital margins are rounded. The frontal and parietal tubercles are not prominent. The coronal and sagittal sutures are not fully obliterated on the external surface of the skull but are almost invisible on the internal surface. This specimen belongs to a male slightly older than 30 years of age, according to X. Zhang et al. (1991).

Traces of cutting seem to exist along the fronto-nasal suture, zygomatic process of the frontal bone, margin of the temporal squama, and the lambdoidal suture. Symmetrically positioned holes occur near the left and right margins of the frontal. Zhang et al. inferred that they are drilled from both the internal and external sides, and they suggested that this is evidence of cannibalism in this population.

Skull II

Only parts of two parietals are preserved. Their black and brown color may be the result of burning.

Skull III

Only the glabellar part of the frontal bone and part of the occipital are preserved. The brow ridge is well developed.

Skull IV

Only part of the frontal is preserved. The brow ridge is weak.

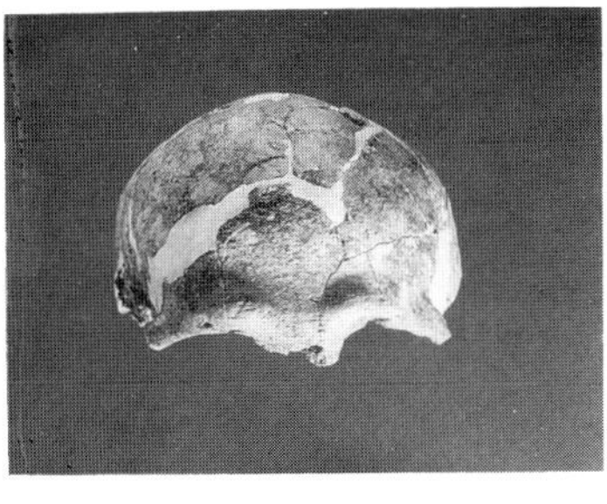

FIG. 4.13 *Homo sapiens sapiens* skull from Mengzi (Courtesy of Xinyun Zhang.)

Mandible I

The specimen includes the right mandibular ramus, condyloid process, and the mandibular body behind the first molar.

Mandible II

This fragmentary juvenile specimen of the left side contains DM2.

Teeth

Dental remains include a right upper third molar and a right upper second premolar. The crown and root of the former are intact; the root of the latter is broken.

Femur

This 18-cm-long upper part of a right femur is without the femoral head.

Tibia

This 15-cm-long lower part of a left tibia is without the articular part.

Zhang et al. (1991) estimated that these specimens represent at least five individuals.

Geology

Maludong cave is about 15 m above ground level. The original cave roof was weathered away, exposing the deposits to the open air. The deposits consist of three layers; from the surface downward they are as follows:

1 Red sandy clay with a few animal fossils, 4.5 m.
2 Grey cultural layer, most human fossils were unearthed from its upper part, most animal and plant fossils are in its northern part, and most artifacts are in its southern part. This layer is intercalated with cemented blocks and a thin layer of red clay, 5 m.
3 Big blocks of limestone or the cave floor.

The mammalian fauna unearthed are as follows (X. Zhang et al., 1991):

Primates
- *Homo sapiens*
- *Macaca robustus*

Rodentia
- *Rhizomys* sp.
- *Sciuridae* gen. et sp. indet.
- *Hystrix* sp.

Carnivora
- Canidae gen. et sp. indet.
- *Felis* sp.
- *Viverra* sp.
- *Meles* sp.
- *Sclenarctos thibetanus*

Artiodactyla
Sus sp.
Muntiacus reevesi
Muntiacus muntjak
Axis cf. *yunnanensis*
Cervus unicolor
Cervus nippon
Caprinae gen. et sp. indet.
Bovinae gen. et sp. indet.

All the mammalian species except *Axis yunnanensis* are extant. Zhang et al. suggested that this site dates to the Late Pleistocene or the earlier part of the Holocene, or falls between the two.

Archaeology

Zhang et al. (1991) reported that 89 stone artifacts, 8 antler spades, 6 antler awls, and 46 pieces of raw material for manufacturing antler awls were recovered. The stone artifacts include 31 hammers, 3 nuclei, 3 choppers, and 10 scrapers. The hammers are disc-like, ball-like, or irregularly shaped. Six scrapers are made of quartz. Hearths, carbon particles, burnt bones and stones, as well as red burnt clay were also found.

MAOMAODONG (105°01′ E, 25°12′ N)

Maomaodong Cave is situated about 25 km northeast of Zingyi City in Dingxiao District, Xingyi County, Guizhou Province. The site was found in the winter of 1974 and excavated by Zetian Cao in 1975. Cao (1982) described the materials found at the site.

Human fossils

The human fossils from this cave have not been studied. They include: a male mandible, a female mandible, a fragment of the right side of a female mandible, an anterior fragment of a child's mandible, and three femoral fragments. Two mandibles have double mental foramina on each side of the mandible.

Geology

Nine fossil mammalian species including *Rhinoceros sinensis,* were recovered. No detailed study has been published (Cao, 1982). The uranium date is 14,600 ± 1200 years BP (Zhang, 1983).

Archaeology

Cao (1982) examined 1121 pieces of stone artifacts: 129 nuclei, 317 flakes, 116 hammers and anvils, 66 worked pebbles, and 493 stone tools. The stone tools consist of 345 scrapers, 107 points, 40 choppers, and 1 burin. Cao (1982) reported 5 bone awls, 1 bone knife, and 8 antler spades.

TAOHUA (105°27′ E, 26°12′ N)

Taohua cave is situated 1.2 km northwest of Linzhi City, which is 172 km west of Guiyang, the capital of Guizhou Province. The cave is 1298 m above sea level and 5 m above the present ground level.

Five days of excavation in 1983 yielded a human femur, artifacts, and mammalian fossils in the basal part of the black clay and upper part of the yellow clay. These deposits are 70–150 cm thick. The mammalian fossils include *Megatapirus* sp., *Bison* sp., Cervidae gen. et sp. indet., *Ursus* sp., and *Hystrix* sp., among others.

The total of 240 stone artifacts included hammers, anvils, nuclei, flakes, scrapers, points, and choppers. There are also two bone awls and two perforated shells. An ash layer 40–70 cm thick contains burnt bones and stones, as well as carbon particles. Po Cao (1984) places the materials in the Late Pleistocene. He considers the stone artifacts to be similar in retouching technique and type of implement to those from Chuandong in the neighboring county.

TUBO (109°05′ E, 24°12′ N)

A team from the Guangxi Museum and the Institute of Vertebrate Paleontology and Paleoanthropology, Academia Sinica, excavated Ganjian Cave (Locality 81001) in Tubo District, Liujiang County, Guangxi Zhuang Autonomous Region. The cave, 25 km northwest of the city of Liujiang, is about 30 km from the site yielding the Liujiang fossil human skull. Five human teeth and mammalian fossils were found in situ and four human teeth came from a drug store (Tables 4.15, 4.16). The materials were discovered and studied by Y. H. Li et al. (1984).

Human fossils

Teeth

Incisor (left upper I1 no. KS81001A). The crown of this unerupted incisor has three mammelons on the cutting margin. The labial surface is not bulging. The lingual surface is distinctly shovel-shaped but it has no lingual tubercle.

Molar (left upper M1 no. KS81001B). The molar's unworn crown is intact and the root is not fully formed. The occlusal surface is roughly rhombdoidal and has four cusps. The paracone is separated from the metacone by a groove extending from the buccal to the occlusal surface. The protocone connects with the metacone. Among the cusps, the protocone is largest, the paracone is next largest, then the metacone, and the hypocone is the smallest. There are four marginal ridges along the buccal, lingual, mesial, and distal borders. A ridge with two accessory ridges appears on the sloping surface of the protocone, paracone, and the metacone. A short and thick longitudinal ridge connecting the protocone with the metacone resembles that found in a *Homo erectus* molar from Yunxian. There is a V-shaped

Table 4.15
Measurements of the crowns of the human teeth from Tubo

		Length	*Breadth*	*Height*
Incisor (upper, median)	(KS81001A)	8.2	5.5	9.8
Molar (upper, first)				
	(KS 81001.B)	11.0	11.5	7.7
	(KS 81001.C)	10.7	12.4	—
	(KS 81001.D)	10.7	12.8	—
	(KS 81001.E)	10.1	11.0	—
(upper, third)	(KS 81001.G)	8.8	11.4	—
(lower, first)	(KS 81001.H)	11.6	11.0	—
Molar (lower, deciduous),	(KS 81001.I)	10.7	9.9	6.5

According to Li et al. (1984)

Table 4.16
Measurements of the root of the human teeth from Tubo

			Length	*Breadth*	*Height*
Molar	(upper, first)	(KS 81001.C)	8.8	12.0	12.2
		(KS 81001.D)	8.2	11.3	(10.1)
	(upper, third)	(KS 81001.G)	—	—	11.0

According to Li, et al. (1984)

groove on the mesial half and a transverse groove on the distal half of the occlusal surface. A Carabelli cusp exists on the lingual surface of the protocone. No cingulum exists.

Molar (right upper M1 no. KS81001.C). This molar is heavily worn; the dentine is almost fully exposed on the occlusal surface. The root is short and slender; its lingual branch diverges somewhat from the buccal branch and slightly inclines toward the lingual side.

Molar (right upper M1 no. KS81001.D). The two branches of the root of this heavily worn molar converge toward the top.

Molar (right upper M1 no. KS81001.E). This heavily worn upper right molar is missing a large part of the root. There is a caries fovea on the mesial side of the neck.

Molar (right? upper M2 no. KS81001.F). Attrition on this tooth is so marked that a large part of the crown height is missing. The contour of the crown is elliptical, and the crown resembles that of a third molar because there is no contact facet on the distal surface. On the other hand, the branching of the root favors attribution to the second molar. Both the buccal and lingual branches of the root are slender, but the buccal is more slender. There is a caries fovea on the mesial surface of the crown.

Molar (left upper M3 no. KS81001.G). This slightly worn molar lacks a contact facet on the distal surface. The rectangular occlusal surface has four cusps. The paracone and hypocone are separated by the protocone and metacone, which unite with each other. Except on the hypocone, each cusp has a ridge on the slope, and smaller ridges to each side. There are no anterior or posterior fovea. The neck constricts distinctly. The slender root has no branching and inclines distally.

Molar (right lower M no. KS81001.H). Only the molar's crown is preserved. Because the root has not been formed, the tooth is attributed to a child younger than 6 years old. The mesial surface is flat and the distal surface is rounded. The occlusal surface shows the Y-5 pattern. The protoconid is separated from the entoconid by the metaconid and hypoconid, which unite with each other. The smallest cusp is the hypoconulid situated in the buccal half of the occlusal surface. The other four cusps are approximately similar in size. Each cusp has a main ridge, on the sides of which are found a few accessory ridges. The trigonid and talonid are about equal in breadth, but the former is shorter in length. There is no cingulum.

Li et al. (1984) attributed KS 81001.A, B, and H to the same individual.

Molar (left lower deciduous M2 no. KS81001.I). The crown of this left molar is slightly worn. There are no contact facets on the distal surface. Two deep grooves extending from the upper part of the buccal to the occlusal surface separate the hypoconid from the protoconid and hypoconulid, respectively. Another groove extending from the lingual to the occlusal surface separates the metaconid from the entoconid. The occlusal surface shows a "+" pattern. The entoconid is the largest cusp, the metaconid is the next largest, and the hypoconulid is the smallest. The protoconid and hypoconid are almost equal in size. The talonid is longer and broader than the trigonid. The mesial surface is flat and the distal surface is rounded. Only part of the root is preserved. Its mesial and distal branches constrict at the neck of the tooth. There is no cingulum.

Geology

The human and other mammalian fossils are embeded in reddish-yellow clay at the bottom of the cave. Most fossils are represented only by isolated teeth, and display gnawing by porcupines. The nonhuman mammalian fossils include (Y. H. Li et al., 1984):

Primates
- *Homo sapiens*
- *Pongo* sp.
- *Macaca* sp.

Rodentia
- *Hystrix subcristata*
- *Rhizomys* sp.

Carnivora
- *Ursus thibetanus*
- *Ailuropoda melanoleuca baconi*
- *Arctonyx collaris*
- *Crocuta ultima*
- *Cuon javanicus*
- *Panthera tigris*

Proboscidea
Stegodon sp.
Perissodactyla
Megatapirus augustus
Rhinoceros sinensis
Artiodactyla
Sus scrofa
Cervus sp.
Muntiacus sp.
Ovis sp.
Bovidae gen. et sp. indet.

Five of the species listed are extinct. Li et al. (1984) suggest that this site dates to the late Pleistocene.

DU'AN (108°00′ E, 23°59′ N)

The Du'an fossil human teeth were found in 1977 by geologists from Hydrogeological Team II of the Chengdu College of Geology. The materials are from Cave R5013, located 12 km northwest of Du'an city, Guangxi Zhuang Autonomous Region, southern China. The site and the remains were described by Zhao et al., (1981).

Human fossils (Table 4.17)

Teeth

Molar (left upper M1 or 2). This heavily worn molar has an uncomplicated pattern of wrinkles. The enamel is broken at the lingual surface, and at part of the mesial and distal surfaces. The root is broken off.

Molar (left upper M3). On this heavily worn molar the exposures of dentine at the protocone, metacone, and hypocone unite to form a whole piece. The root has three branches resting side by side.

Molar (right lower M1 or 2). Like the others, this molar is heavily worn, especially at the distal part of the occlusal surface. There is significant dentine exposure. There is no hypoconulid. The grooves separating the protoconid, hypoconid, metaconid, and the entoconid are "+" shaped. Only the root's stem is preserved.

Zhao et al. (1981) are unable to ascertain if the above three molars come from the same individual.

Molar (right lower deciduous M2). The molar has a protoconid, hypoconid, hypoconulid, metaconid, entoconid, and a sixth cusp arranged in a Y pattern. The largest cusp, the metaconid, contacts both the protoconid and hypoconid, and sends a triangular ridge from its tip to the longitudinal groove of the occlusal surface.

Table 4.17

Measurements of Human Molars From Du'an

	Length	*Breadth*	*Height*
Molar (lt. upper M3)	9.6	10.1	(7.0)
Molar (rt. upper M1 or M2)	11.5	11.5	(6.6)
Deciduous molar (rt. lower dM2)	10.7	9.0	5.8

Cited from Zhao et al (1981)

Parenthesis indicates a considerably worn tooth.

Another ridge mesial to this ridge connects with the corresponding ridge on the protoconid at the longitudinal groove. The trigonid and talonid are similar in height. Two vertical grooves on the buccal surface separate the three buccal cusps. The mesial groove is deeper than the distal groove. The buccal surface is more inclined than the lingual surface. The vertical groove separating the metaconid and the entoconid on the lingual surface is indistinct. The neck is very restricted. The stem of the root is very short and its mesial and distal branches are flat and broad. Zhao et al. (1981) considered this deciduous molar to belong to a 7- or 8-year-old-child.

Geology

The cave is 9.5 m above ground level and 166.5 m above sea level. The human teeth were collected from disturbed deposits in the inner part (Chamber II) of the cave. Local farmers dug these deposits for fertilizer.

Zhao et al. (1981) inferred that the human teeth were probably unearthed from the upper part of the fossiliferous layer. This layer is mainly a reddish-yellow sandy clay 70 cm thick and covered by greyish-yellow travertine (3 to 5 cm thick), dark yellow loam (12 cm thick and containing mollusca), a yellow travertine (2 cm thick), and a yellow mild clay (20 cm thick). The deposits and vertebrate fossils were probably transported by water into the cave. Most fossils are represented by teeth without roots. The vertebrate fossils include (Zhao et al., 1991):

Mammals

Primates

Homo sapiens

Hylobates cf. *concolor*

Macaca sp.

Rhinopithecus sp.

Rodentia

Hystrix subcristata

Atherurus sp.

Bandicota indica

Rhizomys sp.

Muridae gen. et sp. indet.

Carnivora

Arctonyx collaris

Viverra sp.

Paguma larvata

Cuon javanicus
Nyctereutes procyonoides
Crocuta sp.
Ursus thibetanus
Ursus sp.
Ailuropoda melanoleuca baconi
Panthera tigris
Panthera pardus
Felis teilhardi
Felis microtis
Felidae gen. et sp. indet.

Proboscidea
Stegodon cf. *hypsilophus*
Elephas maximus

Perissodactyla
Megatapirus augustus
Tapirus sp.
Rhinoceros sinensis

Artiodactyla
Sus scrofa
Rusa unicolor
Cervus sp.
Muntiacus sp.
Bubalus sp.
Budorcas sp.
Capricornis sumatraensis
Ovis sp.

Birds

Phasiandae gen. et sp. indet.

Reptiles

Testudinidae gen. et sp. indet.
?*Alligator* sp.

Deer teeth, totalling more than 1000 specimens dominate the mammalian assemblage. Wild pig teeth are next in abundance and total several hundred specimens. The fauna suggests a Late Pleistocene date (Zhao et al., 1981).

NALAI (105°08′ E, 24°53′ N)

In 1984, Shilin Peng and his colleagues from the Guangxi Museum found two human molars in Nalai cave, located at Hongyanshan (Red Rock Hill) in Xiangbo

District, about 34 km northwest of the city of Longlin, Longlin County, Guangxi Zhuang Autonomous Region, southern China. Peng et al. (1987) described the site and the fossils.

Human fossils

The human teeth include a left upper second molar (Guangxi Museum Cat. no. KL 84701) and a left lower second molar (G.M.- K.L. 84702) (Table 4.18). The occlusal surface of the well-preserved upper molar is nearly rhombdoidal in shape. The buccal and lingual halves are nearly equal. All four cusps are well developed, and the protocone, hypocone, and metacone are slightly worn. The attrition of this molar is between degree I and II. In degree I wear the apex of the cusp or the margin of the occlusal surface is only slightly worn. In degree II wear the apex of the cusps are worn flat or the occlusal surface is worn to form a depression at the center. Based on attrition patterns of modern southern Chinese, this molar may belong to an adult of around 25 years old. The lingual root is largest, the mesio-buccal root is next, and the disto-buccal root is thinnest. All roots incline slightly backward.

The occlusal surface of the well-preserved lower molar is nearly quadrangular and concave. Although worn there is no exposed dentine; however, a depression is formed at the center as a result of wearing. There is a 4+ cusp pattern. Neither tooth has distinct ridges on the cusp surfaces. These two molars supposedly belong to a same individual.

Geology

There are three chambers in this cave. Mammalian fossils are found in the posterior chamber and the passageway between this chamber and the middle chamber. Human fossils are found in the posterior chamber among undisturbed deposits (Peng et al., 1987). The geology from the surface down is as follows:

1. Yellow clay containing human and other mammalian fossils, 10–30 cm
2. Greyish-green sandy clay that is calcified, 10 cm
3. Yellow sandy clay, 5–7 cm
4. Greyish-yellow travertine crust, 3 cm
5. Greyish-yellow sandy clay of unknown thickness

No fossils were found in Layers 2–5

Peng et al. (1987) suggested a late Pleistocene date for the site based on faunal correlations. Mammalian fossils from the uppermost layer include the following (Peng et al., 1987):

Table 4.18

Measurements of the human molars from Nalai Cave

	Length	*Breadth*
Upper second molar	8.2	10.2
Lower second molar	10.6	10.0

According to Peng et al. (1987)

Primates
Homo sapiens
Macaca sp.
Rodentia
Hystrix subscristata
Carnivora
Cuon javanicus antiquus
Ailuropoda melanoleuca baconi
Panthera tigris
Ursus thibetanus
Proboscidea
Stegodon sp.
Perissodactyla
Rhinoceros sinensis
Artiodactyla
Sus scrofa
Cervus sp.
Muntiacus sp.
Ovis sp.
Bubalus sp.
Capricornis sumatraensis

BAOJIYAN (110°18′ E, 25°12′ N)

In May 1979, students of Affiliated Middle School of the Normal College of Guangxi found a human tooth and other mammalian fossils in Baojiyan cave located on the campus. This school is situated in the city of Quilin, Guangxi Zhuang Autonomous Region. The fossils were sent to the Guangxi Museum for scientific diagnosis. In July Linghong Wang and local archaeologists surveyed the cave and found another human tooth, some stone artifacts, and a large number of other fossils. The following description relies on Wang et al. (1982).

Human fossils

Teeth

Molar (right lower M2) (Guangxi Museum Cat. no. KP79501). Four cusps are discernible on this heavily worn tooth. The protoconid contacts with the entoconid. The mesial and distal branches of the root fuse at the buccal part.

Molar (left lower M3) (Guangxi Museum Cat. no. KP79712). This tooth is much less worn than the other molar. The protoconid is the largest and the hypoconulid is the smallest cusp. The wrinkles are rather complex compared to late *Homo sapiens*. The mesial and distal branches of the root are separated from each other only at the tip by longitudinal grooves on the lingual and buccal sides (Table 4.19).

Table 4.19

Measurements of the human molars from Baojiyan

Specimen	*KP 79501* *Right lower M2*	*KP 79712* *Left lower M3*
Crown		
Length (M-D d.)	10.6	10.4
Breadth (B-L d.)	10.7	10.2
Height	—	8.1
Root		
Length	8.8	9.0
Breadth	9.5	8.6
Height	15.4	13.9

Cited from Wang et al (1982)

Geology

The cave has a main chamber, which yielded the fauna and stone artifacts, and several branches. Most of the deposits were dug for fertilizer before the cave was scientifically explored. The section of the remaining deposits in the region of the sites in the main chamber is as follows: Layer 1 is a stalactite crust, 10–25 cm; Layer 2 is a greyish, yellow cemented loam intercalated with several calcified hard crusts, in some places 20–200 cm; and Layer 3 is a yellow loam of 70 cm.

Based on the faunal assemblage listed below, Wang et al. (1982) considered this site to date from the Late Pleistocene. The mammalian fauna of this cave includes (according to Wang et al., 1982):

Primates
- *Homo sapiens*
- *Macaca* sp.
- *Hylobates* sp.

Rodentia
- *Hystrix subcristata*
- *Rhizomys* sp.

Carnivora
- *Ursus thibetanus*
- *Arctonyx collaris*
- *Ailuropoda melanoleuca baconi*
- *Crocuta ultima*

Perissodactyla
- *Megatapirus augustus*
- *Rhinoceros sinensis*

Artiodactyla
- *Sus scrofa*
- *Cervus* sp.
- *Muntiacus* sp.

Bubalus sp.
Ovis sp.
Proboscidea
Stegodon sp.

Archaeology

The 12 pieces of stone artifacts include 7 nuclei, 4 choppers, and 1 scraper.

TIANDONG (106°59′ E, 23°36′ N)

A human tooth was found in Dingmo Cave situated at the foot of Dingmo Hill, Tiandong County, Guangxi Zhuang Autonomous Region. A team from the Institute of Vertebrate Paleontology and Paleoanthropology, Academia Sinica, and the Guangxi Museum found the tooth during a geological survey in December 1980. Li et al. (1985) described the site and its remains.

Human fossil

Tooth

Molar (right upper M1 or 2). The molar is so heavily worn that the occlusal surface has the appearance of a concave smooth plane. A small hole at the center of this plane may have been caused by dental disease. The protocone is probably the largest cusp. The mesio-distal and bucco-lingual diameters are 10.3 mm and 11.0 mm, respectively. Mineralization is rather low. The root was gnawed off by a porcupine.

Geology

The molar comes from the terminal part of the southern branch of the cave, some 85 m from the entrance. The excavated deposit is divided into two layers. The upper layer, a loose brownish-yellow sandy clay about 0.5 m thick, contained the molar and a large part of the other mammalian fossils. A residue of a hard calcium carbonate crust on the higher cave walls was destroyed by local farmers digging for fertilizer. Li et al. (1985) date the site to the Late Pleistocene.

Many mammalian fossils are from the northern branch of the cave. Li et al. (1985) reported that fauna from both the northern and southern branches of the cave are similar. The fauna includes:

Primates
Homo sapiens
Pongo sp.
Macaca sp.
Carnivora
Ailuropoda melanoleuca

Arctonyx collaris
Ursus thibetanus
Proboscidea
Stegodon orientalis
Palaeoloxodon namadicus
Perissodactyla
Megatapirus augustus
Rhinoceros sinensis
Artiodactyla
Sus scrofa
Rusa unicolor
Bubalus sp.
Ovis sp.

LIPU (110°24′ E, 24°31′ N)

Shuiyan Dongdong (East Cave of Water Cave) Cave, in Lipu County, Guangxi Zhuang Autonomous Region, yielded one human tooth. When Wu and his colleagues (X. Wu et al., 1962) visited in 1961 most of the cave deposits were disturbed. The tooth and other mammalian fossils were found in the remaining undisturbed deposits.

Human fossil

Tooth

Premolar (left lower P2). The specimen is intact except for the tip of the root, which was slightly broken on the mesial side. The distal occlusal surface was slightly worn. The buccal cusp is largest, the mesio-lingual cusp next largest, and the disto-lingual cusp is the smallest. Each cusp has a triangular ridge extending from the tip to the center of the occlusal surface. There is a U-shaped groove in the central fovea. The single root turns slightly to the buccal and distal direction. The length, breadth, and height of the crown are 8.7 mm, 10.0 mm, and 8.1 mm, respectively. Similar dimensions of the root are 5.1 mm, 8.0 mm, and 12.4 mm, respectively.

Geology

The fossils were found in a yellow sandy clay below the stalactic columns. The associated fauna are as follows: *Hystrix cf. subcristata, Ailuropoda melanoleuca fovealis, Sus scrofa,* Cervidae gen. et sp. indet., and Bovinae gen. et sp. indet.

Archaeology

No artifacts were recovered. Like the others mentioned in this chapter, the site is probably of Late Pleistocene dating.

SHUICHENG (104°40′ E, 26°45′ N)

One human tooth was found in Xiaohuidong Cave located 25 km northwest of the city of Shuicheng in Shuicheng County, Guizhou Province. The cave was excavated by the Provincial Museum of Guizhou in December 1973 and January 1974. Z. Cao (1978) described the site and the remains.

Human fossils

Tooth

Canine (left upper C). The crown looks robust despite being heavily worn. The remaining crown height is 9.0 mm; the labial surface is smooth and bulging in the transverse direction. There are no marginal triangular prominences on either the mesial or distal surfaces. The basal tubercle is not very developed on the lingual surface. The root is almost complete with a longitudinal shallow fovea on the mesial surface. The mesio-distal and labio-lingual diameters and the height of the root are 9.3 mm, 9.8 mm, and 19.0 mm, respectively. The diameter of the root diminishes gradually toward the tip.

Geology

The cave is about 40 m above the local water surface and about 1700 m above sea level. The undisturbed deposits were divided into three layers.

1 Yellow, grey, and white stalatctic cover, 0.1–0.3 m thick and devoid of fossil and cultural remains
2 Black, grey, red and white ash layer with carbon particles, burnt bones, and burnt stone, 0.05–0.15 m thick and cemented by calcium carbonate
3 A 0.2–0.7 m thick yellow sandy clay and breccia containing fossils and stone artifacts

Most mammalian fossils, represented by isolated teeth, belong to *Stegodon* sp., Bovinae, Ovinae, *Sus* sp., and *Cervus* sp. The thorium date on oxen tooth from the third layer of the deposit is 52,000 ± 3000 years BP (Yuan et al., 1986).

Archaeology

The 53 pieces of stone artifacts include 2 hammers, 33 flakes, and 5 tools with secondary retouch.

LIANHUA (119°24′ E, 32°09′ N)

Lianhua Cave, situated about 6 km southwest of Zhenjiang, Jiangsu Province, was excavated in 1981 by Yucheng Shi and others. It yielded some mammalian fossils, including a human tooth.

Human fossil

Tooth

Molar (lower M). This slightly worn (probably second lower) molar has four cusps. There is evidence of enamel hypoplasia. There is no contact facet on the distal surface.

Geology

Li et al. (1982) considered the site to be of Late Pleistocene dating. The cave is about 35 m above sea level. The test excavation exposed four layers of deposits. From top to bottom they are as follows:

4 Surface deposit, 0.2 m
3 Brown loam with fossils and a few small broken limestone blocks, 1 m
2 Brownish-red loam with fossils and a few large limestone blocks and concretions, about 1.5 m
1 Brownish-yellow sandy loam without fossils. The thickness is unknown.

The vertebrate fossils include (Li et al., 1982):

Primates
  Homo sapiens
  Macaca speciosa
Lagomorpha
  Laporidae gen. et sp. indet.
Rodentia
  Hystrix sp.
Carnivora
  Nyctereutes procyonoides
  Ursus arctos
  Ursus thibetanus
  Arctonyx collaris
  Crocuta ultima
Perissodactyla
  Dicerorhinus sp.
Artiodactyla
  Sus scrofa
  Cervus cf. *unicolor*
  Cervus sp. (1)
  Cervus sp. (2)
  Bubalus cf. *wansjocki*
  Bos sp.

No artifacts were found.

XICHOU (104°42′ E, 23°26′ N)

Xuanrendong Cave, located about 0.3 km southeast of Xichou County, Yunnan Province, was dug in 1965 and 1970 by local archaeologists who sent mammalian fossils to the Provincial Museum of Yunnan. In 1972 a team from the Institute Vertebrate Paleontology and Paleoanthropology, Academia Sinica, found three human teeth and vertebrate fossils in deposits taken from the cave. In 1973 another team from the Institute found two other human teeth in disturbed deposits. Chen and Qi (1978) described the site and the materials.

Human fossils

Teeth

Canine (right lower C Nos. PA 539, and PA 540). Both canines are gracile, and the crowns are narrow and high. There is no cingulum. The enamel on the buccal surface is more extended than on the lingual surface. A distinct lingual tubercle exists at the basal part of the lingual surface. A distinct lingual fovea and a marginal ridge appear in specimen PA 539.

Both canines are heavily worn. The length-breadth indices of the preserved part of the crowns of specimens PA 539 and PA 540 are 118.6 and 116.7, respectively. The roots of both canines are very gracile. The root of specimen PA 540 distinctly turns to the distal side, and the root of PA 539 has longitudinal grooves on both the mesial and distal surfaces.

Premolar (right lower P1 no. PA 541). The high and short crown of this tooth germ is unworn. There is no cingulum. The buccal is higher than the lingual cusp. The cusps are connected by a transverse ridge that divides the occlusal surface into two almost equal parts. The mesial fovea is slightly smaller than the distal one. The buccal cusp is situated closer to the mesial border; the lingual cusp is situated closer to the distal border. The buccal surface is separated from the mesial and distal surfaces by small grooves. The lingual surface is bulging in both vertical and horizontal directions. The root is incompletely formed.

Molar (right lower M1 or 2 No. PA 542). Although the crown is heavily worn, the "+" cusp pattern is recognizable. The mesial border of the occlusal surface is flat, while the distal border is slightly convex. Enamel extends downward slightly more on the buccal than on the lingual surface. The buccal surface bulges at its middle. There is no cingulum. The breadth of the trigonid and talonid are 10.8 mm and 11.0 mm, respectively. The length-breadth index of the crown is 94.74.

The root is rather small and divides into two branches. A longitudinal groove on the distal surface of the mesial root divides it into buccal and lingual parts. Another longitudinal groove on the mesial surface of the distal root divides it into buccal and lingual parts.

Deciduous molar (right lower DM2 no. PA 538). The buccal part has heavier wear than the lingual part. There is no cingulum, but there is a sixth cusp. There is no contact between the protoconid and the metaconid. There is a mesial fovea at the mesial part of the occlusal surface. The trigonid is smaller than the talonid.

Geology

Although many species of mammalian fossils were recovered, most are represented by isolated teeth. The Late Pleistocene fauna includes (Chen and Qi, 1978):

Primates
- *Homo sapiens*
- *Pongo* sp.
- *Hylobates* sp.
- *Macaca* sp.

Chiroptera
- *Myotis* sp.
- Chiroptera gen. et sp. indet.

Rodentia
- *Eothenomys* sp.
- *Rattus* sp.
- *Rhizomys* sp.
- *Hystrix subcristata*

Carnivora
- *Cuon javanicus antiquus*
- *Ursus thibetanus*
- *Ailuropoda melanoleuca baconi*
- *Ailurus fulgens*
- *Arctonyx collaris*
- *Lutra lutra*
- *Viverra zibetha*
- *Viverricula indica*
- *Paguma lavarta*
- *Crocuta ultima*
- *Panthera tigris*
- *Neofelis* sp.
- *Felis microtus*

Proboscidea
- *Stegodon orientalis*

Perissodactyla
- *Equus* sp.
- *Rhinoceros sinensis*
- *Megatapirus augustus*

Artiodactyla
- *Sus scrofa*
- *Rusa* sp.
- *Muntiacus* sp.

Capricornis sumatraensis
Bubalus sp.

No artifacts were recovered.

ZHAOTONG (103°48′ E, 27°30′ N)

In 1982, while surveying cultural relics, members of the Cultural Club of Zhaotong Municipality, Yunnan Province, found Late Pleistocene mammalian fossils, including a human tooth, which came from Guoshandong limestone cave. The cave, situated 15 km north of the city of Zhaotong, is described by Liang Zheng (1985).

Human fossil

Tooth

Molar (left lower M2). The length and breadth of the crown are 13.0 mm and 11.7 mm, respectively. The height of the heavily worn crown is 5.3 mm. The occlusal surface is more or less elliptical and concave at the mesial side. There is no contact facet on the distal surface. An accessory internal tubercle is discernible between the metaconid and entoconid. The protoconid is the largest cusp, followed in size by the hypoconid, metaconid, entoconid, hypoconulid, and an accessory internal tubercle. The protoconid does not connect with the entoconid. The trigonid is slightly broader than the talonid. The buccal part of the occlusal surface is broader than the lingual part.

Geology

Guoshandong is about 2075 m above sea level and about 70 m above the local water surface, which is about 2 km south of the cave. From the entrance the cave floor inclines slightly upward for about 7 m; then it becomes horizontal for about 10 m. This horizontal part is about 3–5 m wide. The geological sections of the deposit follow: Layer 1 is a disturbed surface deposit about 20 cm; Layer 2 is travertine crust about 10–20 cm; Layer 3 is reddish-brown sandy clay about 40 cm; Layer 4 is red clay covered by a thin travertine crust about 30 cm; and Layer 5 is a cemented yellowish-brown sandy clay with human and mammalian fossils and carbon particles covered by a thin travertine crust about 40 cm.

The mammalian fauna include: *Stegodon* sp., *Rhinoceros* sp., *Tapirus* sp., Bovinae, and *Cervus* sp.

Archaeology

No artifacts were found.

DONGZHONGYAN (111°47′ E, 23°30′ N)

A premolar, the first human fossil from Dongzhongyan, was found in 1978 in a cave situated about 200 km west of Guangzhou, the capital of Guangdong Province.

The fossil was recovered by a team organized by the Guangdong Provincial Museum to survey cultural relics. In the winter of 1989 this museum and the Fengkai County Museum excavated in the cave and found a human molar and more mammalian fossils. The site and the fossils were described by Qiu et al. (1986) and Song et al. (1991).

Human fossils

Teeth

Premolar (right upper P2 [L. Qiu et al., 1986]). The occlusal surface is more or less rectangular. The buccal cusp is larger than the lingual cusp, and the former is separated from the mesial and distal margins of the tooth by deep grooves. The tooth was so worn that the dentine was exposed on the buccal cusp. The mesiodistal and bucco-lingual diameters are 7.3 mm and 9.6 mm, respectively. The existing height of the crown is 7 mm. On its mesial and distal surfaces the root has vertical grooves that gradually widen toward the tip of the root.

Molar (right lower M3 [Song et al., 1991]). The occlusal surface of this newly erupted tooth is more or less spherical. Its length and breadth are 11.4 mm and 10.7 mm, respectively. The protoconid, paraconid, metaconid, hypoconid, and the hypoconulid, and the grooves separating them are distinct. An accessory tubercle seems to be present at the antero-buccal side of the protoconid.

Geology

According to Song et al. (1991), the cave is on the northern slope of Huangkoushan Hill, 15 m above the local surface. The entrance faces northeast. The fossils were found in 1985 in the reddish-brown clay of a branch of the cave that is 6 m long, 2–3 m wide, and 0.7–1.5 m high. It begins from a small hall 5 m long and located 4 m from the cave's entrance. The deposits in the hall from the surface downwards are as follows:

1. Greyish-black sandy clay, 20–40 cm
2. Purple-brown travertine crust, 2–4 cm
3. Cemented sandy clay having a small amount of animal fossils, 20–35 cm
4. Light yellow travertine crust, 2–4 cm
5. Reddish-brown clay having human and other animal fossils, 30–70 cm
6. Clay with iron and manganese nodules

The human teeth are considered to date to the early part of the Late Pleistocene (L. Qiu et al., 1986) or the Late Pleistocene (Song et al., 1991) The mammalian fauna include (Song et al., 1991):

Primates
Homo sapiens
Macaca sp.
Rhinopithecus tingianus

Rodentia
- *Rattus* sp.
- *Hystrix subcristata*
- *Atherurus* sp.

Carnivora
- *Ailuropoda melanoleuca fovealis*
- *Ursus thibetanus*
- *Arctonyx* sp.
- Mustelidae gen. et sp. indet.
- *Hyaena* sp.
- *Acinonyx* sp.
- *Panthera* cf. *tigris*

Proboscidea
- *Stegodon orientalis*
- *Palaeoloxodon namadicus*

Perissodactyla
- *Megatapirus augustus*
- *Rhinoceros sinensis*

Artiodactyla
- *Sus australis*
- *Sus* sp.
- *Cervus unicolor*
- *Pseudaxis* sp.
- *Muntiacus muntiak*
- *Bubalus* sp.
- *Bison* sp.
- Caprinae gen. et sp. indet.

Archaeology

No artifacts were recovered.

QINGLIU (117°03′ E, 26°02′ N)

A human tooth from Qingliu was collected in 1988 by Xuechun Fang and others in Hulidong Cave (Fox Cave), Shaowu district, Qingliu County, western Fujian Province, eastern China, during a general survey for cultural relics in Sanmin Prefecture. The site and the fossils were reported by You et al. (1989).

Human fossils

Tooth

Molar (left lower M1). The crown is slightly worn and has bucco-lingual and mesio-distal diameters that are 10.8 mm and 10.2 mm, respectively. Decay exists at the middle of the anterior half of the occlusal surface; however, the protoconid, hypoconid, metaconid, entoconid, and the hypoconulid are discernible. The protoconid is the largest buccal cusp, the metaconid is larger than the entoconid, and the hypoconulid is larger than the hypoconid. The buccal, lingual, mesial, and distal surfaces are all bulging.

The mesial and distal roots unite 2 cm below the neck of the tooth. The mesial root is broken, and the distal root is 8.1 mm high. The mesial root is broader than the distal root. The distance between the tips of these roots is probably shorter than the length of the crown. A more or less horizontal line 1–3 mm below the occlusal surface indicates hypoplasia. The crown surface above this line is distinctly smaller than the part below this line.

Geology

The cave's entrance is about 80 m above the river surface; the cave is 10 m long and 4 m in maximum breadth. The deposits have been disturbed. Several species of mammalian fossils were collected, including *Megatapirus augustus, Hystrix subcristata,* Rodentia, *Ursus* sp., *Sus scrofa,* and *Cervus* sp. They were unearthed from a greyish-yellow sandy clay that includes fine gravels. This layer is 0.6–0.8 m thick and about 2 m below the top surface of the cave deposits.

JIANDE (119°38′ E, 29°32′ N)

The human fossil found in Wuquidong (Turtle) Cave in 1974 in Lijia District, Jiande County, Zhejiang Province, was described by Han and Zhang (1978).

Human fossil

Tooth

Canine (right, upper C. PA 536). The canine's crown is slightly damaged. A shallow groove on the labial surface indicates gnawing by a porcupine. The tip of the root was also gnawed. The mesio-distal, labial-lingual diameters, and the height of the crown are 8.2 mm, 9.5 mm, and 11.6 mm, respectively. Judging from the attrition and size of this canine, Han and Zhang (1978) considered it to belong to a male approximately 30 years of age.

Geology

Wuquidong cave is 15 m above the surface of the river. The original roof was severely damaged. Below a layer of recent deposits about 50 cm thick are two layers of fossil-bearing deposits: an upper layer of purple clay about 35 cm thick and a lower layer of yellowish-red clay about 110 cm thick. The human canine was found in the upper layer. Mammalian fauna from the upper layer include (Han and Zhang, 1978):

Primates
- *Homo sapiens*
- *Macaca* sp.

Carnivora
Crocuta ultima
Arctonyx collaris
Ailuropoda melanoleuca fovealis
Perissodactyla
Rhinoceros sinensis
Artiodactyla
Bubalus sp.
Ovis sp.
Cervus sp.
Muntiacus sp.
Sus scrofa
Proboscidea
Stegodon sp.

Mammalian fauna from the lower yellowish-red clay include:

Primates
Macaca sp.
Rodentia
Hystrix subcristata
Carnivora
Ursus thibetanus
Arctonyx collaris
Ailuropoda melanoleuca fovealis
Perissodactyla
Rhinoceros sinensis
Megatapirus augustus
Artiodactyla
Cervus sp.
Muntiacus sp.
Sus scrofa
Bubalus sp.
Ovis sp.
Proboscidea
Stegodon orientalis
Palaeoloxodon cf. *namadicus*

Han and Zhang considered the upper layer of purple clay to belong to the later Late Pleistocene and the lower layer of yellowish-red clay to the early Late Pleistocene. They suggest that the human canine belongs to the later period. Yuan et al. (1986) published uranium dates from the upper layer. Dates based on cow teeth are 108,000 + 9,000/ −8,000 years ago and 97,000 ± 8,000 years ago.

LONGTANSHAN SITE 1 (102°49′ E, 24°49′ N)

From 1973 to 1977 quarry workers and others found mammalian fossils and stone artifacts at Longtanshan (Dragon Pool Hill), Dayu District, Chengong County, Kunming Municipality, Yunnan Province. In April 1977, Shaojin Hu, an officer of the county Cultural Club, found two human teeth. Excavation in July 1977 yielded many mammalian fossils (Zhang et al., 1978).

Human fossils

Teeth

Premolar (right upper P1 no. YU 1361). The crown is heavily worn. The root turns backward, its cross section is triangular with rounded angles, and the buccal is thicker than the lingual segment.

Molar (left lower M1 no. YU 1362). The nearly square occlusal surface has four differentially worn cusps. The wrinkle pattern is "+" type. The hypoconid was most worn and metaconid was the least worn cusp. An anterior fovea exists in front of the protoconid.

Geology

Longtanshan Hill is about 40 m high. Its many fissures are filled with deposits. One of the fissures, Site 1, is about 5 m above the water surface of nearby Dianchi Lake. The deposits consist of two layers: a red loam and a yellowish-brown sandy clay, with thin and nearly horizontal travertine plates occurring in its upper and basal parts. The former is 2 m thick, slightly cemented, and has a few fossils in its basal part. The latter is 1.3 m thick, well cemented, and also contains fossils.

Judging from the color, degree of mineralization, and deposit adhering to the fossils, X. Zhang et al. (1978) considered the human fossils to have been unearthed from the cemented yellowish-brown sandy clay. The Late Pleistocene-dated mammalian fauna include the following (Zhang et al., 1978):

Primates
- *Homo sapiens*

Rodentia
- *Hystrix* sp.
- *Rhizomys* sp.
- *Rattus* sp.
- *Myospalax* sp.

Lagomorpha
- *Lepus* sp.

Carnivora
- *Crocuta ultima*
- *Cuon* sp.

Perissodactyla
- *Rhinoceros sinensis*

Artiodactyla
Pseudaxis sp.
Cervidae (?*Muntiacus* sp.)
Sus sp.
Bovidae gen. et sp. indet.

Archaeology

No artifacts were reported.

LONGTANSHAN SITE 2 (102°49′ E, 24°49′ N)

This site, situated in Chengong County, Yunnan Province, was found in 1975 by Shaojin Hu, an officer of the Administrative Committee of Cultural Relics of Kunming, Yunnan Province. Zhonglang Qui and Yinyun Zhang of the Academia Sinica and Shaojin Hu excavated the site in 1982 and found a human tooth and other fossils. The site and its contents were reported by Qiu et al. (1985).

Human fossil

Tooth

Molar (left lower DM2). This tooth is very atypical in morphology and its attribution is debatable. The occlusal surface of the complete crown is lightly worn and is oval. The mesio-distal and bucco-lingual diameters and the crown height are 13.1 mm, 11.1 mm, and 6.7 mm, respectively. The cusps are sharp with short wrinkles. In addition to the five main cusps usually seen in human lower molars, there is an accessory internal tubercle between the metaconid and entoconid, and a sixth cusp between the entoconid and hypoconulid. The sixth cusp is situated at the distal side of the occlusal surface and is divided by two shallow grooves into three unequal parts. There is no paraconid. The distal ridge of the anterior fovea connects the protoconid with the metaconid and is cut by narrow sagittal grooves. The anterior fovea is a narrow short pit, the central fovea is a larger shallow basin, and there is no distinct shallow posterior fovea.

The buccal and lingual surfaces bulge both horizontally and vertically. Deep grooves on the buccal surface separating the protoconid and hypoconid, and separating the hypoconid and hypoconulid, widen toward the occlusal surface. The grooves separating cusps on the lingual surface are narrow and short. There is no contact facet on the distal surface.

The trigonid is smaller than the talonid in both mesio-distal and bucco-lingual dimensions, but the former is slightly higher than the latter. Crown height is relatively low compared with crown length and breadth. The roots are short and weak. The mesial root is short. A large part of the distal root was damaged, but it may be similar to the mesial root in general condition. The distance between the tips of the two branches of the root is large. The pulp cavity is large.

Qiu et al. (1985) suggested that this tooth belonged to a child 6 years old or younger. However, because of the atypical morphology of this tooth, this conclusion may be revised.

Geology

The strata from the surface downward are as follows:

1. Brown surface deposit, 1 m
2. Brown sandy clay with stone artifacts and fossils, 1 m
3. Travertine, 0.3 m
4. Brown sandy clay with burnt bones, stone artifacts, and fossils, 2 m

Qiu et al. (1985) considered Layers 2 and 4 to belong to the same period. The fragmentary mammalian fossils include: *Rhinoceros sinensis,* Bovinae, *Cervus* sp., and *Sus* sp.

The date of the carbon particles and burnt bones from Layer 4 is 30,500 + 800 years BP (Qiu et al., 1985).

Archaeology

Qiu et al. (1985) reported 107 pieces of stone artifacts, including 3 nuclei, 57 flakes, and 9 stone tools. There are 4 flake choppers and 5 scrapers, which were probably made by hammering. Certain characteristics of a convex side-scraper resemble some European Mousterian specimens. The technique for making this scraper is seen in the Late Paleolithic Shuidonggou Site.

ZUOZHEN (120°18′ E, 23°12′ N)

The human fossils include two right fragmentary parietals (nos. 200-1 and 200-3), a left fragmentary parietal (no. 200-2), an upper right first or second molar (No. A570), and a lower right first molar (no catalogue number). All were described by Chaomei Lian (1981) and Shikama et al. (1976), and were collected by amateurs from the gravels of the river bed of Chouqu (Chou-chu) segment of Cailiaoxi Creek, Zuozhen district, Tainan County, Taiwan Province. Shikama et al. (1976) considered these fossils to date to 20,000–30,000 years BP on the basis of fluorine content (0.76%) and manganese content (0.25%), and the assumption that the soil conditions of Taiwan and Japan are more or less similar. Lian (1981) considered the fossils to belong to the Late Pleistocene.

Human fossils

Right parietal fragment (220-l)

The fragment comes from the part between the middle segment of the sagittal suture and the parietal tuberosity. The thickness of the bone ranges from 4.1 mm to 5.6 mm. The sagittal suture is visible, the superior temporal line is not marked, and the groove for the middle meningeal artery is shallow. Shikama et al. (1976) considered that the fragment came from a young male.

Left parietal fragment (220-2)

This piece is the central part of the posterior half of the bone. The superior and inferior temporal lines, and the groove for the middle meningeal artery, are visible.

Traces of scraping on both the internal and external surfaces were considered by Shikama et al. to have been human produced. Lian (1981) suggested that the marks might have been caused naturally, for example, during water transportation. Shikama et al. could not sex the fragment.

Right parietal fragment (200-3)

The thickness of this fragment from a small piece of the medial part of the bone ranges from 4.0 mm to 5.5 mm. The sagittal sulcus is very deep. Shikama et al. could not sex the fragment.

Teeth

Molar (right upper M1 or 2) (A 5710). The occlusal surface is nearly rhombodial and heavily worn. Fine wrinkles are visible on the occlusal surface. A mesio-buccal fovea is well depressed and there are no disto-lingual fovea. The four cusps are about equal in size, the protoconid is slightly larger. X-rays show that the root has three branches. The mesio-distal, bucco-lingual diameters, and the height of the crown and the height of the root are 10.5 mm, 12 mm, 6.5 mm, and longer than 8.5 mm, respectively.

Molar (right lower M1) (no cat. no.). The crown of this moderately worn tooth is only partially preserved. There is a Y-5 cusp pattern. The hypoconulid is the smallest cusp. The mesio-distal and bucco-lingual diameters are 11 mm and 10.2 mm, respectively. This tooth is much smaller and less mineralized than specimen A 5710. Lian considered the lower molar to date later than the upper molar.

The geological deposit is unclear, no artifacts or associated fauna have been reported. However, there is a passing mention of an in situ rhinoceros.

SUMMARY

H. sapiens sapiens sites come from all regions of China except for the vast western regions of Xizang (Tibet), Qinghai, and Xinjiang. Although dating has been somewhat of a problem, the earliest of these sites may date to about 67,000 years ago at Liujiang in southern China, to about 52,000 years ago at Shuicheng in southern China, and about 50,000 years ago at Salawusu in Inner Mongolia. Other sites with chronological dates are in the range of 10,000–30,000 years old.

From China *H. sapiens sapiens* populations migrated to other parts of the world. One migratory route eventually found Asian populations settling into North and South America. Exactly when these migrants crossed into the New World is hotly debated, but the route was across Beringea. The Asian ancestors of the Paleo-Indians entered the New World in several waves of migration, an assertion supported by dental, archaeological, linguistic, and mitochondrial DNA evidence.

5

Evolution and Dispersal

Ten anatomical traits seem to link the early *Homo sapiens sapiens* samples from China with earlier Chinese *Homo erectus* populations. This is an indigenous evolutionary event. Archaeological evidence also seems to suggest that the appearance of anatomically modern *Homo sapiens* in China was a local evolutionary event. The Late Paleolithic assemblage in China is not markedly different from the Middle Paleolithic assemblage. This possibility of an indigenous evolutionary event casts doubt on the Out-of-Africa scenario.

China has more than 60 sites containing human fossils and many more sites yielding Paleolithic artifacts. Humans may have inhabited China for 1.7 mya. What were the evolutionary fates of Chinese Pleistocene humans? Are they the direct ancestors of modern *Homo* or did some, like *H. erectus,* became extinct without further issue? Was the appearance of *H. sapiens* in China an indigenous phenomenon or did *H. sapiens* migrate to China from elsewhere, like West Asia? These questions have long been debated by paleoanthropologists.

Weidenreich (1943) found a series of common morphological traits shared by *H. erectus pekinensis* and modern Mongoloid populations, and he proposed an ancestor-descendant relation based on these traits. His evidence did not convince many scientists, partly because of a lack of fossil materials from China to support his viewpoint. However, after 1949 many new human fossils were recovered in China. These materials belong to different periods of human evolution. The majority of the traits mentioned by Weidenreich are found in these newly discovered specimens; one common trait is shovel-shaped incisors. X. Wu (1988a,b, 1990) has described a further suite of shared traits not mentioned by Weidenreich that seem to support the continuity hypothesis that Weidenreich originally favored.

1. Midsagittal elevation

This structure that exists on the frontal and in *H. erectus* from Zhoukoudian also on the parietal, is usually called a *sagittal crest.* In *Homo erectus* from Hexian, the sagittal crest covers the upper three fourths of the frontal and a trace can also be seen on the parietal. Midsagittal elevation, seen in all archaic *H. sapiens* skulls from China (such as Dali, Jinniushan, and Maba), is limited to the middle of the frontal. The skulls from the site of Quyuan River Mouth have no midsagittal elevation.

Midsaggital elevation is absent in most European specimens. However, in specimens such as Arago and Petralona, midsagittal elevation is just discernible. Midsagittal elevation on the Petralona skull has a different morphology from that seen on Chinese skulls. In Petralona the structure is broader at the top and becomes narrower downward. The ratio formed by its thickness (or the degree of prominence of the elevation) compared with the transverse diameter (the breadth of the elevation) is smaller than that in its Chinese contemporaries. In the latter, midsagittal elevations are rodlike or fusiform (with transverse diameters more or less constant on the length of the structure or attenuating towards both upper and lower ends). In African specimens the morphology of the midsagittal elevation is more similar to that of Petralona than to Chinese fossils.

2. *Flatness of the nasal saddle*

All Pleistocene human fossils preserving the nasal saddle show it as a flattened area. This characteristic persisted for at least 1 million years in China, from the Lantian *H. erectus* to modern *Homo*. There was never a Chinese population with a protruding nose, such as seen in the Neandertals.

3. *Orientation of the antero-lateral surface of the frontal process of the zygomatic*

The antero-lateral surface of the frontal process of the zygomatic bone faces more forward in modern and fossil skulls from China than is true with Neandertals, Petralona, and some African fossils such as Bodo and Kabwe. There is no Chinese fossil skull whose antero-lateral surface of the frontal process faces as laterally as it does among Neandertals.

4. *Contour of the lower border of the zygomatic process of the maxilla*

In most Chinese specimens the junction between this border and the maxillary body is rather high (not close to the alveolar border). The lower border of the zygomatic process extends mainly laterally to connect with the zygomatic. In European specimens, like the Neandertals and Petralona and African specimens such as Broken Hill (no. 1), the junction of this border with the body of the maxilla is closer to the alveolar border. The lower border of the zygomatic process of the maxilla extends upward and laterally, forming a more or less oblique straight line instead of a curve, as in most Chinese specimens. In no Pleistocene specimen from China is the contour of this border close to the condition seen in the Neandertals. However, in the Lantian *H. erectus* maxillary fragment, the lower border of the zygomatic process of the maxilla arises from the maxillary body at a lower level.

5. *Lower upper face*

The upper facial index of Chinese Pleistocene skulls is around 50. The one exception, the Maba archaic *H. sapiens* specimen, might have a higher upper facial index, judging from its high orbit. Many European and African specimens have a higher upper face.

6. *Shovel-shaped incisor*

All Pleistocene human upper incisors from China are shovel-shaped. Although shovel-shaped incisors are reported in European and African Pleistocene specimens, the frequencies are much lower than in China.

7. *Position of the maximum breadth of the skull*

The maximum breadth of all Chinese Pleistocene skulls is at the middle third of its maximum length. In European and African specimens, the position of the maximum breadth is more variable. For example, skulls from Eliye Springs and Laetoli 18 in Africa, and some typical European Neandertals, have their maximum breadth around or posterior to the junction between the middle and posterior thirds of the maximum length of the skull.

8. Orbital margins

Orbital margins are rounded in all Chinese Pleistocene skulls except Maba, where the infero-lateral margin of the orbit has a sharp edge. The orbital margins of specimens from other continents are more variable.

9. Sutures between the frontal and the nasal and maxillary bones

These sutures form a more or less horizontal curve in Pleistocene skulls from China.

10. Lambdoidal ossicle

Among Chinese human fossils, this small bone was first reported by Weidenreich in his study of "Peking Man". He called these small bones *Inca bones,* and documented their existence in four of five skulls from Zhoukoudian.[1] Among archaic Chinese *H. sapiens* there was probably a high occurrence of Inca bones. We previously described the probable presence of Inca bones in the Dali and Dingcun specimens and the two Xujiayao parietals. There is no evidence of Inca bones in Africa and Europe. Only Ndutu, Petralona, Saccopastore, and Atapuerca have an irregularly shaped small bone in the lambdoidal region. Inca bones are less frequent in Pleistocene *H. sapiens sapiens* samples from China, and they are only reported in the Chuandong skull.

There is a continuity of the 10 above-mentioned morphological features in Chinese hominid evolution. Features such as the flat nose, orientation of the frontal process of the zygomatic, the incidence of shovel-shaped incisors, and the position of the maximum breadth of the skull, are uniformly expressed in Chinese specimens. Other features, such as the midsagittal elevation on the frontal bone, the lower upper face, the contour of the lower margin of the zygomatic, the rounded orbital margin, and the horizontal course of the sutures between the frontal and the nasal and maxillary bones, occur in most Chinese Pleistocene human specimens. Exceptions are few. The lambdoidal ossicle had a high frequency in *H. erectus* and archaic *H. sapiens* samples in China and was relatively rare in African and European specimens. All these features are commonly seen in Pleistocene China specimens, but irregularly occur in specimens outside of China.

In the course of human evolution changes occur in these morphological features. The position of the maximum breadth of the skull is raised and midsagittal elevation is decreased. The supraorbital torus, occipital torus, and angular torus weakened. The cranial wall thinned and postorbital constriction lessened. The mandibular fossa became wider and shallower. In "Peking Man" the dimension of the cerebral fossa of the occipital bone divided by the dimension of the cerebellar fossa of the occipital bone is 2:1. It is 3:2 in Dali and 1:2 in modern humans. The inclination of the frontal squama lessens. In earlier specimens the occipital plane of the occipital

[1] During Weidenreich's time only five skullcaps were available. Therefore, he documented the existence of Inca bones in four of five skulls. In 1966 an additional occipital part was found for Skull V, from Locus H, so we now have six skullcaps. In "Peking Man" the lambdoidal bone is bilateral and triangular shaped, so it is called an *Inca bone.* In Ndutu, Saccopastore, and Atapuerca, the lambdoidal bones are not bilateral. In Petralona the shape is irregular so it should be called a *lambdoidal* rather than an *Inca* bone.

joined the nuchal plane by an angular transition, but a rounded transition appeared in later forms.

Changes occurred at different points in time. For example, thinning of the cranial bones occurred slowly from *H. erectus* to *H. sapiens,* but occurred more rapidly from archaic *H. sapiens* to *H. sapiens sapiens.* The relative height of the temporal squama, as expressed by its length-height index, increased from the Zhoukoudian to the Hexian *H. erectus* samples. However, it changed a little or not at all from late *H. erectus* to *H. sapiens.*

Some features that Wood (1984) suggested to be derived characters of *H. erectus* are not concordant with the evidence for human evolution in China. Among those features suggested by Wood are thick cranial bones, a midsagittal crest, angular turn of the occipital region, distinct angular torus, exaggerated postorbital constriction, and a low temporal squama. The average thicknesses of the cranial bones from Dali (dating ca. 200,000 years ago) and Xujiayo (dating ca. 120,000 years ago) are very close to those of the Zhoukoudian *H. erectus.* In parts of the skull, the Dali and Xujiayao cranial bones are even thicker than those of the Zhoukoudian *H. erectus.* Three among seven Zhoukoudian *H. erectus* specimens are thinner than Dali. A midsagittal crest existed on all archaic *H. sapiens.* The Dali and Jinniushan archaic *H. sapiens* specimens have an angular turn between the occipital and nuchal plane. The angular torus is present in *H. erectus* and *H. sapiens* from Dali, Ziyang, and Chuandong. Postorbital constriction in the Hexian *H. erectus* skull was much weaker than that in the Maba *H. sapiens.* Clearly, thick bones, a midsagittal crest, an angular torus, and an angular turn of the posterior part of the skull are not unique to *H. erectus.* These traits cannot be used to differentiate *H. erectus* and *H. sapiens.* It is difficult to define these two species in China on the basis of the so-called derived traits of *H. erectus.* A mosaic of some "derived characters" of *H. erectus* and advanced characters usually seen in *H. sapiens,* in late *H. erectus,* and in early *H. sapiens* specimens from China may be the result of different rates of change in different features in different local populations and may indicate that *H. sapiens* in China evolved from an indigenous *H. erectus* population.

The origin of *H. sapiens sapiens* in China was probably an indigenous event rather than resulting from a migration from Africa. Archaeological evidence provides no support for an African origin of Chinese *H. sapiens sapiens.* If early populations of this subspecies migrated from Africa to China and replaced indigenous inhabitants, they should have brought along their paleoculture. However, the Late Paleolithic assemblage from China is definitely "inherited" from the Middle Paleolithic of China (Zhang, 1987). Very few Chinese late Paleolithic sites exhibit the Mousterian technique. According to the Out-of-Africa model, populations like those at Skhūl and Qafzeh have been suggested to be ancestral to *H. sapiens sapiens* in China. If descendants of these two populations migrated to China and replaced the local population, the Mousterian techinique should appear (perhaps dominate) in the Late Paleolithic of China, but, in fact, it does not.

If Chinese Pleistocene human populations were replaced by migrations from Africa, the latter migrants must have had some advantage(s) strong enough to wipe out the indigenous people and "Africanize" the Late Paleolithic Chinese cultural assemblage. If a natural (ecological) disaster eliminated the indigenous Chinese populations, then the African migrants could easily have occupied China. However, no such disasters seem to have occurred. For example, the Pleistocene fauna in China, especially in southern China, developed continuously without any dramatic change (Han and Xu, 1989; Qi, 1989).

The morphological features mentioned above are not unique to human fossils from China. Each trait existed in populations in other parts of the world, but in lower or much lower frequencies. Two factors may be responsible for the shared occurrence of such features in geographically diverse regions. Perhaps the common origin of humanity led to the sharing of some traits, and perhaps variation in the frequency of expression of traits is tied to genetic drift in geographically isolated populations. For example, facial flatness and shovel-shaped incisors might be pleisiomorphic characteristics existing in high frequencies in China but in lower frequencies in Europe. As examples, we cite the flat face in Steinheim and shovel-shaped incisors in Arago and Krapina.

The second possibility is that gene flow may explain the distribution and variation in frequency of traits. If genetic drift resulted in the absence of some traits, then perhaps gene flow resulted in the presence of some traits in some areas. If a new feature occurred in one area, gene flow may account for its presence in other areas. For example, the bulging of the bone surface supero-lateral to the piriforme orifice is a very common feature of Neandertals and also occurs in Petralona. However, the trait is seldom seen in Chinese skulls except Dali. The eye orbit of the Maba specimen is high and spherical; this is quite different from other Chinese Pleistocene skulls, but resembles the Neandertal condition. Many Late Pleistocene or Early Holocene *H. sapiens sapiens* skulls from China, especially specimens from southern China, have weak occipital buns. These features, spherical orbit in Mapa, paranasal bulging in Dali, and an occipital bun in many of the late Paleolithic specimens from south China, shared by Chinese and European specimens, are probably the results of gene flow. High rates of gene flow should result in a similar frequency of numerous shared traits. Since the difference between frequencies of shared traits between Chinese and non-Chinese *H. sapiens* is obvious, it can be assumed that gene flow was not extensive.

Gene flow among *H. sapiens sapiens* populations seems to have been more extensive. Recently, a number of attempts have been made to assess human migration and dispersal in East and Southeast Asia (X. Wu, 1987a,b, 1988a,b, 1992a,b). The coefficients of divergence between various skulls from eastern and southeastern Asia are given in Table 5.1.

Table 5.1 illustrates that the coefficient of divergence of two skulls of the same sex and from the same site are 0.029 and 0.033 in the Upper Cave sample and Minatogawa, Okinawa, Japan, respectively. If a figure of around 0.030 approximates the amount of normal intrapopulation variability, cranial variation between the Liujiang and Minatogawa specimens and variation between Liujiang and Niah could be considered intrapopulation variation. The coefficient of divergence between specimens from Liujiang and specimens from Niah in Borneo and Wajak in Java seems to increase as the geographical distance increases.[2] The Niah Cave specimen is morphologically closer to Liujiang in China than to Keilor in Australia and Wajak in Java. On the other hand, the latter two are morphologically very close.

[2] The coefficient of divergence between Liujiang and Niah is 0.033, so they are close. The coefficient of divergence between Liujiang and Minatogawa is 0.029, which is close. The coefficient of divergence between Wajak and Keilor is 0.028, which is also close. The geographic distance between Liujiang and Wajak is greater than that between Liujiang and Niah. The distance between Liujiang and Keilor is greater than that between Liujiang and Wajak. The geographical distance between the Upper Cave and Liujiang is not great, but their coefficient of divergence is large. The reason for this is not yet known, but it is the subject of a future study.

Table 5.1

Coefficients of divergence between various skulls from the West Pacific Region

Liujiang (male)	—	Upper Cave 101 (male)	0.056
Upper Cave102 (female)	—	Upper Cave 103 (female)	0.030
Liujiang	—	Minatogawa I (male)	0.029
Minatogawa II (female)	—	Minatogawa IV (female)	0.033
Liujiang	—	Niah	0.033
Liujiang	—	Wajak	0.057
Liujiang	—	Keilor	0.074
Niah	—	Minatogawa I	0.046
Niah	—	Wajak I	0.068
Niah	—	Keilor	0.077
Wajak	—	Keilor	0.028
Wajak	—	Minatogawa I	0.071

The formula employed in making these measurements is:

$$\text{Coefficient of Divergence (CD)} = \sqrt{\frac{\sum (a_1 - b_1)^2}{k}}$$

$$a_1 = \frac{A_1}{A_1 + B_1};\ b_1 = \frac{B_1}{A_1 + B_1}, \text{etc.}$$

A_1 is the measurement of one item of one skull (e.g. for the maximum cranial length of Liujiang A_1=189.3 mm)

B_1 is the measurement of the same item of another skull (e.g. for the maximum cranial length of Upper Cave Skull 101, B_1=204 mm).

K is the number of items measured that are used for comparing two skulls. If there are 12 measurements, then k = 12.

Therefore, CD indicates the average divergence between the two skulls.

According to Wu (1987, 1988, 1992a)

The Niah Cave specimen has long been considered as having close affinities to Australian Aborigines. Brothwell (1960) calculated the coefficient of divergence between the Niah skull and modern Australian Aborigines, Tasmanians, and prehistoric Chinese. The figures are 0.042, 0.028, and 0.070, respectively. Brothwell's "prehistoric Chinese" sample was a pooled sample of Aeneolithic (the period between the Stone Age and the Bronze Age) age. Results based on comparisons between a Paleolithic specimen such as Niah, on the one hand, and a modern or Aeneolithic sample, on the other, would be less conclusive than a comparison between two late Paleolithic specimens, such as Liujiang and Niah. Furthermore, the cranial index of the Niah Cave skull is 77.8, the indices of the Liujiang, Ziyang, and Chuandong skulls from China are 75.1, 77.4, and 76.8, respectively. Indices of Australian Pleistocene or Early Holocene skulls are not higher than 73 and most are lower than 70. Niah is closer to southern Chinese contemporaries in its cranial index than it is to Australian contemporaries. The broad palate of the Niah skull also favors a close connection to Chinese specimens.

Human remains (mandibles, frontal, and nasals) from Tabon, a late Paleolithic site in Palawan, Phillippines, have features suggesting an affinity to continental Asians. A narrow longitudinal ridge along the midsagittal line of the nasal saddle of Tabon also appears in specimens from Dali, Maba, and on Upper Cave no. 101. The congenital absence of a mandibular third molar in Tabon suggests affinities between Tabon and Mongoloids rather than between Tabon and aboriginal Australians. The occurrence of dental agenesis of the third molar could be traced to Lantian *H. erectus*. The frequencies of occurrence of dental agenesis of the mandibular third molar are 5.14–40.2% in modern Mongoloids and 1.5% in Australian Aborigines.

Xinzhi Wu is of the opinion that the traits of Niah and Tabon may well represent interbreeding between Mongoloids and Australoids. Australia was influenced by migrations from China in the Late Paleolithic. Thorne (1980), who separated Pleistocene skulls from Australia into a robust and a gracile group, considered that the latter originated in China. Xinzhi Wu indicates that these skulls of Pleistocene Australia belong to one highly heterogenous population instead of belonging to two (a gracile and a robust) populations. This population exhibited robust features originating in Java and gracile features originating in China (X. Wu, 1987a,b).

The estimated statures of Liujiang, materials from the Upper Cave, and specimens from Minatogawa, Okinawa are 157.36 cm, 174 cm, and 156 cm for males' respectively. These estimates of statures and the coefficients of divergence indicate that Liujiang is closer to Minatogawa than to the Upper Cave. The physical difference between modern southern and northern Chinese might probably be traced to the late Pleistocene in age.

The dispersal of human populations from east Asia to the Americas has been recognized for a long time. The pinched nose of Upper Cave Skull no. 101 and Minatogawa add more morphological evidence to support this dispersal.

SUMMARY

The 10 morphological traits that seem to link Chinese *H. erectus* and *H. sapiens sapiens* samples have been reviewed. Weidenreich, followed by Coon and others, originally proposed an indigenous origin of Asian *H. sapiens sapiens* populations. Chinese and some Western researchers have expanded Weidenreich's original concept. Xinzhi Wu favors an indigenous evolutionary sequence, and his evidence has been presented here.

6

Important Fossil Hominoids in China

Although perhaps not so rich either in variety or quantity of hominoid remains as in other parts of the Old World, China has yielded important fossil apes, and only further research will reveal their exact number and variety. *Gigantopithecus* and *Lufengpithecus* are arguably the most important Chinese hominoids, and both have at one time or another been touted as human ancestors.

The fossil apes found in China, *Gigantopithecus* and *Dryopithecus*, are found in other parts of the world. The Indian variety of *Gigantopithecus*, which dates to 9–5 mya, is older than the Chinese representatives, which date to about 500,000–250,000 years ago. *Gigantopithecus* from India does not overlap with hominids, but *Gigantopithecus* in China seems to overlap with *Homo erectus*. In China, *Gigantopithecus* was first found in 1935 among teeth purchased in a Chinese drug store.

The widest array of fossil apes from China and the largest number of remains come from Yunnan Province and belong to a form usually called *Lufengpithecus*. By 1986 mandibles, skulls, teeth, and fragments of the upper limb skeleton had been found. A third hominoid, also from Yunnan, is known as *Dryopithecus kaiyuanensis*.

LUFENGPITHECUS

In May 1975 Zhengju Wang, an employee of the Cultural Club of Lufeng County, Yunnan Province, southwestern China, found a lower molar of a Tertiary ape and other mammalian fossils in a small lignite mine at Shihuiba, 9 km north of Lufeng City [102°04′ E, 25°11′ N]. Lufeng is 100 km west of Kunming, the capital of Yunnan. In the winter of 1975 and the spring and winter of 1976, a joint team led by Qinghua Xu and organized by the IVPP and the Yunnan Provincial Museum, excavated the site, and found two mandibles and more than 100 teeth. In 1978 the first hominoid skull was found. The Lufeng specimens were attibuted to *Sivapithecus* and *Ramapithecus* in 1979 (Xu and Lu, 1979). In 1980, R. Wu and his colleagues (1986) reported that the site had yielded 5 skulls, 10 mandibles, 6 cranial fragments, 41 jaw fragments, 29 dental rows, 650 isolated teeth, the fragment of 1 scapula, 1 clavicle, and 2 digits. Rukang Wu (1987) attributed the Lufeng fossils to a new genus and species, *Lufengpithecus lufengensis*. Conroy (1990), among others, agrees with this designation. Specimens previously attributed to *Sivapithecus* may represent males, while those attributed to *Ramapithecus* may be females.

Fossils

The following description relies on the work of R. Wu et al. (1983, 1985, 1986) and R. Wu (1987).

Female skull (Fig. 6.1)

The only well-preserved specimen (PA 677) shows that the temporal ridge starts at the lateral end of the supraorbital ridge. It passes medially and almost parallel to the supraorbital ridge, and turns more backward at a point behind (or above) the middle of the supraorbital ridge. The temporal ridges converge at the vertex of the skull and then they gradually diverge and become thinner. Specimen YV 652 shows both temporal ridges joining the nuchal ridge at the occipital. The measurements of the face, maxilla, and mandible are given in Table 6.1.

There is no sagittal crest or the crest is very weak. On specimen YV 652 it begins at the external occipital protuberance as a slightly prominent ridge. It extends toward the place where the temporal ridges converge and disappears nearby. Specimen YV 652 shows that the external occipital protuberance is very small. The

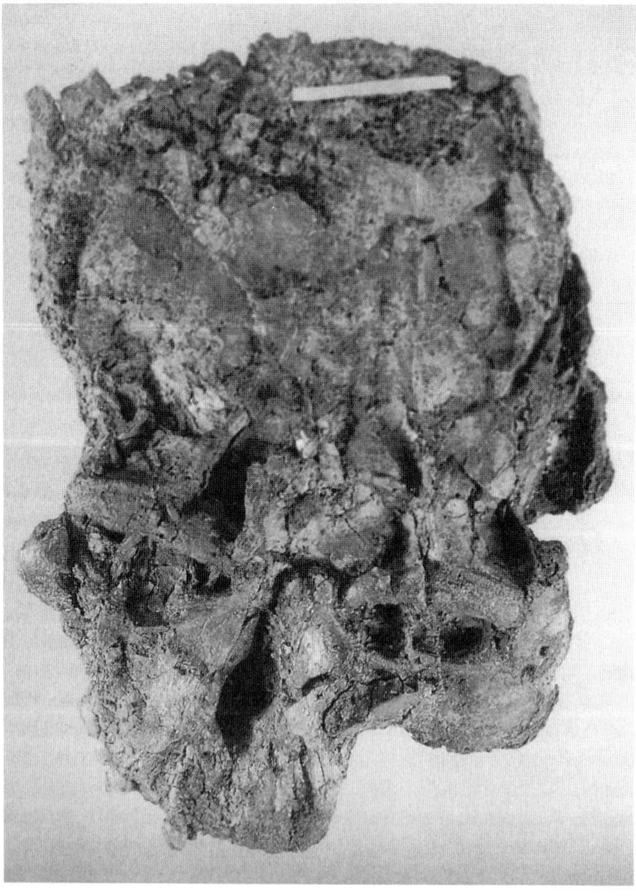

FIG. 6.1 Female skull of *Lufengpithecus lufengensis*. (Courtesy of Rukang Wu.)

Table 6.1

Facial, maxillary and mandibular measurements of *Lufengpithecus lufengensis*

Nasal height (n-ns)	(58.0)*
Bizygomatic breadth (zy-zy)	172.0
Alveolare to nasospinale (alv-ns)	22.0
Nasal aperature height	(30.0)
Nasal aperature breadth	20.0
Nasal bone length	(32.0)
Orbital height	33.0
Orbital breadth	42.0
Zygomatic depth	(25.0)
Orbitale-alveolar height (or-alv)	(45.0)
Minimum biorbital breadth	27.0
Biorbital breadth across lacrymal crests	34.0
Palatal length	70.0
Palatal breadth at canine	38.0
Palatal breadth at upper M2	49.0
Alveolar breadth at canine	54.0
Alveolar breadth at upper M2	65.0
Symphysis depth of mandible	52.0
Symphysis thickness of mandible	27.0
Depth of mandibular corpus at second premolar	44.0
Depth of mandibular corpus at third molar	40.0
Thickness of mandibular corpus at third molar	31.0
Condylar height	105.0
Condylar length	63.8
Ramus notch height	94.0

*Figures in parentheses are estimated.

Cited from Wu (1987)

external surface of the occipital is smooth and has very weak external occipital ridges.

The face is rather narrow; the distance between the external margin of both orbits is 106.6 mm. The weakly developed supraorbital ridges are separated by a broad and depressed glabellar region. The orbital part of specimen PA 677 was broken and deformed. However, its well-preserved left lower orbital margin indicates that the orbit is probably quadrangular instead of spherical. The interorbital breadth is 23.5 mm.

Specimen PA 677 shows that the fronto-sphenoidal process of the zygomatic is narrow. Two zygomatico-facial foramina appear on the left zygomatic. The anterior nasal aperture is probably of a long pear shape, and the upper margin is slightly higher than the lower margin of the orbit. The zygomatic process of the maxilla is rather anteriorly located so that the alveolar process is not very protruding. The naso-alveolar slope is short and concave, and gently continues with the nasal floor.

The cranial base of PA 677 was severely deformed. R. Wu et al. (1986) supposed its foramen magnum to be heart-shaped. Rukang Wu and his colleagues supposed that the foramen magnum of *Lufengpithecus* might be more anteriorly located than that in extant apes. The occipital condyles are narrow, long, and kidney-shaped. A short and broad mandibular fossa and articular surface are preserved on the left side of skull PA 677. There is a prominent postglenoid process and a small mastoid process. The palate is not very depressed. The alveolar process rises from the anterior rearward. There is a greater palatal foramen above the third molar of both sides.

On the basis of the dental arches preserved in PA 677 and in other specimens, the anterior dental arch could be rounded, and because the posterior part on both sides of the dental arch diverge, the whole arch is V-shaped. The canine is low; only one third of the crown height exceeds the dental plane. There is a narrow diastema between the canine and the second incisor. The canine is small. The canine jugum is not distinct and the canine fossa is shallow.

Male skull

Specimen PA 644 is the best of the male skulls, and R. Wu et al. (1986) used it as the type specimen of male skulls of *Lufengpithecus*. It is generally larger and more robust than the female skulls, and there is a very prominent V-shaped temporal ridge and a sagittal ridge. The temporal ridge begins at approximately the same place as on the female skull. After turning backward and medially behind the midpoint of the supraorbital ridge, its course is straighter and oblique. The temporal ridges of both sides join at the midpoint of the vertex, and continue with the sagittal ridge, resulting in a large triangular flat area at the forehead.

The face is short and broad; the distance between the external margin of both orbits is 146.5 mm. The facial profile is concave, and the zygomatic arches flare laterally. The supraorbital ridges, only slightly more robust than in the female skull, are separated by a depressed glabellar region. There is no supraorbital groove. The interorbital breadth is about 30.5 mm; the interorbital region is depressed and continues with the depressed glabellar region. The orbits were deformed by the weight of the covering strata. Judging from the slightly angular shape of the infero-lateral orbital margin, the orbit was probably spherical with somewhat angular corners. The fronto-sphenoidal process of the zygomatic is broad and flat, and has two zygomatico-facial foramina at its basal part. Wu and his colleagues supposed that the anterior nasal aperture is long and pear-shaped, based on its well-preserved lower part. The upper margin of the nasal aperture is slightly higher than the infra-orbital margin. The lower part of the lateral margin of the nasal aperture is rounded. The naso-alveolar slope, which is short, narrow, and concave in sagittal cross section, continues with the nasal floor. The anterior nasal spine is weak. The zygomatic process of the maxilla is rather anteriorly located, the alveolar process is slightly protruding (prognathous).

The canine is large and the canine jugum is prominent. The frontal view shows that the canines on both sides diverge in the shape of an upside down V. The canine fossa is deep. The jugum of the median incisor is prominent. There is a wide and deep sagittal furrow between the incisal juga of both sides. Although somewhat deformed, the dental arch may have been U-shaped with the posterior end slightly diverged. Almost two thirds of the height of the canine crown exceeds the dental plane. There is a wide diastema between the canine and lateral incisor.

Both sexes of *Lufengpithecus* share a series of morphological features (Table 6.2). They are (R. Wu et al., 1986) as follows:

Table 6.2

Sexual differences between the skulls of *Lufengpithecus*

Skull feature	*Female*	*Male*
Temporal ridge	Weak and separated	Strong and join together at vertex
Sagittal crest	Absent or weak	Strong
Mid-facial part	Relatively narrow	Broad
Zygomatic arch	Less expanding	Quite expanding
Canine juga	Weak	Strong
Canine fossa	Shallow	Very deep
Canine	Low crown and small	High crown and big
Max dental arch	Approximately V-shaped	Approximately U-shaped

According to Rukang Wu et. al (1986)

Notes: Quite expanding means further from the neurocranium and less expanding means closer to the neurocranium.

1. A weak supraobital ridge separated by a wide and depressed glabellar region lacking a supraorbital groove
2. A wide interorbital distance
3. A somewhat quadrangular orbit
4. A narrow anterior nasal aperture that is long and pear-shaped
5. A low face, anteriorly located zygomatic process of the maxilla, and a slightly protruding alveolar process
6. A short nasal alveolar slope that continues with the nasal floor.

Differences between the male and female skulls are shown in Table 6.2

Female mandible (Fig. 6.2)

The upper part of the anterior surface of the symphyseal region is more or less vertically oriented. The lower part inclines backward, while the median part has a Y-shaped crestlike prominence extending downward to the diagastric spine (Specimen PA 895). The anterior teeth are vertically oriented. The division of the superior and inferior branches of the lateral eminence is not obvious because of deformation (Specimen YV 710). The single mental foramen, located at the middle of the mandibular height below the second premolar, opens upward and forward (Specimens YV 678 and YV 710). On the internal surface of the symphyseal part, the alveolar plane slopes at about a 60° angle. The superior transverse torus is not distinct. The moderately broad genioglossal fossa extends backward for a considerable distance (Specimens PA 580 and PA 895). The inferior transverse torus is thick. There is a diagastric fossa on the anterior of the basal margin of the symphyseal part. A diagastric spine on the anterior margin of the middle of this fossa extends backward to form a midsagittal crest to the genioglossal fossa (Specimen PA 895). The anterior dental arch is a rounded curve. The cheek tooth rows gradually diverge, as they extend backward, forming a dental arch that approximates a V shape (Specimen PA 580). A small part of the mandibular ramus preserved on Specimen PA 580 shows that the anterior margin of the ramus corresponds to the middle of the third molar.

Male mandible (Fig. 6.2)

The anterior teeth are vertically oriented as in the female mandible. The upper part of the anterior surface of the symphyseal region is vertical, with its lower part inclining backward. A rounded eminence at the center of the anterior surface of the symphyseal region narrows as it extends downward and terminates at the diagastric spine (Specimen PA 548). The lateral eminence starts at the anterior margin of the lateral surface of the ramus of the mandible and divides into an upper and a lower branch. The upper branch extends forward to the jugum of the second premolar, while the lower branch extends along the inferior margin of the mandible and continues with the canine jugum. The single mental foramen is located at the lower third of the height of the mandibular body below P1–P2. The foramen opens forward and upward (Specimens PA 548 and PA 896). On the internal surface of the symphyseal aspect, the alveolar plane inclines 50°. The superior transverse torus is not distinct. The moderately wide genioglossal fossa extends slightly further backward than in the female specimen. The inferior transverse torus extends further backward than that in female specimens LC 102 and YV 711. Because of its backward and upward inclination, the posterior end of the inferior transverse torus is not as close to the lower margin as in female specimen LC 102. The inferior transverse torus is thick (Specimens LC 102 and YV 711). The anterior of the basal margin of the symphysis has a rough bony ridge, behind which is located the wide and rough diagastric fossa (Specimen YV 711). There is a prominent diagastric spine at the middle of this ridge. Behind the spine is a thick and prominent midsagittal ridge, which passes the posterior margin of the inferior transverse torus and disappears at the genioglossal fossa (Specimen PA 548). The cutting edge of the anterior teeth is slightly protruding (Specimens PA5 48 and LC 102). The cheek tooth rows are almost parallel and diverge slightly posteriorly, and the dental arch is almost U-shaped (Specimens PA 548 and LC 102). The mandibular ramus is wide and high, and its steep anterior margin corresponds to the mesial margin of the third molar (R. Wu et al., 1986).

Female teeth

Upper median incisor. The tooth has a lingual tubercle and one or two fingerlike processes lateral to which are enamel wrinkles in the lingual fossa. The mesial and distal marginal ridges start at the lingual tubercle and extend halfway up the lingual fossa. The central part of the labial surface of the crown is bulging, and the lower part inclines obviously toward the lingual side. The labio-lingual diameter is long. The mesial segment of the cutting edge is horizontal, while the distal segment is curved. The tip of the single root inclines slightly toward the distal and lingual sides.

Upper lateral incisor. This tooth is much smaller than the upper median incisor. The labio-lingual diameter is much greater than the mesio-distal diameter. The lingual tubercle and fingerlike processes are weaker than those of the median incisor. The cutting edge is V-shaped, and its distal segment is more oblique than its mesial segment. The tip of the single root turns slightly to the distal side.

Upper canine. This tooth is low and small; its apex is not very sharp. The mesial and distal marginal ridges, which are not sharp, join the well-developed cingulum at the base of the crown to form small processes. On the anterior part of the lingual surface is a main ridge that is separated from the mesial marginal ridge

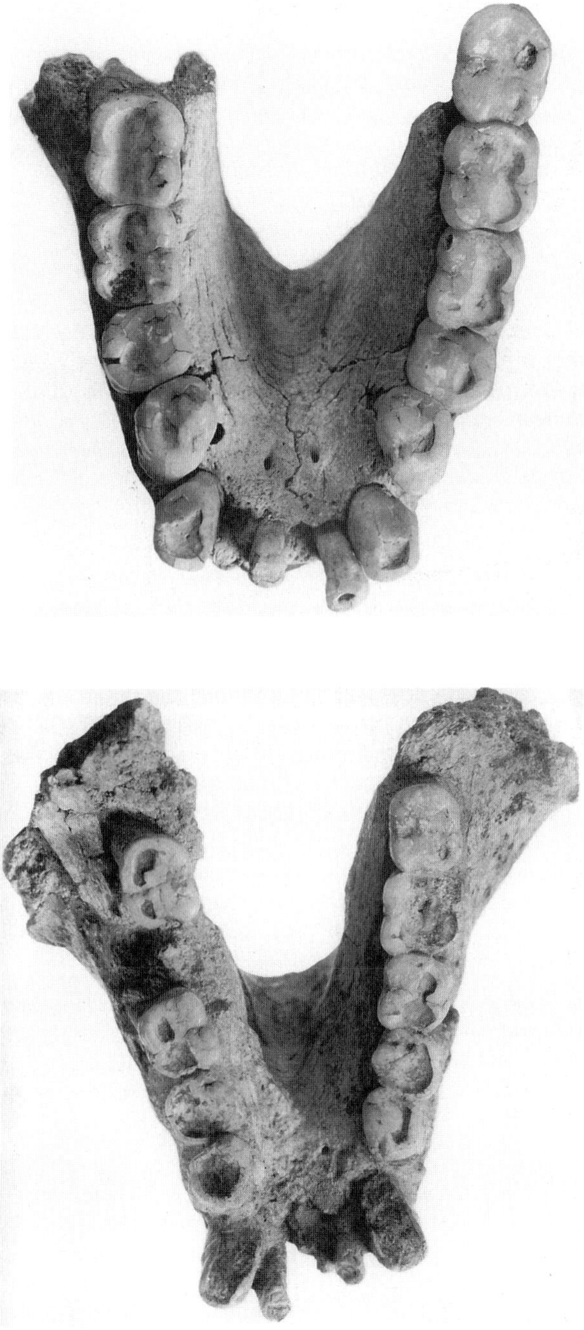

a

b

FIG. 6.2 Male mandible (a) of *Lufengpithecus* and (b) female mandible. (Courtesy of Rukang Wu, IVPP.)

by a deep groove. Vertical enamel wrinkles cover the lingual surface. The root is weak, and its tip inclines laterally and backward.

Upper first premolar. The upper first premolar has two cusps, and because the paracone is higher and bigger than the protocone, the mesio-buccal corner protrudes forward. The contour of the occlusal surface is close to a triangular shape. Two ridges radiating from each cusp on the occlusal surface join at a longitudinal groove in the middle of the surface, resulting in anterior, middle, and posterior foveae. The buccal surface of the crown is large, high, and bulging, and has a buccal anterior cingulum. The lingual surface is low and small. Of the three branches of the root, the lingual branch is the thickest, and the anterior buccal branch is bigger than the posterior buccal branch.

Upper second premolar. The contour of the occlusal surface is ovoid. The paracone is slightly higher, shorter, and located more anteriorly than the protocone. A main ridge on both cusps has two accessory ridges extending from both sides that form the anterior, middle, and posterior foveae. Both the buccal and lingual surfaces are curved in horizontal section. The root has three branches; the two buccal branches unite at the lower part, and the lingual branch is thicker.

Upper first molar. The occlusal surface is square or somewhat rhombdoidal. The buccal cusps are higher and sharper than the lingual cups. The protocone is largest and the hypocone is the smallest cusp. The paracone, almost equal in size to the metacone, is connected to the protocone by a transverse ridge. An oblique ridge connects the protocone and metacone. These ridges form the anterior and posterior foveae and the trigonid fovea on the occlusal surface. The anterior fovea is slender and long, the trigonid fovea is large and deep, and the posterior fovea is very small. The wrinkle pattern is relatively simple. The protocone and paracone have a main ridge and anterior and posterior accessory ridges. The main ridge on the metacone is well developed and has fine wrinkles appearing on the slopes of both sides. The main ridge of the hypocone is very short, with one or two wrinkles. The buccal surface of the crown is steeper than the lingual surface. The Carabelli's pit exists at the mesio-lingual corner of the crown in a few specimens. The worn tooth shows that the molar's enamel is thicker than the enamel on the anterior teeth. Caries are occasionally visible. The root has three branches; the two buccal branches are smaller than the lingual branch.

Upper second molar. This is the largest of the upper molars. The occlusal surface is rhombdoidal or trapezoid. Like the first molar, it possesses four cusps, two ridges, and three foveae. The protocone is especially large. Wrinkles are more abundant on some specimens than on others. Both Carabelli's pit and cingulum are absent. The root has three branches.

Upper third molar. This tooth is more variable in size and morphology. The posterior part of the crown is very narrow because of a reduced metacone. The contour of the occlusal surface is trapezoidal. The paracone, metacone, and protocone are present in most specimens. The hypocone is usually divided into small processes. The existence of transverse and oblique ridges define three foveae on the occlusal surface. Cusps on the third molar are usually lower than those on the other molars. Enamel wrinkles are abundant, and there is no cingulum. The root has three branches.

Lower median incisor. The labial surface of this chisel-like tooth is oblique, and the lingual surface is concave in vertical section. The mesial and distal marginal ridges join at the base of the crown to form a V-shape. The horizontal cutting edge forms right angles with both the mesial and distal surfaces. The labio-lingual diameter of the crown greatly exceeds the mesio-distal diameter. The single root is very flat. Its labio-lingual diameter is almost twice that of the mesio-distal diameter.

Lower lateral incisor. Other than the fact that the cutting edge of the lateral incisor forms an obtuse angle with the distal surface, the lower incisors are almost similar in size and morphology.

Lower canine. The lower canine is low and small. The single cusp is not sharp. The mesial marginal ridge is short, it extends upward obliquely and terminates at the apex of the triangular mesial surface. The distal marginal ridge is longer and extends steeply toward the base of the crown to join the cingulum on the lingual side to form a small tubercle. The labial and mesial surfaces of the crown form a labial anterior cingulum. On the lingual surface, a main ridge extending from the apex of the crown to the basal cingulum divides the crown into a flat anterior and a depressed posterior part. A prominent lingual cingulum extending from the apex of the mesial surface of the crown goes obliquely downward and terminates at a small tubercle at the base of the mesial marginal ridge. The single flat root has a labio-lingual diameter that exceeds its mesio-distal diameter.

Lower first premolar. This tooth has a large and high protoconid, and a small and low metaconid. A longitudinal groove separates these two cusps. The contour of the occlusal surface is more or less triangular. The transverse diameter of the talonid fovea is larger than its sagittal diameter. The lingual surface is obliquely oriented with a prominent cingulum at its base. The buccal surface is bulging and has an anterior marginal cingulum. The distal surface is low. The root has anterior and posterior branches.

Lower second premolar. The occlusal surface is flattened and rhombdoidal. Roughly equally sized cusps are separated by a longitudinal groove. Each of the cusps radiates an anterior transverse ridge, which, together with the anterior marginal ridge, surrounds a small anterior fovea. The transverse ridge, buccal, lingual, and posterior margins of the crown surround a large and deep talonid fovea within which there are enamel wrinkles. The root divides into two branches.

Lower first molar. This is the smallest of the lower molars. The protoconid protrudes forward and the entoconid protrudes inward. The five main cusps are arranged in a *Dryopithecus* Y-5 pattern. The metaconid and entoconid are slightly higher and sharper than the three buccal cusps. The apex of the three buccal cusps is more centrally located than the apex of the lingual cusps. The lingual surface of the crown is steeper than the buccal surface. The metaconid connects to the protoconid by a transverse ridge, and the entoconid connects with the hypoconulid by a short oblique ridge. These ridges divide the anterior, middle, and posterior foveae on the occlusal surface. The anterior fovea is slender, the middle fovea is large and deep, and the posterior fovea is very small. Fine wrinkles on the slope of the five cusps face the center of the middle fovea. The root is divided into a vertical anterior branch and an oblique posterior branch.

Lower second molar. On this, the largest of the lower molars, the contour of the crown is nearly quadrangular and has a rounded posterior margin. The main structures on the occlusal surface are similar to those on the first lower molar. The wrinkles are somewhat more abundant in some specimens. The enamel is thick and there is no cingulum. The anterior branch of the root is more vertically oriented than the posterior branch.

Lower third molar. The crown is larger than that of the first molar and slightly smaller than that of the second molar. The lower third molar is more variable in morphology. The posterior part of the crown is narrower. In addition to the five cusps, there are two accessory cusps in some specimens, one posterior and medial to the metaconid and another corresponding to the position of the posterior fovea. The enamel is thick and wrinkles are abundant. There is no buccal cingulum. The anterior branch is the broadest of the two branches of the root.

Male teeth

Upper median incisor. Other than being more robust, this tooth is similar to that of female specimens. A large labio-lingual diameter typifies the upper median incisor of both sexes. On the cutting edge of the unworn specimen there are four or five mammilary processes.

Upper lateral incisor. This tooth is morphologically similar to female specimens. It is much smaller than the median incisor. The cutting edge is V-shaped.

Upper canine. This upper male canine, which is quite different from the female canine, has a crown that is almost twice the height. The tooth looks like a curved knife, the distal margin of the crown is concave, and the mesial margin is convex. Both margins are very sharp. A thick and blunt main ridge on the anterior of the lingual surface is separated from the mesial ridge by a very deep groove. Behind this ridge is extensive vertical wrinkling. A cingulum is generally present at the base of the crown. Vertical wrinkles on the labial surface of the crown are weaker than those on the lingual surface. A longitudinal groove exists slightly behind the middle line of the labial surface. There is a small process at the anterior and posterior extremities of the base of the crown. The anterior process is larger. The robust root inclines backward and laterally.

Upper first premolar. The upper first premolar is similar to that characteristic of females in having two cusps, two ridges, and three foveae. However, its crown is more robust, the mesio-buccal corner protrudes further forward, and the buccal anterior cingulum is more developed. There is a central main ridge on the mesial oblique surface, in addition to the two transverse ridges. The main ridge is not distinct in female specimens.

Upper second premolar. Other than being larger and more robust, this tooth is morphologically similar to female specimens.

Upper first molar. This tooth is morphologically similar to the females' first upper molar and is similar to the females' upper second molar in size. The occurrence of Carabelli's pit is slightly higher in male upper first molars than in female teeth. Enamel on the occlusal surface is thick. Caries are also seen.

Upper second molar. This is the largest of the three upper molars. Like the female upper second molar, it has four cusps, two ridges (transverse and oblique), and three foveae. Wrinkles are slightly more abundant than in the first molar, and vertical wrinkles are usually present on the lingual surface. Both buccal and lingual surfaces are obliquely oriented. There is no Carabelli's pit or cingulum. Enamel on the occlusal surface is thick. The buccal branch of the root bifurcates, and the lingual branch is robust.

Upper third molar. Sizewise, this tooth is intermediate between the first and the second molars. The posterior two cusps are reduced. The posterior margin is rounded. The cusps of the third molar are lower than those of the other molars of the same individual, and the wrinkles of the former are more abundant than on other molars. The hypocone is especially small and is usually divided into several small processes. The anterior, middle, and posterior foveae are clearly seen. There is no cingulum. The root divides into three branches.

Lower median incisor. This tooth is morphologically similar to its female counterpart, but its crown is higher and the root is more robust. The cutting edge had two indentations before wearing; the worn edge is a horizontal line.

Lower lateral incisor. The sexual difference between the lower lateral molars of *Lufengpithecus* is mainly in size. The mesio-distal diameter of the lower lateral incisor is slightly larger than that of the median incisor.

Lower canine. The male lower canine has a higher crown than its female counterpart. Its sharp mesial and distal borders extend to the base of the crown. The mesial surface is not triangular, as it is in female specimens. The main ridge on the lingual surface is sharp and located posteriorly. It is separated from the distal marginal ridge by a very deep groove. There are many vertical wrinkles on the lingual surface. The lingual cingulum is well developed and extends from the base of the mesial marginal ridge to the base of the distal marginal ridge, where it forms a small tubercle. The root is robust.

Lower first premolar. This tooth is quite different from female specimens. It is sectorial; a sharp main ridge extends from the apex of the cusp to the base of the disto-lingual corner. A small tubercle exists at the middle of the main ridge. In a very few specimens this small tubercle develops into a small metaconid, as is true in female specimens. The buccal anterior cingulum and the cingulum at the base of the mesio-lingual surface are more prominent in males, but the shape is similar in both sexes. The root has two branches.

Lower second premolar. Morphologically, this tooth is more or less similar to that characterizing females. However, the disto-lingual corner in male specimens is more protruding and the male tooth is larger.

Lower first molar. This is the smallest among the three lower molars. The male lower first molar is larger than its female counterpart.

Lower second molar. This is the largest of the three lower molars. Five cusps, two ridges, and three foveae are present on the occlusal surface. The cusps, arranged in a Y-5 pattern, are robust, and their ridges are thick and blunt and have fine wrinkles. The occlusal enamel is thick. There is no buccal cingulum. The root is divided into a vertical anterior and an obliquely oriented posterior branch.

Lower third molar. This tooth is usually smaller than the second lower molar. Its narrow posterior makes the posterior end somewhat pointed in appearance. Generally, the third lower molar has five cusps, but sometimes there is a sixth cusp. The metaconid is higher than the other four main cusps. The thick enamel of the occlusal surface is abundantly wrinkled. The root has anterior and posterior branches.

The foregoing dental descriptions are based on the work of R. Wu, et al. (1986).

Scapula (YV 716-1)

This fragment of a left scapula includes part of the acromian, corocoid process, and the glenoid cavity, as well as anterior and inferior parts of the supraspinous fossa. The root of the acromian is robust. Two broad and smooth prominences are visible on the lateral margin of the scapular spine, which is nearly horizontally oriented. The rough surface of the root of the coracoid process suggests robusticity. The shallow glenoid cavity is pear-shaped and the margin is moderately rounded. The supraglenoid tuberosity is pointed and protrudes forward and laterally. The surface of the infraglenoid tuberosity is big, rough, and triangular. The scapular notch is deep and narrow, and the bone is thick at the base.

Clavicle (YV 7l6-2)

This fragment represents part of the lateral third of a left clavicle. The bone is roughly prismatic in shape. The shaft is especially thick medial to the conoid tubercle. The acromian extremity was lost, but its broken end shows that the clavicle had a short sagittal diameter. The anterior surface of the clavicle is rounded and becomes thinner posteriorly. The conoid tubercle is small. A ridge parallel to the posterior border of the bone extends to the acromian end. Both surfaces of this ridge are rough. A slender ridge appears parallel and anterior to this ridge. Both ridges form a shallow preconoid fossa. The clavicle is especially robust medial to the conoid tubercle. A thick ridge, possibly a deltoid ridge, starting at the place corresponding to the conoid tubercle, extends along the anterior border of the clavicle.

The information for the clavicle and scapula is from R. Wu et al. (1986).

Phalanges (PA 1056 and PA 1057)

This information is from R. Wu et al. (1986). These two specimens are similar in morphology and size. A longitudinal prominence along the ulnar part of the palmar surface extends almost the whole length of the bone. Specimen PA 1056 probably represents the proximal phalange of a middle or fourth finger of the right hand. Specimen PA 1057 probably represents the proximal phalange of an index finger of the left hand.

The long axis of PA 1056 is very curved. The palmar surface has ulnar and radial ridges for the attachment of the tendon sheath. The ulnar ridge is on the distal part of the bone; the radial ridge is more proximally located. In the volar view the radial margin of the trochlear articular surface exceeds the proximal margin. The dorsal surface is rounded in cross section.

There are small differences between these two phalanges. The proximal part of the ulnar prominence of PA 1057 is slightly larger than in specimen PA 1056. The ridge for the tendon sheath is more developed in the former. PA 1057 is shorter and slightly more slender.

On the basis of a preliminary study of the five skulls, more than 40 jaw fragments, and more than 1000 teeth of the specimens previously called *Sivapithecus* and *Ramapithecus* found at Shihuiba, Lufeng, Yunnan; R. Wu (1987) proposed that they all belong to a new taxon: *Lufengpithecus lufengensis*. This taxon shares the following features with *Sivapithecus meteai* from Turkey and *S. indicus* from Pakistan:

1. The lateral view of the middle facial part shows a concave line, the upper jaw bends upward.

2. The left and right supraorbital ridges are thin and separate.
3. The pyriforme aperture is long and narrow.
4. Both canines are shaped like an upside down V.
5. The root of the upper incisor is obviously curved, and the incisor juga are prominent.
6. The mandibles are robust.
7. The molars have thick enamel.
8. Sexual dimorphism is obvious.

The Lufeng specimens differ from hominoids from Turkey and Pakistan as follows:

1. The interorbital distance is broad.
2. The middle face is broad and short.
3. The palate is short, broad, and shallow.
4. The orbits are ovoid and the long axis of the orbit is horizontal.
5. The cheek tooth rows of both sides diverge posteriorly rather than parallelling each other.
6. Higher cusps on the molar have more abundant wrinkles on the occlusal surface.

Geology

The Shihuiba fossil site (IVPP site no. 75033) is at the northern part of the Lufeng basin. The average height above sea level is 1560 m. Guoqin Qi (1985) described the geological section that was exposed in 1983 excavation. From the surface downward the section includes:

1. Yellow sandy clay with lenses of gravel, 0.5–2 m
2. Interbedding of blackish-brown carbonaceous clay and grayish-white fine sand, rich with molluscs, 0.7–3 m
3. Blackish-brown massive lignite with a compact and solid texture, partially intercalated with sands and sandy clay, 0.3–1.4 m
4. Interbedding of black carbonaceous clay and gray fine sand, 0.2–1.8 m
5. Grey, uniform, and pure fine sand with intercalations or lenses of black carbonaceous clay and lignite containing considerable molluscs. The first *Lufengpithecus* skull was found in this layer in 1980, 1.5–2.5 m
6. Lignite consisting of two to three laminated layers, intercalated with fine sands and carbonaceous clay, 0.5 m
7. Greyish-white, blackish-grey clay; the thickness is 1.6 m in the test pit
8. Purplish-red, orange-red, yellowish-brown clay; the exposed thickness is 0.8 m

Qi (1985) reported the following vertebrates from Layers 1–6.

Mammals

Insectivora

Tupaiidae gen. et sp. indet. (Tree shrew)

Galerix sp. (Hedge hog)

Lanthanotherium sp. (Hedge hog)
Scalopini gen. et sp. indet. (Mole)
Heterosoricinae gen. et sp. indet. (Shrew)
cf. *Crocidura* sp. (White toothed shrew)
Blarinella sp. (Asiatic short tailed shrew)
Soricinae gen. et sp. indet. (Shrew)
Anourosorex sp. (Shrew)

Chiroptera
Pteropidae gen. et sp. indet. (Bat)
Hipposideridae gen. et sp. indet. (Bat)
Myotis sp. (Little brown bat)
Epthsicus sp. (Bat)
Pipistrellus sp. (Pipistrelles)
Plecotus sp. (Long eared bat)

Rodentia
Sciurotamias sp. (Squirrel)
cf. *Callosciurus* sp. (Beautiful squirrel)
cf. *Dremomys* sp. (Red cheeked squirrel)
cf. *Tamlops* sp. (Asiatic stripped squirrel)
Albanensia sp. (Flying rat)
? *Forsythia* sp. (Flying rat)
cf. *Hylopetes* sp. (Flying rat)
cf. *Monosaulax* sp. (Beaver)
Platacanthomys sp. (Rat)
Typhlomys sp. (Larger rat)
Typhlomys sp. (Smaller mouse)
Leptodontomys sp. (Mouse)
Brachyrhizomys nagrii (Bamboo rat)
Brachyrhizomys sp.* (Bamboo rat)
cf. *B. pilgrimi* (Bamboo rat)
B. tetracharax (Bamboo rat)
Kowalskia sp. (Hamster)
cf. *Kowalskia* sp. (Hamster)
Mus sp. (Mouse)
Parapodemus sp. (Mouse)
Parapelomys sp. (Mouse)
Muridae gen. et sp. indet. (Mouse)
Hystrix sp. (Porcupine)

Lagomorpha
Alilepus sp. (Hare)
*Alilepus longisinuosis*** (Hare)

Primates
Sinoadapis carnosus (Chinese adapis)
Sinoadapis shihuibaensis *** (Chinese adapis)
Laccopithecus robustus (Lesser ape)
Ramapithecus sp.#
Sivapithecus sp.#
Carnivora
Ursavus depereti (Bear)
Indarctos sinensis (Bear)
Indarctos sp. (Bear)
Ursinae gen. et sp. indet. (Bear)
cf. *Martes palaeosinensis* (Weasel)
Martes sp. (Weasel)
Mustelinae gen. et sp. indet. (Weasel)
cf. *Eomellivora wimani* (Badger)
Melinae gen. et sp. indet. (Badger)
Proputorius lufengensis (Marten)
Proputorius sp. (Marten)
Sivaonyx bathygnathus (Otter)
Lutra sp. (Otter)
Lutrinae gen. et sp. indet. (Otter)
Mustelidae gen. et sp. indet. (1) (Weasel)
Mustelidae gen. et sp. indet. (2) (Weasel)
Viverra sp. (Civet)
Viverrinae gen. et sp. indet. (1) (Civet)
Viverrinae gen. et sp. indet. (2) (Civet)
Paradoxurinae gen. et sp. indet. (Civet)
Viverridae gen. et sp. indet. (1) (Civet)
Viverridae gen. et sp. indet. (2) (Civet)
Ictitherium gaudryi (Hyena)
Ictitherium sp. (Hyena)
Epimachairodus fires (Saber-toothed cat)
Pseudaelurus sp. (Cat)
Felis sp. (Cat)
Proboscidea
Gomphotherium sp. (Elephant)
? *Serridentinus* sp. (Serrated toothed elephant)
Zygolophodon lufengensis (Short maxilla elephant)
Perissodactyla
Hipparion sp. (1) (Three toed horse)
Hipparion sp. (2) (Three toed horse)

Macrotherium salinum (Odd toed ungulate with clawed digit)
Macrotherium sp. (Odd toed ungulate with clawed digit)
Tapirus sp. (Tapir)
Chilotherium sp. (big lip rhinoceros)
Aceratherium sp. (Hornless rhinoceros)

Artiodactyla

Hyotherium sp. (Wild pig)
cf. *H. palaeochoerus* (Wild pig)
Lophochoerus lufengensis (Wild pig)
Potamochoerus sp. (1) (River pig)
Potamochoerus sp. (2) (River pig)
Potamochoerus sp. (3) (River pig)
Suidae gen. et sp. indet. (Wild pig)
Dorcabune progresus## (Small deer)
Dorcabune (Small deer)
Dorcatherium sp. ### (Small deer)
Moschus sp.(Musk deer)
Dicroceus sp. (Deer)
cf. *Metacervulus simplex* (Muntijac)
Metacervulus sp. (1) (Muntijac)
Metacervulus sp. (2) (Muntijac)
Muntiacus cf. *nanus* (Small muntijac)
Muntiacus sp. (1) (Muntijac)
Muntiacus sp. (2) (Muntijac)
Muntiacus sp. (3) (Muntijac)
Cervidae gen. et sp. indet. (1) (Deer)
Cervidae gen. et sp. indet. (2) (Deer)
Selenoportax sp. (Antelope)
Bovidae gen. et sp. indet. (Ox)

Birds

Anas sp. (Duck)
Aythya shihuibas (Duck)
Bambusicola sp. (Chicken)
Diangollus mious (Chicken)
Phasinaus lufengia (Pheasant)
Yunnanus gaoyuansis (Sparrow)
Bird fam. et order indet.

Reptile

Cuora pitheca (Turtle)

Fish

Cyprinus sp. (Carp)

[* Added in Qi (1986). ** Species name identified by Qiu and Han (1986). *** Added in Pan and Wu (1986). # Considered *Lufengpithecus lufengensis* by R. Wu (1987). ## Species name identified by Han (1986). ### Changed to *Yunnanotherium simplex* by Han (1986).]

OTHER APES ALLIED TO *LUFENGPITHECUS*

Specimens found in Kaiyuan, Yunnan

In Febuary 1956 Taimao Wang and Wenshan Lin of the Geological Bureau of Southwestern China found five ape teeth in the Xiaolongtan (Small Dragon Pool) lignite mine [101°15′ E, 23°50′ N], Kaiyuan County, Yunnan Province. The teeth include a left and right lower second premolar, a left and right lower second molar, and a right lower third molar (Fig. 6.3). J. Woo (1957) suggested that the teeth belonged to a new species, *Dryopithecus kaiyuanensis*. In 1958, he reported five more ape cheek teeth from the right lower dentition collected from the same site by the Yunnan Provincial Museum. J. Woo (1958) suggested that the former specimens belonged to females and the latter specimens to males of the same species. In the same mine Jinliang Sun and colleagues found three lower molars in 1980 and an upper jaw fragment with 12 teeth in 1982 (Y. Zhang, 1987) (Table 6.3).

Fossils

J. Woo (1957) described the teeth as follows. Transverse ridges connect the buccal and lingual cusps of the premolar. The anterior fovea is small and shallow, while the posterior fovea remains only as a small pit. Among the five main cusps of the second lower molar, the metaconid is the largest and highest, the protoconid and hypoconid are about equal in size, the entoconid is slightly smaller, and the mesoconid is the smallest. Cusp height decreases as follows: metaconid > entoconid > hypoconid > protoconid > mesoconid. The cingulum on the buccal surface is more obvious than that on the lingual surface. A sulcus between the protoconid and the hypoconid terminates on the buccal surface at the buccal cingulum. A sulcus between the metaconid and entoconid is shallower and terminates at the extremely reduced lingual cingulum.

The lower third molar is a little longer than M2. Its talonid is relatively more reduced with respect to the trigonid than that in M2. There is a trace of an anterior fovea but the posterior fovea was obliterated by wear. As the crown narrows poste-

FIG. 6.3 Lower dentition of hominoid from Kaiyuan: Right P1-M3. (Courtesy of IVPP.)

Table 6.3

Measurements of female and male ape teeth from Kaiyuan

	Sex	*Length*	*Breadth*	*Trigonid breadth*	*Talonid breadth*	*Author*
Upper I2	F	5.9	6.3	—	—	Zhang (1987)
Upper C*	F	9.9	8.5	—	—	Zhang (1987)
Upper P1*	F	8.1	9.7	—	—	Zhang (1987)
Upper P2*	F	6.4	10.5	—	—	Zhang (1987)
Upper M1*	F	10.0	10.5	—	—	Zhang (1987)
Upper M2*	F	10.5	11.2	—	—	Zhang (1987)
Upper M3*	F	10.3	10.7	—	—	Zhang (1987)
Lower P3	M	10.9	9.0	—	—	Woo (1958b)
Lower P4	M	8.1	9.3	—	—	Woo (1958b)
	F	9.0	9.5	—	—	Woo (1957)
Lower M1	M	11.8	10.5	10.5	10.3	Woo (1958b)
	F	11.0	9.3	9.3	9.2	Zhang (1987)
Lower M2	M	13.5	12.1	12.1	12.0	Woo (1958b)
	F	11.8	10.0	10.0	9.8	Woo (1957)
	F	11.8	10.2	10.2	10.0	Woo (1957)
	F	12.7	11.5	11.5	10.7	Zhang (1987)
Lower M3	M	14.2	11.9	11.9	10.9	Woo (1958b)
	F	12.0	10.0	10.0	9.6	Woo (1957)
	F	13.0	11.5	11.5	9.8	Zhang (1987)

* Two teeth are measured

riorly, the mesoconid occupies a lateral position. As with M2, the buccal cusps of this tooth show a greater degree of wear than the lingual cusps. The metaconid is the largest and highest cusp. The description of the size and the height of various cusps of M2 applies equally well to the third molar.

Unlike the second molar, the lower third molar has a sixth cusp between the mesoconid and entoconid, and a small accessory cusp between the metaconid and entoconid. The cingulum is a broad ridge that is more marked on the buccal than on the lingual surface. The furrow separating the protoconid and the hypoconid on the buccal surface terminates at the cingulum, while the furrow separating the lingual cusps on the lingual surface is less marked. The maximum breadth of this tooth is near the mesial end. The buccal and lingual crown surfaces converge backward. J. Woo (1957) described the general shape of the occlusal surface as roughly subtriangular.

J. Woo (1958) described the teeth attributed to the male specimen found in 1958. The lower first premolar is compressed and elongated. Its main axis extends obliquely from the mesio-buccal to the disto-lingual corner of the crown. The junction between the mesio-lingual and buccal surfaces is angular. The oblique convex slope of the mesiobuccal surface of the crown was worn by the upper canine. The buccal cusp is large, and the lingual cusp is vestigial. The base of the crown is surrounded on most sides by a projecting band, which forms triangular prominences on the anterior and posterior sides. The talonid fossa is incipient. The root has a mesial and a distal branch.

The lower second premolar has a hypoconid and entoconid on the talonid. The metaconid is slightly higher than the protoconid, and is separated from the latter by a well-marked sulcus. The anterior fovea and talonid fossa are well developed. A weak cingulum exists on the buccal surface. The root branches on the buccal side but is united on the other sides. The lingual side of the root is convex, but there is a median longitudinal depression on both its mesial and distal surfaces.

The lower first molar is generally similar to the lower second molar, but the former is smaller and its mesoconid is about the same size as the entoconid. The mesoconid of the lower first molar lies in a more central position than its counterpart on the lower second molar. The root of the lower first molar divides into mesial and distal branches. The mesial branch is nearly straight and projects only slightly mesially on the buccal side. The distal branch projects markedly backward, especially on the buccal side. Except for size and the position of the mesoconid, the lower second and third molars of the male specimens are generally similar to those of a female.

Xingyun Zhang (1987) described three isolated left lower molars (Cat. no. YVO 734) and a maxillary fragment (YVO 720) found in the same lignite mine by Jinliang Sun of the local Fossil Preservation Group. Zhang suggested that these specimens belonged to the same species as the 1956 specimens found in the same mine. He identifies them as *Ramapithecus kaiyuanensis*.

The lower first molar has a weak cingulum, which is most obvious on the buccal side. The metaconid is the largest cusp, the protoconid and hypoconid are about equal in size, the entoconid is smaller, and the hypoconulid is the smallest. The vertical sulcus on the buccal surface terminates at the cingulum and is deeper than that on the lingual surface. The anterior fovea is a transverse sulcus, and the posterior fovea is a small pit. The second molar found in 1980 is similar to that described by J. Woo (1957). The third molar has a sixth cusp and an accessory cusp between the metaconid and entoconid.

The maxilla was deformed post mortem. Although 12 teeth are preserved, the shape of the dental arch is uncertain. There is an obvious diastema between the canine and the lateral incisor. The attached teeth include a left lateral incisor, canine, two premolars, and three molars, as well as a right canine, two premolars, and the first and second molars (Fig. 6.4). Y. Zhang (1987) described the teeth attached to this maxilla.

Lateral incisor. A thick ridge on the well developed lingual tubercle extends along the median axis of the tooth to the cutting edge of the crown and divides the lingual surface into a mesial and a distal fovea.

Canine. The crown and the root are well preserved. The canine is triangular in top view, and the angles are rounded. The crown is nearly a bilateral triangle in lateral view. The worn surface of the left canine extends obliquely distally. The heavily worn surface of the right canine extends obliquely to both the mesial and distal sides. The buccal surface of the crown is smooth. The buccal and lingual surfaces meet at the disto-lingual corner and form a posterior marginal ridge. The anterior marginal ridge on the lingual surface is the main ridge of the crown. A triangular deep sulcus intervenes between the anterior marginal ridge and the buccal surface. There are some vertical ridges and furrows on the lingual surface. A wave-like cingulum on the base of the lingual surface intersects with both the mesial and distal marginal ridges to form the mesial and distal accessory cusps. The crown's height slightly exceeds that of the tooth row and corresponds to the lateral incisor. The tip of the robust root inclines slightly distally.

FIG. 6.4 Hard palate with upper dentition of hominoid from Kaiyuan. (Courtesy of Xingyun Zhang.)

First premolar. The crown and root are well preserved. The top view of the tooth shows a triangular shape with rounded angles. A longitudinal sulcus divides the occlusal surface into a much larger buccal segment with a high cusp and a smaller lingual segment with a lower cusp. The buccal is broader than the lingual surface. The buccal surface is smooth, triangular, and bulging. In front and behind the buccal surface is a shallow fovea. The anterior fovea is more evident than the posterior one. Lateral to the fovea there is a thick ridge, which extends to the occlusal surface and forms two small tubercles on the buccal margin of the occlusal surface. The lingual surface is also smooth and more rounded than the buccal surface.

Second premolar. The well-preserved right second premolar is slightly worn. A longitudinal sulcus divides the occlusal surface into a smaller buccal and a slightly larger lingual part. The lingual margin is a rounded curve, but because the other sides are nearly straight the contour of the tooth is squarish. The buccal cusp is higher than the lingual cusp, and the buccal surface is slightly smaller than the lingual surface. The enamel is more extensive on the buccal than the lingual surface. At the base of the buccal surface is a middle triangular eminence, on both sides of which are the anterior and posterior foveae. Neither fovea is as distinct as that on the first premolar, and the anterior fovea is smaller. There are small tubercles on the two ends of the buccal margin of the occlusal surface. There are two marginal ridges on the occlusal surface. The lingual part of the occlusal surface is triangular; the buccal part is pentagonal. The transverse ridge on the sloping surface of both cusps extends from the apex of the cusp to the longitudinal sulcus. The root has a buccal and a lingual cusp.

Upper first molar. The lingual cusps of this tooth are slightly worn, while the buccal cusps are even less worn. The mesio-buccal and disto-lingual corners are at a right angle; the other two corners are blunter. Cusp size reduces in the order of protocone > metacone > hypocone > paracone. The latter two cusps are about equally sized. Cusp height reduces in the order of paracone > metacone > protocone > hypocone. The former two cusps are about equally sized. The vertical sulcus separating the cusps terminates at the middle of the buccal surface. The sulcus on the lingual surface terminates at the upper middle part of this surface. There is an anterior accessory cusp at the mesio-buccal corner of the crown. The anterior fovea is anterior of the paracone; the posterior fovea is shallower than the anterior fovea. The cingulum is distinct on the mesio-lingual part of the protocone. The root has a mesio-lingual, mesio-buccal, and a disto-buccal branch.

Upper second molar. Because the mesio-buccal and disto-lingual corners are less rounded than the other two corners of the crown, the tooth is rhomboidally shaped. The order of height and size reduction of the four cusps is similar to that of the first molar. There is a weak cingulum on the buccal and lingual surfaces. There are accessory cusps anterior to the protocone and located mesio-buccally to the paracone. The anterior is more distinct than the posterior fovea. The oblique ridge connecting the paracone and the protocone, and that connecting the protocone and metacone constitutes a V-shaped ridge.

Upper third molar. The exceedingly large protocone occupies three fourths of the lingual half of the occlusal surface, and the remaining one fourth is occupied by the reduced hypocone. The posterior part of the crown is reduced. The fifth cusp is located between the metacone and the hypocone. Small accessory cusps are located anteriorly and posteriorly to the protocone and in front of the paracone.

Xingyun Zhang suggested the name *Ramapithecus kaiyuanensis* for the Kaiyuan specimens he described in 1987 and those described in 1957 by J. Woo.

Geology

Fauna found in the Xiaolongtan lignite mine include the following (Zhang, 1987):

Primates
- *Ramapithecus kaiyuanensis*
- *Sivapithecus* sp.

Rodentia
- Rodentia gen. et sp. indet.

Carnivora
- Mustelidae gen. et sp. indet.

Proboscidea
- *Gomphotherium xiaolongtanensis*
- *Gomphotherium cf. macrognathus*
- *Tetralophodon* cf. *sinensis*
- *Zygolophodon chinjiensis*

Perissodactyla
Tapirus cf. *yunnanensis*
Artiodactyla
Hexaprotodon sp.
Propotamochoerus parvulus
Listriodon sp.
Cervidae gen. et sp. indet.

The geological age is late Miocene.

Specimens found at Yuanmou, Yunnan

In 1986 Zhang and his colleagues found Neogene ape fossils in Baozidongqin (Leopard Cave Valley) near Zhupeng village, Yuanmou, Yunnan. In 1987 they found another Neogene ape fossil site at Hudieliangzi (Butterfly Hill) near Xiaohe (Little River) village in the same county [101°46′ E, 25°55′ N]. About 2 km separates the villages.

X. Zhang et al. (1987a) proposed the names *Homo orientalis* and *Ramapithecus hudiensis* for the higher primates from Baozidongqin and Hudieliangzi, respectively. In 1988 they reported a facial part of the so-called *Ramapithecus hudiensis* from Hudieliangzi (X. Zhang et al., 1988) and more ape teeth were subsequently found at nearby Gaipailiangzi Hill about 1 km southwest of Zhupeng.

In 1991 Chu Jiang and his colleagues found seven hominoid teeth dating from the Neogene near Leilao village, about 8 km south of Zhupeng. Jiang et al. (1993) attributed the teeth to *Sivapithecus*.

Fossils

Based on a preliminary analysis of the teeth from Baozidongqin and Hudieliangzi, Xinzhi Wu (1989) suggested dropping the name *Homo orientalis* for the Baozidongqin specimens and putting the ape fossils from Baozidongqin and Hudieliangzi into a single species. The morphological features that Zhang et al. used as the basis for erecting the new species *Ramapithecus hudiensis* are not significantly different from those of *Lufengpithecus*.

X. Zhang et al. (1988) gave a preliminary description of the facial bone (YV 0999) from Hudieliangzi (Fig. 6.5). The slightly deformed specimen of a youngster consists of the lower part of the frontal squama, both orbits, nasal bones, maxillae (Fig. 6.6), zygomatics, and eight teeth. The intact teeth include the first permanent molar and deciduous dentition (except the incisors) of both sides.

The face is short and broad. The interorbital distance is large compared to that of extant great apes of the same age. The interorbital region is not depressed, the zygomatic arch does not flare, and the supraorbital ridge is broad and flat. The supraorbital margins of both sides are connected by a very low ridge. The metopic suture is discernible only below glabella. The orbit is spherical and the superolateral angle is somewhat angulated. The vertical diameter is slightly smaller than the horizontal diameter. The medial and superior orbital margins are more rounded than the lateral and lower margins. The zygomatico-orbital foramen has a long orifice. The zygomatico-frontal and zygomatico-maxillary sutures are serrated. Because the malar surface of the zygomatic is smooth, broad, and very convex, the face looks wide.

FIG. 6.5 Frontal view of hominoid facial bone from Hudieliangzi, Yuanmou. (Courtesy of Xingyun Zhang.)

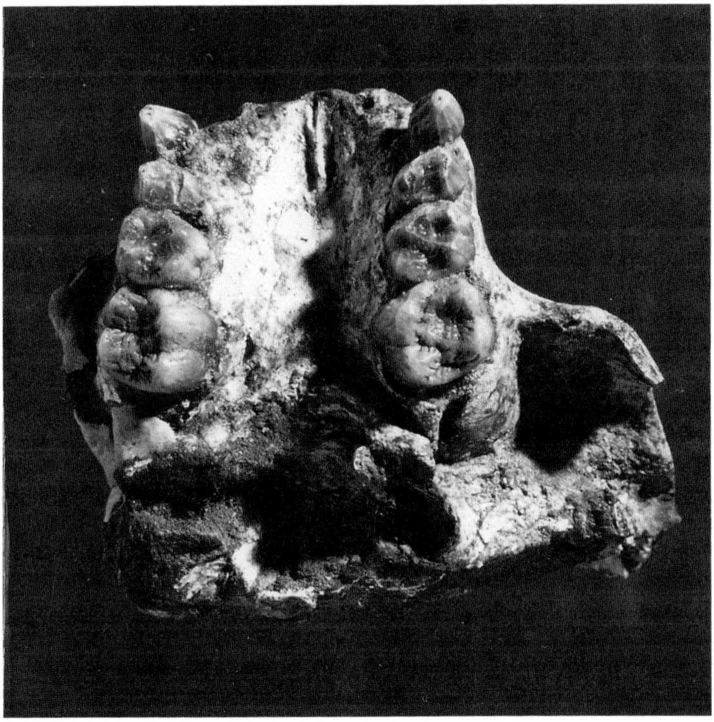

FIG. 6.6 Basal view of hominoid maxilla from Hudieliangzi, Yuanmou. (Courtesy of Xingyun Zhang.)

The lower part of the nasal bone is wider than the upper part. The middle segment of the fronto-nasal suture is deeply wavy, the lateral segments are serrated and extend obliquely down and laterally. The internasal suture seems to be obliterated. The lower part of the nasal bone is slightly elevated. The well preserved left maxilla is without deformation. The pyriform aperture is short, broad, and pear-shaped. Its height slightly exceeds its breadth. The upper part of the lateral margin of the pyriform aperture is sharp; the lower part is rounded. The upper margin of the aperture is slightly more forwardly located than its lateral margins. The lower margin of the pyriform aperture is rounded.

The alveolar process is low, and its anterior surface is somewhat vertical. Beneath the distinct infraorbital foramen is a shallow canine fossa. The zygomatic process of the maxilla arises from above the septum between the decidous and permanent molars. The canine jugum is prominent and vertically oriented. The juga of the median incisors are prominent and separated by a groove. The juga of the lateral incisors are not prominent. The alveolar plane is convex in lateral view. The alveola of the incisor and permanent molar are higher than those of the intervening teeth. The hard palate is well preserved. The interpalatal suture is distinct. The palatal surface is concave, and the palate becomes shallower in its anterior part. There is a single incisive foramen. The dental arch as a whole is V-shaped and parabolic, but posteriorly the arch diverges slightly. The incisor portion of the anterior dental arch is curved. The breadth of the dental arch at the level of the canine is shorter than that at the posterior end of the arch. The broken bone shows the existence of the maxillary sinus.

The alveolum of the median incisor is nearly spherical with the mesio-distal diameter slightly longer than the labio-lingual diameter. The alveolum of the lateral incisor is elliptical with the labio-lingual diameter longer than its mesio-distal one. The labio-lingual diameter of the lateral incisor is about half that of the median incisor. In basal view the incisors are arranged in a curve instead of in a straight line.

The crown of the decidous canine is low and small; its mesio-distal diameter exceeds the labio-lingual diameter. The labial surface is smooth and triangular. From the apex of the crown, two oblique ridges extend to the base on the mesial and distal parts of the labial surface. On the lingual surface a vertical ridge extends from the apex and connects with the cingulum. It forms a small triangular fovea with the mesial ridge at the base of the crown. The apex of the tooth is a small flat surface with an exposure of dentine. There seems to be a small diastema between the incisor and the deciduous canine.

The first deciduous molar has buccal and lingual cusps. Ridges that connect these two cusps separate the anterior, middle, and posterior foveae. Dentine is exposed at the apex of both cusps. The contour of the crown is triangular with the labial side longer than the lingual side.

The second deciduous molar is slightly worn, square, and has rounded angles. The buccal cusps are higher than the lingual ones. Cusp size decreases in the order of: protocone > metacone > paracone > hypocone. The protocone connects the paracone with transverse ridges and connects the metacone by an oblique ridge, which separates the protocone and hypocone. These ridges form anterior, trigonid, and posterior foveae with the anterior and posterior marginal ridges. The trigonid fovea is broader than the others. The apex of the cusps is close to the margin of the occlusal surface. There is no cingulum. The enamel of the deciduous molars is thin and smooth, and lacks wrinkles.

The first permanent molar is square and, although slightly worn, has no exposed dentine. Morphologically somewhat similar to the second deciduous molar, the first permanent molar is larger, more robust, and has slightly more wrinkles.

Geology

Zhang et al. (1988) described the geological section yielding the ape facial bone from Hudieliangzi. From the surface down the layers are as follows:

1. Disturbed dark red sandy clay, 0–1.7 m thick
2. Red sandy clay with concretions, pebbles, and a few fossils, 0.9–2.5 m thick
3. Yellow and purplish-red sands with lenses of various size or intercalated gravels (3–20 cm thick), many mammalian fossils, 1.3–2 m thick

Mammalian fauna from this section include the following (X. Zhang et al., 1988):

Primates
Ramapithecus hudiensis
Pliopithecus sp.
Hylobates sp.
Rodentia
Hystrix sp.
Rhizomys sp.
Carnivora
Ursus sp.
Hyaena sp.
Cynailurus sp.
Machairodontinae gen. et sp. indet.
Probosdcidea
Gomphotherium sp.
Perissodactyla
Hipparion sp.
Chilotherium sp.
Nestoritherium sp.
Artiodactyla
Cervidae gen. et sp. indet.
Suidae gen. et sp. indet.

Guanfu Zong et al. (1991) described the geology and mammalian fauna of the ape site at Dashuqinliangzi, near Leilao Village in Bango Basin, Yuanmou County. From the surface down the section includes:

Pleistocene:
13. Deep brownish-red sandy clay with stone artifacts, 1–5 m thick Pliocene (early stage):

———Erosion surface———

12. Brown sandy claystone, rich with concretions, 5–10 m thick
11. Bluish-gray sandy claystone, slightly cemented, 1 m thick
10. Brown sandy claystone with a grayish-white layer of fine sands, slightly cemented, 5 m thick
9. Grayish-green, grayish-yellow, or brown sandy claystone, intercalated with silty layers, slightly cemented, contains ape fossils, 2–6 m thick

8. Brownish-red sandy claystone, slightly cemented, 2–6 m thick
7. Yellow sands or silt intercalated with grayish-yellow and grayish-green layers of fine sands, with *Chilotherium*, 3–5 m thick
6. Fine brown sands, slightly cemented, 2–4 m thick
5. Hard gravels, with small fragments of mammalian fossils, cemented, 1–2 m thick
4. Hard brown gravels, 1–2 m thick
3. Brown sandy claystone, appears as lens 1 m thick, slightly cemented
2. Hard brown gravels with claystone lens, cemented 2–3 m thick

———Erosion surface———

Miocene or Early Pliocene:

1. Brown, grayish-white or grayish-green claystone, sandy stone, slightly cemented, 3–10 m thick

Mammalian fossils come from the upper Pliocene levels of this section. Zong et al. (1991) suggested that this set of strata were rapidly deposited, so the fossils are considered to be the same geological age. Mammalian fossils associated with the apes are as follows:

Rodentia
- *Brachyrhizomys nagrii*
- *Monosaulax* sp.

Carnivora
- Hyaenidae gen. et sp. indet.

Proboscidea
- *Stegolophodon* aff. *banguoensis*
- *Stegotetrabelodon primitium*

Perissodactyla
- *Hipparion platyodus*
- *Chilotherium yunnanensis*
- *Macrotherium salinum*

Artiodactyla
- *Potamochoerus* sp.
- Suidae gen. et sp. indet.
- *Yunnanotherium* sp.
- *Dorcabune* sp.
- *Muntiacus* sp.
- *Muntiacus fenhoensis*
- *Cervavitus* sp.
- Cervinae gen. et sp. indet.

Zong et al. (1991) suggested that the Leilao, Hudieliangzi, Baozidongqin, and Gaipailiangzi sites date to 5 to 6 mya.

GIGANTOPITHECUS

The original *Gigantopithecus* fossils became known in 1935, when G.H.R. von Koenigswald proposed a new taxon, *Gigantopithecus blacki,* for a right lower third molar he found in a Hong Kong drug store (von Koenigswald, 1935). In 1937 F. Weidenreich suggested that this material belonged to an orang-utan. In successive years von Koenigswald collected a left lower third molar in another Hong Kong drug store and a right upper molar in a Guangzhou (Canton) drug store. After study of these three teeth, *Pithecanthropus* Skull IV, and two other mandibles from Java, Weidenreich changed the genus name *Gigantopithecus* to that of *Gigantanthropus* because he thought it bore more hominid than anthropoid traits. Weidenreich (1945, 1946) considered that this giant creature was the direct ancestor of *Pithecanthropus* and *Sinanthropus.*

In 1952, von Koenigswald restudied all the *Gigantopithecus* teeth obtained from drug stores that he had kept in his possession. He agreed with Weidenreich's hominid attribution of the giant primate, but he considered it to be an extinct, aberrant branch of hominid evolution (von Koenigswald, 1952).

In the winter of 1955 a field team from the Laboratory of Vertebrate Paleontology, Academia Sinica, led by Wenzhong Pei, went to Guangxi in the southern part of China close to Vietnam's border. The team sought to find the locale and determine the geological age of *Gigantopithecus.* In the spring of 1956 the team found an upper molar, a lower third molar, and a first lower premolar *in situ* in the Heidong Cave (Black Cave), Niushuishan Hill (Cow Sleep Hill) in Daxin County, southwestern Guangxi. The cave is located at 107°17′ E and 22°43′ N. The associated mammalian fossils belong to the *Ailuropoda-Stegodon* fauna common in south China caves. In the same year, the team collected 47 more teeth from stores of Supply Cooperatives and Export Companies. Based on the new materials, Pei and Woo (1956) concluded that *Gigantopithecus* may belong to an anthropoid side branch whose divergence from the hominid-pongid line was closer to the hominid than the pongid line.

In the winter of 1956 Pei and his team returned to Guangxi to search for more *Gigantopithecus* materials and other fossil mammals in the caves. A local farmer, Xiuhuai Qin, had dug cave deposits for fertilizer and discovered many fossils in a cave on Lengzhaishan (Lengzhai Hill) (109°14′ E., 24°30′ N) near the village of Xin Shechuncun (New Shechun Village). Persuaded by a government employee, he presented the fossils to Pei's team through government sources. One *Gigantopithecus* mandible with 14 teeth attached was identified among these materials. Guided by Qin, the team immediately excavated the cave and found another mandible, several isolated teeth, and other mammalian fossils. The excavation of 1957–1958 yielded another large mandible (no. 3) and more teeth.

In the winter of 1965 another field team from the Institute of Vertebrate Paleontology and Paleoanthropology led by Yinyun Zhang found 12 *Gigantopithecus* teeth in a cave on Bulishan Hill (108°13′ E, 22°59′ N) near Ganxu District, Wuming County, Guangxi Zhuang Autonomous Region. In 1973 they found a fourth *Gigantopithecus* site in Guangxi in a cave on Lonmoushan, near Nahe Village, Bama County. One right lower third molar was found here. By 1973, 1,070 isolated teeth and 37 attached mandibular teeth had been found in Guangxi and Guangdong.

In the spring of 1968 an IVPP team led by Chunhua Xu collected almost 200 *Gigantopithecus* teeth in Hubei Province, central China. In 1970 this team found five *Gigantopithecus* teeth in Longgudong (Dragon Bone Cave) (110°04′ E, 30°38′ N) near Gaopin Village, Jianshi County, Hubei Province. They also collected dozens of *Gigantopithecus* teeth from drug stores throughout Hubei. Hubei yielded 292 *Gigantopithecus* teeth: 281 from a drug store, 5 from Longgudong, and 6 probably from Xiniudong cave (Rhinoceros Cave) near Luojiaba Village, Jiangshi County. A new *Gigantopithecus* site located near Hubei is Longgupo (Dragon Bone Slope) (109°35′ E, 30°48′ N), near Damiao town (Big Temple town), Wushan County, Sichuan Province. Twelve *Gigantopithecus* teeth associated with other mammalian fossils were found here. Two were found in 1985, 2 in 1986, 4 in 1987, and 4 in 1988 (Gu and Fang, 1991b).

Fossils

The following description of the *Gigantopithecus* mandibles and teeth is according to J. Woo (1962a,b).

Mandibles (Table 6.4)

Mandible I (Fig. 6.7). Except for the parts corresponding to and behind the second molar, a large part of the mandibular body was preserved. The teeth are attached except for the third molars. The crowns of the right lateral incisor and left median incisor are broken. The mandible is attributed to an elderly female.

Mandible II. Except for the parts corresponding to and behind the second molar, a large part of the mandibular body was preserved. All the teeth are attached, except the two median incisors, canines, and the left first premolar. The slightly worn teeth indicate that the mandible belongs to an adolescent.

Mandible III. This is the largest of the three mandibles. The mandibular rami are almost completely lost. Except for two median incisors and a right second molar, all the teeth are attached. The teeth are heavily worn and the mandible is attributed to an old male. The basal border is well preserved and symmetrical. The right side of the dentition is more anteriorly positioned than the left side.

Lateral side of the mandibles. The weak lateral eminence of the right side of Mandible I divides into two branches: the superior lateral torus and the marginal torus behind the second molar. The former continues with the jugum of the second premolar above the mental foramen. The latter extends along the basal margin of the mandible to a point below the mental foramen, and then continues with the jugum of the canine. There is no jugum for the first premolar.

The distinct lateral eminence of the right side of Mandible II divides into two weak branches below the second molar. The superior branch, positioned slightly higher than on Mandible I, continues with the jugum of the second premolar. The inferior branch turns downward and forward below the mental foramen. It bifurcates in front of the mental foramen to continue with the jugum of the canine and the first premolar. The remaining part of the left half of the mandibular body shows a similar condition in the marginal torus. The canine jugum is prominent. The jugum of the first premolar is slightly obvious; the jugum of the second premolar is indistinct. The lateral eminences are most prominently seen on Mandible III. The lateral eminence of Mandible III divides into two branches. The superior branch is horizontal and disappears at the level of the second premolar, the inferior branch disap-

Table 6.4

Measurements of *Gigantopithecus* mandibles

Specimens			*I*	*II*	*III*
At symphysis:	Height		64	60.5	85
	Thickness		35	41	39
	Circumference		187	196	247
At foramen mentale:	Height:	left	58.5	54	66
		right	58	53	66
	Thickness:	left	—	—	33
		right	28	—	31
	Circumference:	right	161	—	—
Behind P2:	Height:	left	55	57	81
		right	59	56	—
	Thickness:	left	27.5	30	34
		right	27	31	32
	Circumference:	right	159	—	—
Between M1 and M2:	Height:	left	—	—	83
		right	—	—	81
	Thickness:	left	—	32	34
		right	—	32	35
At the middle of M2:	Height:	left	—	—	84
		right	—	—	81
	Thickness:	left	—	—	32
		right	30	—	34
Between M2 and M3:	Height:	left	—	—	85
		right	—	—	82
	Circumference:	left	—	—	218
		right	—	—	192
Incision-gnathion			76	77	108.5
Length from incision to the foot of perpendicular line of fora. supraspinosum			21	21	33
Breadth between mental foramina (left)			55	54	55
Height of mental foramen:	left		16	18	17
	right		17	18	19
Length of three molars:	left		—	—	60
	right		—	—	59
Angle of inclination			59°	55°	51.5°
Length of I1-M2			82	90	96
Length of P1-M2			65	74	72
Length of alveolar arch			99	112	117
Breadth of alveolar arch			65	72	75
Length of anterior alveolar arch			45	50	58.5
Breadth of anterior alveolar arch			56	56	56
Total length of alveolar arch (M3-I1-M3) :	int.		—	—	244
	ext.		—	—	283
Breadth of basal arch at M2			—	—	85
Breadth of alveolar arch at M2			78	84	83
Angle of molar rows			12°	13°	20°
Angle formed by id-gn line with alveolar plane			59°	55°	51.5°

According to J. Woo (1962a, b).

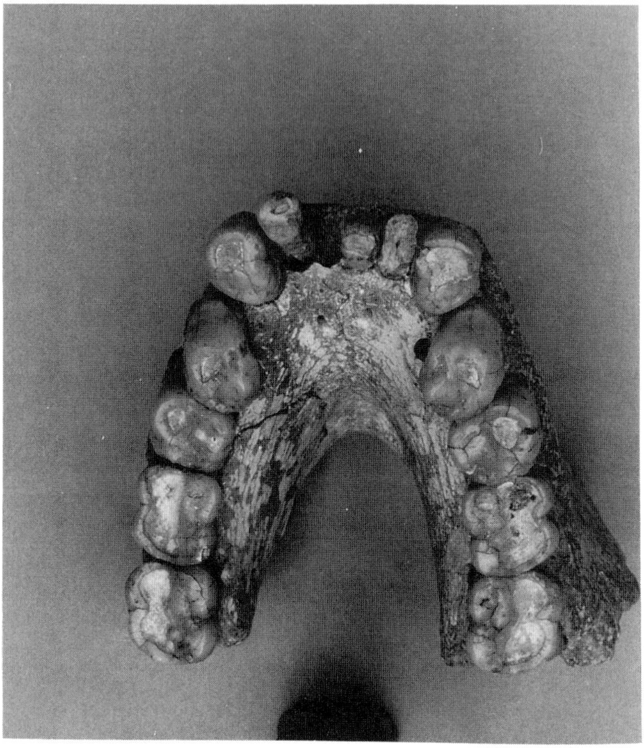

FIG. 6.7 *Gigantopithecus* mandible I (female). (Courtesy of IVPP.)

pears at the basal margin of the mandible at the level of the first molar. The surface is slightly depressed between the two branches.

There are no anterior or posterior marginal tubercles or stria platysmatica on any of the mandibles.

Mental foramen

Mandible I. This foramen on the right side is level with the posterior root of the second premolar and opens forward and slightly upward. The remaining anterior half of the left mental foramen is level with the anterior root of the second premolar and opens forward.

Mandible II. The mental foramen is a small depression level with the anterior root of the second premolar on the right side. The left mental foramen is located corresponding to the anterior root of the second premolar. The foramen opens forward and upward.

Mandible III. The right mental foramen is level with the interalveolar septum of the premolars and opens forward. The position and orientation of the left foramen is similar to that of the right.

Anterior surface of the frontal part. The upper margin projects more than the lower margin. There is no incurvatio mandibulae and there is no mentum osseum. The central part of the anterior surface bulges forward. The bulging, which is especially marked in Mandibles I and II, is situated in a position lower than in modern apes.

Internal surface of the frontal part. The upper part of this surface is slightly concave. Below, it bulges gradually to form the superior transverse torus, which is more marked in Mandibles I and III than in Mandible II. Located further down is the shallow and wide fossa genioglossi, which appears in all specimens. At the bottom of the fossa in Mandible I is a fairly large opening, to the lower right of which is another miniscule opening. There are only indications of the presence of this opening in Mandibles II and III. The inferior transverse torus is more marked in Mandibles I and II than Mandible III. Below this torus is a concavity corresponding to the digastric fossa. There is no simian shelf. As a whole these mandibles are closer to hominids than to apes in the outlines of the median section through the symphysis.

Internal surface of the lateral part. All three *Gigantopithecus* mandibles lack a definite linea mylohyoidea and torus mandibularis.

Lower margin of the mandibular body. This margin increases in thickness from the posterior and to the level of M2, where the thickness decreases. From the level of P2 the margin becomes thicker again; it maintains a uniform thickness in the anterior part of the basal margin. Mandible III has the lower margin almost intact. The lower margin of the symphyseal region is higher than that of the other parts. A distinct incisura submentalis is anterior to the first molar.

Alveolar arch. In Mandibles II and III the alveolar arch turns sharply at the level of the canine. The cheek teeth are arranged in a nearly straight line. Mandible I lacks this sharp turn and the cheek teeth are arranged in a curve. The larger the canine, the more sharply is the turn of the alveolar arch. The tooth rows of both sides of all three mandibles are arranged in a gentle curve and are slightly diverged backward. In Mandible III the second molars of both sides bulge slightly outward and the third molars are situated inward. The length of the lacteal portion of the alveolar arch is one half of the total length of the alveolar arch.

In Mandible I the canine does not contact with the premolar, there is a very narrow slit instead of a diastema. In Mandible III there is a narrow diastema. The posterior half of the crown of the lower canine is well worn. Only a trace of wear is visible on the mesial crown surface of the lower first premolar of Mandible III. This suggests that during biting the upper canine inserts between the lower canine and the premolar. The worn surfaces of the isolated teeth show that the worn surface of the lower canine is more on the labial than on the lingual side, so the upper canine overlaps the lower canine on the lateral side.

The basal arches of all three mandibles lie behind the alveolar arches. In Mandibles I and II, the basal arch lies before the postlacteal line; it is behind the line in Mandible III. The basal arches are as wide as those in hominids.

Dentition

Among the 1094 teeth studied by J. Woo (1962b), 168 are just erupting and are without attrition. Size variation among the *Gigantopithecus* teeth is quite large. J. Woo divided the teeth into large and small types, representing male and female specimens, respectively.

According to J. Woo's analysis, teeth collected from the Liucheng and Daxin sites and from drug stores in the Guangdong and Guangxi regions represented at least 88 individuals, 41 belong to males and 47 to females. Based on tooth wear, J. Woo considered that the sample contained adults, youngsters, and infants. Dental measurements appear in Tables 6.5 and 6.6.

Permanent dentition

Upper median incisor. The labial surface of the crown bulges in both the vertical and horizontal direction. A pronounced lingual tubercle on the basal part of the lingual surface sends several fingerlike projections toward the shovel-shaped central part. Some of these projections extend to the cutting edge. The labial and lingual surfaces form a 45° angle at the cutting edge which has 3 to 4 small mammilary processes. The mesial border is usually shorter than the distal border.

A large part of the root of specimen PA 1 is conical. The cross section of the root is elliptical. The labio-lingual diameter (12.1 mm) is shorter than the mesio-distal diameter (14.3 mm). The estimated height of the root is 25 mm. The cervical part is slightly constricted. The junction between the crown and the root is slightly higher on the labial than on the lingual surface.

Upper lateral incisor. The labio-lingual diameter of the crown is much larger than the mesio-distal diameter. The labial surface of the crown bulges obviously in the horizontal direction and is more flat in the vertical direction. The cutting edge and the mesial surface form a rounded right angle. The distal margin continues with the cutting edge and forms a curve. There are three mammilary processes on the edge. The mesial and distal margins of the lingual surface are thickened. The thick borders join at the basal part of the crown and form an obvious lingual tubercle extending towards the center of the lingual surface to form a median ridge. The enamel extends to a higher position on the lingual than on the labial surface. The lateral incisor shows more variation in structure and size than does the medial incisor.

Lower median incisor. The labial surface is slightly convex in both vertical and horizontal directions. The lingual surface is shallow and shovel-shaped. The lingual tubercle is distinct. The mesial border forms a right angle with the cutting edge, which continues with the distal border to form a curve. The enamel extends to a much lower position on the labial and lingual surfaces than on the mesial and distal surfaces.

Lower lateral incisor. This tooth is shaped similarly to the lower median incisor. However, its labial surface is much more curved in both the sagittal and transverse directions. The lingual surface is shovel-shaped. The lingual tubercle is distinct on the basal part. The cutting edge forms a rounded right angle with the mesial border of the crown and forms a curve with the distal border. The enamel extends more on the lingual and labial surfaces than on the mesial and distal surfaces.

Upper canine. The crown is low and conical. A cingulum that surrounds the base of the crown is especially distinct on the lingual surface of the female specimen. The cingulum appears as a triangular prominence on both the mesial and distal surfaces, but is especially obvious on the distal surface. The labial surface is very convex in the horizontal direction. Many transverse riblike ridges exist on some specimens. There is a sagittal, deep groove along the anterior margin of the crown.

The lingual surface is flatter, its anterior half is slightly bulging, and its posterior half is slightly depressed. Vertical striae are more prevalent on the posterior half of

Table 6.5

Measurements of *Gigantopithecus* teeth from Liucheng ,Guangxi

		Length			Breadth		
		Average	*(no.)*	*Range*	*Average*	*(no.)*	*Range*
Probable Male							
Upper dentition							
Incisor:	median	15.2	(1)	—	15.2	(3)	14.0-16.8
	lateral	—	(0)	—	—	—	
Canine:		19.0	(8)	16.8-20.8	21.7	(11)	18.8-23.9
Premolar:	first	15.5	(23)	14.1-17.4	19.9	(23)	18.0-22.6
	second	14.2	(30)	13.0-15.6	20.6	(30)	18.8-24.0
Molar:	first and second	20.5	(50)	18.8-23.0	21.1	(50)	18.5-23.6
	third	19.6	(47)	18.0-22.2	20.4	(47)	18.3-24.2
Lower dentition:							
Incisor:	median	7.2	(5)	6.9-7.7	9.5	(6)	8.7-10.8
	lateral	8.2	(6)	8.0-8.3	10.3	(8)	9.6-11.3
Canine:		14.3	(9)	12.9-15.0	19.1	(7)	18.0-21.0
Premolar:	first	16.9	(33)	15.5-18.8	17.1	(33)	15.7-21.0
	second	16.7	(33)	15.4-18.0	17.1	(33)	16.0-18.9
Molar:	first and second	21.3	(47)	19.3-24.1	18.1	(47)	16.0-20.7
	third	21.5	(32)	19.2-26.4	17.6	(33)	15.5-22.3
Probable Female							
Upper dentition:							
Incisor:	median	12.4	(6)	11.8-13.2	12.7	(6)	11.2-14.0
	lateral	8.6	(6)	7.9-9.0	11.2	(5)	10.2-12.8
Canine:		15.3	(5)	14.1-16.7	14.7	(5)	13.0-16.0
Premolar:	first	13.5	(16)	12.5-14.6	17.8	(16)	16.3-19.9
	second	12.4	(35)	10.4-13.1	18.4	(35)	17.0-20.1
Molar:	first and second	17.8	(59)	15.1-20.0	18.5	(57)	15.7-20.4
	third	16.3	(42)	14.0-18.5	17.6	(42)	15.0-20.5
Lower dentition:							
Incisor:	median	6.3	(2)	6.0-6.5	8.2	(2)	8.0-8.4
	lateral	7.6	(2)	7.5-7.7	10.2	(2)	9.8-10.5
Canine:		10.9	(6)	10.4-11.7	15.1	(7)	14.5-16.2
Premolar:	first	14.8	(34)	12.9-16.3	14.9	(34)	12.9-17.1
	second	14.4	(41)	12.5-16.5	14.9	(41)	13.2-16.3
Molar:	first and second	18.2	(53)	16.4-20.0	15.4	(53)	13.1-17.0
	third	17.8	(27)	14.6-19.2	14.8	(27)	13.2-16.1

Figures in parentheses are the number of teeth.

According to J. Woo (1962a, b).

Table 6.6

Measurements of *Gigantopithecus* teeth from the Late Period

Specimens		*Length*			*Breadth*	
	Average	*(no.)*	*Range*	*Average*	*(no.)*	*Range*
Upper Dentition						
I1	12.9	(4)	11.1-14.1	13.0	(3)	11.4-14.9
I2	11.0	(1)	—	11.5	(1)	—
C	18.3	(11)	14.7-21.6	17.5	(11)	14.5-20.0
P1	15.7	(14)	13.7- 17.9	21.3	(14)	18.6-24.1
P2	15.3	(10)	13.6-17.0	23.1	(10)	20.7-26.2
M1	20.3	(15)	18.0-23.5	22.5	(15)	19.6-25.7
M2	20.8	(10)	18.1-24.0	23.1	(10)	19.6-27.8
M3	17.6	(17)	15.5-21.8	21.2	(17)	17.6-25.6
Lower Dentition						
C	14.3	(5)	13.3-16.0	19.6	(5)	18.9-20.4
P1	16.7	(15)	14.6-19.6	18.1	(15)	15.6-20.8
P2	16.0	(19)	14.4-18.8	16.8	(19)	14.6-19.8
M1	21.0	(17)	17.5-24.6	18.0	(17)	14.3-21.3
M2	21.6	(14)	18.0-27.3	18.8	(14)	15.3-23.4
M3	22.3	(28)	19.0-26.7	18.5	(28)	14.9-22.5

According to Zhang (1983). The teeth in this sample come from Jianshi, Daxin, Wuming, Bama, and drug stores in Guangxi, Guandong, and Hubei Provinces.

Figures in parentheses are the number of teeth.

the lingual surface. The slightly concave distal margin is sharper than the slightly convex mesial margin. The apex is situated behind the center of the crown.

Upper canine no. 11 is almost intact. The root is high, robust, and conical. Its mesial surface is rounded and its distal surface is flatter. There is no obvious constricture at the neck of the tooth. The long axes of the crown and the root form a 160° angle. The extension of enamel on the mesial and distal surfaces is less evident than on the labial and lingual surfaces.

Lower canine. Specimen PA 27 is a tooth germ. The mesial and distal sides of the crown are wedge-shaped with a sharp margin. The labial surface bulges in both the sagittal and transverse directions, the lingual surface is only slightly elevated. The cingulum appears triangular at the mesial and distal sides. A shallow fovea exists at the center of the basal part of the triangle. The apex of this triangle continues upwards with the cutting edge. Weak longitudinal and transverse thin ribs exist on the labial surface, except at its apex.

On the basal part of the lingual surface, above the cingulum, is a prominent lingual tubercle, which sends several linear ridges to the apex of the crown. The main ridge, which extends from the anterior part of the lingual tubercle to the apex of the crown, divides the lingual surface into a small anterior and a larger posterior section. The cingulum is less distinct in fully developed specimens.

The root, preserved only in a few of the isolated lower canines, is less robust than that of the upper canine. The labio-lingual diameter of the root of the lower canine is much longer than the mesio-distal diameter. The cervix of the lower canine is not constricted as in the upper canine. The enamel extends downward more on the mesial and distal surfaces than on the labial and lingual surfaces.

Upper first premolar. The mesio-buccal corner of the upper first premolar protrudes more than other corners. Therefore, the mesio-distal diameter of the buccal part of the crown is larger than its lingual part. The lingual cusp is slightly closer to the mesial border of the crown and is lower than the buccal cusp.

A deep longitudinal groove divides the occlusal surface into a large buccal and a small lingual part. The mesio-distal diameter is slightly larger than the height of the buccal surface of the crown in big specimens (males), while the proportion is reversed in small specimens (females). The enamel extends upward more on the buccal surface than on the lingual surface. A cingulum, which surrounds the base of the crown extends upwards at the mesio-buccal corner to form an distinct triangular prominence (tuberculum molare). The cingulum also extends upwards at the bucco-distal corner as a thickened part.

The lingual surface of the crown is bulging. The occlusal surface has a central longitudinal groove dividing the buccal and lingual cusps. Each cusp has a primary ridge, which broadens and divides into two or three branches towards the median longitudinal groove of the occlusal surface. On either side of the primary ridge is a large depression bordered by the marginal major ridge. Several smaller ridges arranged in a mesio-distal direction diverge from the main ridge.

The root is short and broad, and its mesio-distal diameter is shorter than the bucco-lingual diameter. The buccal root divides into a larger mesial and a smaller distal branch. The mesial branch deflects slightly to the mesial side, and the distal branch deflects markedly to the distal side. The two branches form an angle of about 35°. The lingual root deflects toward the lingual side. The lingual branch is the thickest and the distal branch the thinnest. In cross section the branches are ovoid. The neck of the tooth constricts slightly.

Upper second premolar. The buccal and lingual parts of the occlusal surface are almost equal in breadth. Enamel extends upwards more on the buccal and lingual surfaces than on the mesial and distal surfaces. The buccal cusp is slightly higher and bigger than the lingual cusp. Both cusps are situated in the anterior half of the occlusal surface; therefore, the occlusal surface of the buccal part is larger than the lingual part, and the distal part is larger than the mesial one. The mesial and distal parts of the occlusal surface are separated by a groove extending between two cusps. The wrinkles are similar to those on the first premolar but are less distinct than on the first premolar. The cingulum is similar to that on the first premolar but is less distinct. The cingulum of the big types (males) is more distinct than that of small types (females).

The root is similar to that found on the first premolar. The bucco-lingual diameter is much larger than the mesio-distal diameter. The buccal root divides into mesial and distal branches about 5–6 mm above the neck of the tooth. Both branches of the buccal root are more or less equal in thickness. The lingual root is the largest. The tooth is somewhat constricted at the cervix.

Lower first premolar. Its long axis is oblique and extends from the mesio-buccal to the disto-lingual side. The tooth crown is pyramidlike (triangular in cross section) and formed by the buccal, distal, and mesio-lingual surfaces. In some cases there is no clear demarcation between the buccal and distal surface, and the buccal surface gradually merges with the distal surface. The buccal cusp is higher than the lingual cusp. On the occlusal surface the lingual cusp is at the center of the lingual part; the buccal cusp is slightly more anteriorly located on the buccal part. The basal part of the crown is markedly bulging on the buccal surface. A prominent crest, which is especially distinct in large (male) specimens, extends along the mesio-buccal corner to the occlusal surface.

The buccal surface of the crown bulges. A shallow groove extending from above downward is somewhat anterior to the disto-buccal corner of the crown and behind the buccal cusp. The mesio-lingual surface joins the buccal surface to form the mesial border of the crown. In top view the buccal cusp is much larger than the lingual cusp. Both cusps are separated by a longitudinal groove. The posterior part of the crown is much larger than the anterior part. The anterior fovea is smaller than the posterior fovea, and accessory ridges are more common in the latter.

The root divides into two branches. The anterior root in specimen PA 35 is 25.5 mm high and turns slightly backward and toward the lingual side. The posterior branch is slightly flatter, with its bucco-lingual diameter exceeding the mesio-distal diameter. There is a vertical groove on the distal surface of the posterior branch. The tooth is constricted at the cervix.

Lower second premolar. The tooth's shape is more or less quadrangular, with its breadth greater than the length. There is a considerable amount of molarization. The tooth is generally larger than the first premolar belonging to the same individual. The buccal and lingual cusps occupy the anterior two thirds of the occlusal surface, and the remaining part is a talonid. The anterior part of the occlusal surface is much larger than the talonid and is separated from the latter by vertical grooves on the buccal and lingual surfaces. The buccal part of the talonid is higher than the lingual part. Some specimens have two accessory cusps on the talonid; the buccal accessory cusp is the largest and highest. In other specimens, the separation of these two small accessory cusps is unclear. The basal part of the crown especially bulges on the buccal surface. The enamel extends downward more on the buccal surface than elsewhere. The buccal is generally smaller than the lingual cusp, but both are equal in height. A transverse groove on the occlusal surface separates the mesial and distal parts; a sagittal groove separates the buccal and lingual cusps. A primary ridge extending between the apex of both cusps and crossing the sagittal groove radiates small accessory ridges from its anterior and posterior sides. The posterior fovea is much larger than the anterior fovea. There are small accessory ridges on the distal marginal ridges.

The root divides into two branches that are flat and broad almost in their entirety. The root becomes narrower near the tip. Vertical furrows on the mesial surface of the anterior branch and the distal surface of the posterior branch separate each branch into a buccal and a lingual part. The posterior branch is broader and higher than the anterior branch. There is no distinct constricture at the neck of the tooth.

Upper first and second molars. Because it is difficult to differentiate first and second upper molars, J. Woo (1962b) lumped them together for descriptive purposes. In maxillary molars, the paracone and protocone are generally higher than the metacone and hypocone. A vertical furrow separates the paracone and metacone on the buccal surface, and another vertical furrow separates the protocone and hypocone on the lingual surface. The height (or breadth) of the cingulum equals about one third to one quarter of the height of the crown. The cingulum is more prominent on the buccal than on the lingual surface. The mesio-lingual corner (angle) of the crown is least sharp, and the mesio-buccal angle is sharpest. The paracone protrudes more than the protocone, and the hypocone protrudes more than the metacone. The apex of each cusp is close to the margin of the occlusal surface.

A sagittal furrow intervenes between the buccal and the lingual cusps. The transverse groove separating the paracone and metacone does not continue with the groove separating the protocone and hypocone. The grooves are deeper near the center of the occlusal surface. The protocone and metacone contact with each other directly. There is no anterior or posterior fovea in the strictest sense. Instead, there is a small fovea between the paracone, protocone, and the mesial marginal ridge. Another small fovea between the metacone, hypocone, and the distal marginal ridge is more distinct.

In 57% of the cases, the protocone is the largest cusp, the paracone is the second largest, and the hypocone is the smallest. In 21% of the cases, the hypocone is larger than the metacone; in 18% of the cases the paracone and metacone are almost equal in size; and in 4% of the cases, the paracone is larger than protocone. The rounded cusps have complex surface wrinkles. On the sloping surface of the paracone, a robust uneven ridge extends from the apex of the cusp to the Y-shaped furrow of the occlusal surface. The ridge becomes thicker as it extends downward. There are two marginal ridges along both sides of this ridge. Accessory ridges radiate from a ridge extending from the apex of the metacone to the Y-shaped furrow of the occlusal surface. Distinct marginal ridges surround the central eminent part on three sides of the main ridge. A main ridge extending from the apex of the protocone to the longitudinal groove of the occlusal surface divides into two or three branches with marginal ridges on both sides. The ridges on the hypocone are more or less similar to that on the protocone, but ridges on the hypocone are smaller. A wave-shaped ridge extending from the mesio-buccal corner of the paracone to the mesial border or mesio-lingual corner of the protocone suddenly terminates at the lingual end to form a notch. In slightly worn teeth, a triangular accessory tubercle sits between the paracone and protocone.

A triangular scaly structure exists at the mesio-lingual corner of the protocone of specimen PA 20, which is probably the second upper molar. This structure is recognized as Carabelli's cusp. J. Woo considered this cusp an extension of the cingulum of the protocone. This is the only case seen among the available specimens. The root has a lingual branch and two buccal branches that, in cross section, are ovoid with the bucco-lingual diameter larger than the mesio-distal diameter. The mesio-distal diameter of the lingual branch is larger than its bucco-lingual diameter.

The neck of the tooth constricts because of the presence of a prominent cingulum. Enamel extends more on the lingual surface than on the other surfaces. The size of the pulp cavity varies.

Upper third molar. The general pattern of this molar is similar to the first and second molars. However, it is generally smaller, especially in its length. The metacone is the most reduced of the four cusps. In 49% of the sample, the protocone is the largest cusp, followed in size by the paracone and hypocone; the metacone is the smallest. In a few specimens the paracone is not reduced, but the metacone and hypocone are much reduced. In specimen PA 10 the distal part of the crown appears as a distal marginal ridge behind the paracone and protocone. The wrinkles on the slope of the cusps are basically similar to the first and second molars, but they vary according to the different degree of reduction of the various cusps. The transverse ridge is represented by a distal marginal ridge. An indentation of the marginal ridge at the mesio-lingual corner of the protocone separates it from the lingual surface in most specimens.

The root divides into three or four branches. In the latter case, the anterior buccal branch divides into two branches. The neck is slightly constricted. The enamel extends more on the lingual surface than on the other surfaces.

Lower first and second molar. This tooth is more or less quadrangular. The anterior and posterior surfaces are flatter, and the buccal and lingual surfaces are rounded and bulging. The slightly prominent cingulum surrounds the basal part of the crown, and is most obvious on the buccal surface. Vertical furrows on the buccal and lingual surfaces separate the protoconid, hypoconid, metaconid, and the entoconid. Where there is a sixth cusp, a short furrow separates the metaconid and the entoconid. In addition to these cusps, there is a mesoconid at the buccal or central part of the distal portion of the crown. The metaconid is the largest and highest cusp in 46% of the sample. In decreasing order of size are the protoconid, hypoconid, entoconid, and the mesoconid. In 17% of the specimens the metaconid is equal to or smaller than the protoconid. In 17% of the specimens, the metaconid is equal to or smaller than the hypoconid in size. In two specimens the hypoconid equals or is smaller than the entoconid. Because of the large metaconid, the mesial half of the longitudinal furrow has shifted more buccally beyond the midline of the occlusal surface. It shows a typical *Dryopithecus* cusp pattern. The mesial cusps are slightly higher than the distal cusps. With few exceptions, the trigonid is broader than the talonid.

The "tubercle six," including M1 and M2, is present in 26% of the sample. In 34% of the cases there is an accessory internal tubercle between the metaconid and the entoconid. They are less likely to exist on the first than on the second molar. The anterior fovea is narrow and shallow, and the posterior fovea is even thinner and smaller.

On each main cusp there are generally three rounded ridges, extending from the apex to the furrow separating the cusps. The middle ridge is larger than the marginal ridges on either side. Accessory ridges radiate from the main ridge. Both the mesial margin of the occlusal surface and the lingual half of the distal margin are crenulated in unworn specimens. Enamel extends downward more on the buccal than on the other surfaces. The root divides into mesial and distal branches. There is no obvious narrowing at the neck of the tooth.

Lower third molar. In general shape this tooth is not obviously different from the lower first and second molars. Crown length exceeds its breadth. The lingual surface is steeper than the buccal surface because of the presence of the prominent remnant of the cingulum on the basal part of the buccal surface. On the buccal and lingual surfaces, vertical grooves separate the respective cusps. The groove on the buccal surface is narrow and deep, while that on the lingual surface is wide, shallow, and shorter. The former extends upward to two thirds of the height of the buccal surface, the latter extends to only half the height of the lingual surface.

The metaconid is the largest of the five main cusps. In decreasing size it is followed by the protoconid, hypoconid, entoconid, and the mesoconid is the smallest. In 28% of the sample a small tubercle six exists between the entoconid and the mesoconid. In 48 specimens, or 44% of the sample, there is a very small accessory internal tubercle between the metaconid and the entoconid. The ridge pattern on the cusps is similar to that on the first and second molars.

The root divides into mesial and distal branches. The bucco-lingual diameter of the mesial branch is much larger than its mesio-distal diameter. The bucco-lingual diameter of the distal branch is smaller than that of the mesial branch. The mesial branch is more perpendicular, and the distal branch inclines slightly backward and laterally. The neck is slightly constricted. The enamel extends downward much more on the buccal surface.

The decreasing size sequence of the molars in Mandible III is M2, M1, and M3. In the other two mandibles M2 is also larger than M1. The length-breadth index is 89.9 for M1, 90.6 for M2, and 86.1 for M3. The trigonid index is 88.3 for M1, 90.1 for M2, and 84.4 for M3. The talonid index is 87.3 for M1 and M2, and 80.2 for M3.

In both the upper and lower molars, the pulp cavity is confined to the basal part of the crown. It extends to the undivided part of the root or even to the branches of the root. *Gigantopithecus* molars show marked taurodontism.

Deciduous teeth

Upper second deciduous molar. Differences that normally exist between the permanent and deciduous molars in other primates also exist in *Gigantopithecus.* The crown of the upper deciduous molar of *Gigantopithecus* is more or less square. The apex of the deciduous cusps is closer to the margin of the occlusal surface than in permanent molars. The bucco-lingual diameter of the occlusal surface is much smaller than that in the basal part of the tooth's crown. The pattern of the main ridge on the occlusal surface is similar to that in permanent molars, but thin accessory ridges are more abundant.

Lower second decidous molar. Five main cusps occupy the occlusal deciduous surface. In two large specimens and one small specimen, there is a distinct internal accessory tubercle or its trace. There is no tubercle six. In a small specimen the marginal ridge is cut by several small grooves to form a paraconidlike structure. The metaconid is the largest cusp, followed by the protoconid or hypoconid. The smallest of the five main cusps is the mesoconid or entoconid. The talonid is broader than the trigonid. Wrinkles on the occlusal surface are more complex than on the permanent molars.

A large part of the root is preserved only in one large specimen. The root divides into anterior and posterior branches. A furrow on the posterior surface of the anterior root and the anterior surface of the posterior root divides it into two parts connected by a thin bony plate. The posterior root is the largest. The buccal part of both roots is larger than the lingual part. Judging from the dentition of Mandible II, which belongs to a young individual, the canine probably erupted before the third molar.

Zhang (1983) divided the *Gigantopithecus* teeth from various sites into two groups: an early group from Liucheng (dated to the Early Pleistocene) and a late group, including specimens from Daxin, Wuming, Bama, and Gaopin (dating from the late period of the Early Pleistocene to the Middle Pleistocene) and teeth from drug stores in Guangxi, Guangdong, and Hubei. The length and breadth of the cheek teeth of later specimens are generally greater than those of earlier specimens from the same cave (Table 6.6). The relative length of the late group is shorter than in the early group. There are no lower incisors from the late group. The lower canine shows no significant difference in its dimensions between the early and late group. Dental differences between the late and early groups appear in the posterior dentition. In its evolutionary history, the anterior teeth of *Gigantopithecus* remained constant in size, while the posterior teeth become larger.

Zhang found that the frequencies of chipping are 3.0% and 3.2% in the early and late groups, respectively. Given that the environment in south China remained stable in the Pleistocene, the diet of *Gigantopithecus* may not have varied over time (Zhang, 1983). Ciochon (1991) concluded that the diet was mostly of seeds, fruits, and bamboo.

Xuand Han (1992) compared *Gigantopithecus* specimens from Gaopin and Liucheng. The cheek teeth from Gaopin are slightly longer, are obviously broader, and have an obviously larger length-breadth index than those from Liucheng. T-tests show that while the difference in length is insignificant, differences in breadth and the length-breadth index are significant.

Geology

Daxin. *Gigantopithecus* teeth were unearthed from a layer of red-purple clay intercalated with sands below the cemented yellow breccia in Heidong (Black cave) of Niushuishan Hill, Daxin County. Many mammalian fossils, including *Pongo, Ailuropoda, Stegodon,* and *Megatapirus,* were found in the latter layer. *Gigantopithecus* from this cave is dated to the Middle or Early Pleistocene.

Liucheng. *Gigantopithecus* mandibles and nearly 1000 teeth were found in Xiaohuidong cave (IVPP field no. 5704) near Xinshechong Village. This cave, located high on an isolated limestone hill named Lengzhai Hill, is about 90 m above the ground. The cave has at least three upper and lower entrances. The excavated area covers about 180 m^2. Deposits in different parts of the cave exhibit some variation. Pei and Li (1958) described the section that included the three *Gigantopithecus* mandibles. From the bottom layer up there are four layers:

1. Small blocks of red cave loam and violet sands that are slightly cemented; bone is rare; 30 cm
2. Yellow breccia, yellow sand and clay, slightly cemented; rich in fossils; 100 cm thick; Mandibles I and II are from this layer
3. Yellow breccia, mingling into the second layer, less consolidated; few fossils; 150 cm thick; Mandible III is from this layer
4. Hard layer of stagmitic crust, 75 cm

Fauna unearthed from this cave include (Han and Xu, 1989):

Primates
- *Gigantopithecus blacki*
- *Pongo* sp. ?

Rodentia
- *Hystrix magna*
- *Hystrix subcristata*
- *Atherurus* sp.

Carnivora
- *Cuon dubius*
- *Ailuropoda microta*
- *Ursus* aff. *thibetanus*
- *Arctonyx minor*

Arctonyx collaris
Viverra sp.
Paguma larvata
Hyaena brevirostris licenti
Felis teilhardi
Felis sp.
Panthera pardus
Panthera sp.
Cynailurus pleistocaenicus

Proboscidea
Gomphotherium serridentoides
Stegodon preorientalis

Perissodactyla
Equus yunnanensis
Nestoritherium praesinensis
Tapirus peii
Rhinoceros chaii

Artiodactyla
Dicoryphochoerus ultima
Potamochoerus nodosarius
Sus xiaozhu
Sus liuchengensis
Sus australis
Sus peii
Suidae gen. et sp. indet.
Dorcabune liuchengensis
Muntiacus lacustris
Cervocerus fenqii
Rusa yunnanensis
Megalovis guangxiensis
Caprinae gen. et sp. indet.
Bibos sp.

Most fossils from this site are isolated teeth. Many had lost the root and show gnawing by rodents. Some fossils were embedded in loose deposits, perhaps resulting from the roots of plants that made the deposit looser and eroded the fossils. Youheng Li (1960), who excavated this cave, suggested that several agencies may have carried the animal bones into the cave: (1) Cave animals that lived and died in the cave. (2) The prey of carnivores that were carried into the cave. (3) Geological forces. Because the composition of the cave deposits is similar to those outside the cave, big animals such as elephants and rhinoceros could not enter the cave before dismemberment. Their remains might have been introduced into the cave by geological forces such as water flow. (4) Perhaps the remains are the prey of

Gigantopithecus. Li considered many of the bones and teeth of juvenile animals in the assemblage to be the prey of *Gigantopithecus*. This site is of an Early Pleistocene date.

Wuming. The cave is located about 30 m above ground level. It was disturbed prior to scientific investigation. The fossils are embedded in the cemented hard yellow sandy clay. The associated mammalian assemblage includes (Han and Xu, 1989):

Primates
- *Gigantopithecus blacki*
- *Macaca* sp.

Rodentia
- *Rhizomys* sp.
- *Rattus* cf. *edwardsi*
- *Hystrix* cf. *subcristata*

Carnivora
- *Ailuropoda melanoleuca fovealis*
- *Ursus* sp.
- *Lutra* sp.
- *Hyaena* sp.

Proboscidea
- *Stegodon orientalis*

Perissodactyla
- *Rhinoceros sinensis*

Artiodactyla
- *Sus* sp.
- *Capricornis* cf. *sumatraensis*
- Bovinae gen. et sp. indet.

Zhang et al. (1973) considered that this site is most probably of an Middle Pleistocene date.

Bama. The cave (IVPP field no. 73133) is 600 m above sea level and 80 m above the valley floor. The cave was disturbed prior to scientific investigation. *Gigantopithecus*, and other mammalian fossils are embedded in a yellow sandy clay that is about 2 m thick. Most fossils are isolated teeth. They represent (according to Han and Xu, 1989):

Primates
- *Gigantopithecus blacki*
- *Pongo* sp.
- *Hylobates* sp.
- *Macaca* sp.

Rodentia
- *Hystrix subcristata*
- *Atherurus* sp.

Carnivora
- *Cuon javanicus*
- *Ailuropoda melanoleuca baconi*
- *Ursus* sp.
- *Felis* sp.

Proboscidea
- *Stegodon* sp.

Perissodactyla
- *Tapirus* sp.
- *Rhinoceros sinensis*

Artiodactyla
- *Sus scrofa* ?
- Cervidae gen. et sp. indet.
- Ovinae gen. et sp. indet.
- Bovinae gen. et sp. indet.

Zhang et al. (1987a, 1988) concluded that the Middle Pleistocene geological age of the Bama *Gigantopithecus* is later than specimens from Liucheng, but more or less similar to material from Wuming.

Gaopin. The *Gigantopithecus* teeth were unearthed from Longgudong (Dragon Bone Cave, IVPP field no. 70016), some 1 km southwest of Gaopin town. The cave's entrance is about 740 m above sea level and about 85 m above the water surface of the nearby Longdong (Dragon Cave) River. Longgudong Cave is a tunnel-like structure with many branches and two entrances at both ends.

Xu and Lui (1979) excavated this cave in 1970 and studied the *Gigantopithecus* and other materials. *Gigantopithecus* was found with *Homo erectus* in this cave. The geological and faunal data were presented in Chapter 2 in the discussion of the Jianshi *H. erectus* remains.

Wushan. *Gigantopithecus* and hominid teeth were found at Wushan, Sichuan Province. The geological information was presented in Chapter 2 on *Homo erectus*.

SUMMARY

Extensive lignite deposits in Yunnan Province in southern China yielded an array of remains referred to as *Sivapithecus, Ramapithecus,* and now most often as *Lufengpithecus*. Some of the taxonomic disagreement probably stems from the fact that both sexes are being sampled. *Dryopithecus kaiyuanensis* was first found in 1956 in Yunnan. Another ape from Yunnan, sometimes referred to as *Ramapithecus hudiensis,* may not differ from *Lufengpithecus*.

Gigantopithecus from China is related to its Indian predecessor. Who the ancestor is and whether there are descendants of *Gigantopithecus* are two of the intriguing problems yet to be answered. Despite sporadic attempts to relate *Gigantopithecus* to some hominid line, beginning with von Koenigswald's early attempt to tie *Gigantopithecus* to *Homo erectus,* there is little evidence to support a *Gigantopithecus* to hominid relationship.

APPENDIX I

Some Commonly Mentioned Fauna

Insectivora	**Insect-Eating Mammals**
Anourosorex	Short-tailed Shrew
Blarinella	Asiatic short-tailed shrew
Chodsigoa	East Asian shrew
Crocidura	White-toothed shrew
Erinaceus	Hedgehog
Heterosoricinae	Shrew
Neomys	Old World water shrew
Scalopini	Mole, Shrew
Scaptochirus	Mole
Sorex	Long-tailed shrew
Soricinae	Shrew
Talpidae	Mole, Shrew
Tupaiidae	Tree shrew
Chiroptera	**Bats**
Hipposideros	Old World leaf-nosed bat
La io	Great evening bat
Miniopterus	Long-winged bat
Murina	Bat
Myotis	Little brown bat
Pipistrellus	Pipistrelle
Plecotus	Lump-nosed bat, long-eared bat, lappet-eared bat
Pteropidae	Bat
Rhinolophus	Horse-shoe bat
Primates	**Primates**
Gigantopithecus	Giant ape
Hylobates	Gibbon
Macaca	Macaque
Pongo	Orang-utan
Rhinopithecus	Golden monkey
Lagomorpha	**Hares**
Laporidae	Hare
Lepus	Hare

Ochotona	Pikas, Mouse, Hares or Conies
Ochotonoides	Pikas, Mouse, Hares or Conies
Rodentia	**Rodents**
Allactaga	Jerboa
Allocricetulus	Rat
Allophaimys	Vole
Alticola	High mountain vole
Apodemys	Field mouse
Arvicola	Water mole
Atherurus	Bush-tailed porcupine
Bahomys	Rat
Bandicata	Bandicoot rat
Belomys	Hairy-footed flying squirrel
Callosciurus	Tricolored squirrel
Castoridae	Beaver
Chuanocricetulus	Hamster
Citellus	Ground squirrel, Suslik
Clethrionomys	Red-backed mouse
Cricetinus	Rat-like hamster
Cricetulus	Rat-like hamster
Dipus	Rough-legged or Northern three-toed jerboa
Dremomys	Red-cheeked squirrel
Eothenomys	Pratt's vole
Eutamias	Siberian and western American chipmunk
Gerbillus	Northern pygmy gerbil
Hapalomys	Asiatic climbing rat, Mormoset rat
Hystrix	Porcupine
Marmota	Marmot
Meriones	Jird
Microtus	Vole, Meadow mouse
Mimomys	Vole
Muridae	Rat, Mouse, Hamster, Vole, Gerbil
Mus	Mouse
Myospalax	Mole-rat, Zokor
Niviventer	White-bellied rat
Phaiomys	Vole
Pheromys	Flying squirrel
Pitaurista	Flying squirrel
Pitymus	Vole
Pteromys	Old World flying squirrel
Rattus	Rat
Rhizomys	Bamboo rat

Sciuridae	Squirrel, Chipmunk, Marmot, Prairie dog
Sciurotamias	Rock squirrel
Sciurotamias	Squirrel
Sciurus	Squirrel
Scurotamias	Squirrel
Sinocaster	Chinese beaver
Tamias	Eastern American chipmunk
Tamiops	Asiatic striped squirrel
Trogontherium	Large beaver
Wushanomys	Rat
Carnivora	**Carnivores**
Acinonyx	Cheetah
Ailuropoda	Giant panda
Arctonyx	Hog badger
Canidae	Dog, Wolf, Coyote, Jackal, Fox
Canis	Dog, Wolf, Coyote, Jackal
Crocuta	Spotted hyena
Cuon	Jackal, Dhole
Cynailurus	Cheetah
Felidae	Cat
Felis	Small cat, Lynx, Cougar
Gulo	Wolverine
Homotherium	Saber-toothed cat
Lutra	River otter
Lynx	Lynx
Machairodontinae	Saber-toothed cat
Martes	Marten, Fisher, Sable
Megantereon	Saber-toothed cat
Megaviverra	Large civet
Meles	Old World badger
Mustela	Weasel, Ermin, Stoat, Mink, Ferret, Polecat
Mustelidae	Weasel, Badger, Skunk, Otter
Neofelis	Clouded leopard
Nyctereutes	Raccoon dog
Pachycrocuta	Spotted hyena
Paguma	Masked-palm civet
Panthera pardus	Leopard
Panthera tigris	Tiger
Sclenarctos	Black bear
Ursus	Bear
Viverra	Civet
Viverricula	Lesser oriental civet
Vulpes	Fox

Proboscidea	**Elephants**
Elephantidae	Elephant
Elephas	Elephant
Gomphotherium	Mastodon
Mammuthus	Mammoth
Palaeoloxodon	Elephant
Sinomastodon	Mastodon
Stegodon	Elephant
Perissodactyla	**Odd Toed Ungulate**
Coelodonta	Wooly rhinoceros
Dicerorhinus	Two-horned rhinoceros, Sumatran rhinoceros, Hairy rhinoceros
Equus	Horse
Equus hemionus	Donkey
Megatapirus	Giant tapir
Nestoritherium	Odd-toed ungulate with clawed digits
Rhinoceros	Rhinoceros
Tapirus	Tapir
Artiodactyla	**Even Toed Ungulate**
Alces	Moose, Elk
Axis	Axis deer
Bison	Bison
Bos	Ox
Bovidae	Antelope, Cattle, Bison, Buffalo, Goat, Sheep
Bubalus	Asian water buffalo, Anoas
Camelus	Camel
Capreolus	Roe deer
Capricornis	Serow
Caprinae	Goat
Cervavitus	Deer
Cervus	Red deer, Wapiti
Dicoryphoerus	Pig
Elaphodus	Tufted deer
Elaphurus	Père David's deer
Gazella	Gazelle
Hydropotes	Chinese water deer
Leptobos	Cattle
Megaloceros pachyosteus	Thick jaw bone deer
Metacerculus	Muntjac
Moschus	Musk deer
Muntiacus	Muntjac
Naemorhaedus	Goral
Ovibovinae	Musk ox
Ovis	Sheep

Paracamelus	Camel
Pseudaxis	Deer
Rusa	Sambar
Spirocerus	Spiral horn antelope
Sus	Pig

APPENDIX II

Mammals of Locality 1 at Zhoukoudian Grouped According to Climate

Layers	1–3	4	5	6	7	8–9	10	11	12	13
Animals living near water bodies										
Castor sp.				+?						
Trogontherium cuvieri			+	+	+	+				
Lutra melina					+					
Animals living in desert and grassland										
Struhio anderssoni	+	+				+	+			
Paracamelus gigas								+?		
Animals living in cool or cold environments										
Coelondonta sp.						+				
Megaloceros pachyosteus	+	+	+	+	+	+	+			
Megaloceros flabellatus							+	+		+
Hyaena brevirostris sinensis			+	+	+	+	+	+		
Felis teilhardi		+				+	+	+		
Ursus cf. *spelaeus*						+	+	+		
Marmota complicidens					+		+			
Marmota bobak				+		+?		+		
Gulo sp.						+				
Crocuta crocuta ultima	+									
Animals living in warm or hot environments										
Ursus thibetanus kokeni	+	+	+	+		+				
Hystrix cf. *subcristata*						+				
Bubalus teilhardi					+	+				
Palaeoloxodon cf. *hamadicus*		+			+					
Acinonyx sp.			+							
Dicerorhinus mercki	+	+		+	+	+	+	+		
Macaca robusta	+	+	+	+	+	+	+			
Rhizomys sp.	+									

Source: From Jia (1978).

References

Aigner, J. (1981). *Archaeological remains in Pleistocene China.* Verlag C.H. Beck, Munich.

An, Z. (1972). Discussion on the dates of some primitive cultures in China. *Kaogu Xuebao,* 1:111–124 (in Chinese).

An, Z., Gao, W., Zhu, Y., Kan, X., Wang, J., Sun, J., and Wei, M. (1990). Magnetostratigraphic dates of Lantian *Homo erectus. Acta Anthropologica Sinica,* 9:1–7 (in Chinese with English abstract).

Andrews, P. (1984a). An alternative interpretation of the characters used to define *Homo erectus.* In *The Early Evolution of Man.* P. Andrews and J. Franzen, eds. Senckenberg Museum, Frankfurt.

Andrews, P. (1984b). On the characters that define *Homo erectus. Courier Forschungsinstitut Senckenberg,* 69:167–178.

Binford, L. and Ho, C. (1985). Taphonomy at a distance: Zhoukoudian, the cave home of Beijing man? *Current Anthropology,* 26:413–442.

Binford, L. and Stone, N. (1986). Zhoukoudian: A closer look. *Current Anthropology,* 27:453–475.

Black, D. (1927). The lower molar hominid tooth from the Chou Kou Tien deposit. *Palaeontologia Sinica Series D,* Vol. 7, Fascicle 1.

Black, D., Teilhard de Chardin, P., Young, C. C., and Pei, W. C. (1933). *Fossil Man in China. Memoirs of the Geological Survey of China.* Series A, no. 11.

Bräuer, G. (1984). The Afro-European sapiens hypothesis and hominid evolution in Asia during the Middle and Upper Pleistocene. In *The Early Evolution of Man.* P. Andrews and J. Franzen, eds. Senckenberg Museum, Frankfurt.

Brothwell, D. (1960). Upper Pleistocene human skull from Niah Cave, Sarawak. *Sarawak Museum Journal Kuching, Borneo,* 9:323–349.

Cao, P. (1984). A paleoanthropological and paleolithic site found in Guizhou. *Acta Anthropologica Sincia,* 3:182–183 (in Chinese).

Cao, Z. (1978). Xiaohui Cave, Shuicheng, Guizhou, A Paleolithic site. *Vertebrata PalAsiatica,* 16:67–72 (in Chinese).

Cao, Z. (1982). The preliminary study of bone tools and antler spades from the rock shelter site of Maomaodong. *Acta Anthropologica Sinica,* 1:36–41 (in Chinese with English abstract).

Chang, K. C. (1977). Chinese paleoanthropology. *Annual Review of Anthropology,* 6:137–159.

Chao, Z. (Cao, Zetian). (1982). On the Palaeolithic artefacts from Maomaodong (The rock shelter site), Guizhou Province. *Vertebrata PalAsiatica,* 20:155–164 (in Chinese).

Chen, C., Lin, S., Yu, Y., and Gao, S. (1991). Spores and Pollens. In *The Late Pleistocene Environment of North China.* Chongqing Publishing House, Chongqing, pp. 53–78 (in Chinese with English abstract, pp. 183–185).

Chen, D. and Qi, G. (1978). Human fossils and associated mammalian fauna from Xichou, Yunnan. *Vertebrata PalAsiatica,* 16:33–46 (in Chinese).

Chen, T., Yuan, S., and Gao, S. (1984). The study of uranium series dating of fossil bones and an absolute age sequence for the main Paleolithic sites of north China. *Acta Anthropologica Sinica,* 3:259–268 (in Chinese with English abstract).

Chen, T., Yuan, S., Guo, S., and Hu, Y. (1987). Uranium series dating of fossil bones from the Hexian and Chaoxian human fossil sites. *Acta Anthropologica Sinica,* 6:249–254 (in Chinese with English abstract).

Chen, T., Hedges, R. E. M., and Yuan, Z. (1992). The second batch of accelerator radiocarbon dates for the Upper Cave site of Zhoukoudian. *Acta Anthropologica Sinica,* 11:112–116 (in Chinese with English abstract).

Chen, T., Yang, Q., and Wu, E. (1994). Antiquity of *Homo sapiens* in China. *Nature* 368:55–56

Chen, Y. and Du, N. (1991). Palynological analysis. In *Wushan Hominid Site.* Haiyang Press, Beijing, pp. 150–155 (in Chinese with English abstract).

Chia (Jia), L. and Wei, Q. (1976). A Paleolithic site at Hsu-chia-yao in Yangkao County, Shanxi Province. *Kaogu Xuebao,* (2):97–115 (in Chinese with English abstract).

Chia (Jia), L. and Woo, J. (Wu, Rukang), (1959). Fossils of a human skull base of the late Paleolithic stage from Chilinshan, Leipin district, Kwangsi, China. *Vertebrata PalAsiatica,* 3:37–40.

Ciochon, R. (1984/1985). Paleontological field research in Vietnam. *Newsletter, Institute of Human Origins,* 4:7–8.

Ciochon, R., (1991). The ape that was. *Natural History,* November, 54–62.

Conroy, G. (1990). *Primate Evolution.* W.W. Norton and Company, New York.

Coon, C. (1962). T*he Origin of Races.* Alfred A. Knopf, New York.

Daegling, D. (1993). Functional morphology of the human chin. *Evolutionary Biology,* 1:170–176.

Daegling, D. and Grine, F. (1987). Tooth wear, gnathodental scaling and diet of *Gigantopithecus blacki. American Journal of Physical Anthropology,* 72:191–192.

Dai, E. (1966). The paleoliths found at Lantian Man locality of Gongwangling and its vicinity. *Vertebrata PalAsiatica,* 10:30–34 (in Chinese with English abstract).

Dong, G., Gao, S., and Li, B. (1981). New discovery of fossils of Ordos Man. *Kexue Tongbao,* 26:1192–1194. (in Chinese).

Eckhardt, R. B. (1973). *Gigantopithecus* as a hominoid ancestor. *Anthropol. Anz.,* 34:1–8.

Etler, D. (1984). The fossil hominoids of Lufeng, Yunnan Province, The People's Republic of China: A series of translations. *Yearbook of Physical Anthropology,* 27:1–56.

Etler, D. and Li, T. (1994). New archaic human fossil discoveries in China and their bearing on hominid species definition during the Middle Pleistocene. In *Integrative Paths to the Past.* R. Corrucini and R. Ciochan, eds. Prentice Hall, Englewood Cliffs, N.J., pp. 639–676.

Freeman, L. (1977). Paleolithic archaeology and paleoanthropology in China. In *Paleoanthropolgy in the People's Republic of China.* W. Howells and P. Tsuchitani, eds. National Academy of Sciences, Washington, D.C., pp. 79–113.

Gao, J. (1975). Australopithecine teeth associated with *Gigantopithecus. Vertebrata PalAsiatica,* 13:81–88 (in Chinese with English abstract).

Gerasimov, M. M. (1949). *Basis of the Reconstruction of the Face on the Basis of the Skull.* Soviet Academy of Science, Moscow (in Russian).

Greenwell, R. and Poirier, F. (1989). Further investigation into the reported Yeren. *Cryptozoology,* 8:47–57.

Gu, Y. (1978). New Cave Man of Zhoukoudian and his living environment. In *Collected Papers of Paleoanthropology,* Institute of Vertebrate Paleontology and Paleoanthropology, ed. Science Press, Beijing, pp. 158–174 (in Chinese).

Gu, Y. and Fang, Q. (1991a). Human fossils. In *Wushan Hominid Site.* Haiyang (Ocean) Press, Beijing, pp. 16–19 (in Chinese with English abstract).

Gu, Y. and Fang, Q. (1991b). Primate fossils. In *Wushan Hominid Site.* Haiyang (Ocean) Press, Beijing, pp. 86–92 (in Chinese).

Gu, Y. and Jablonski, N. (1989). A reassessment of *Megamacaca lantianensis* of Gongwangling, Shaanxi Province. *Acta Anthropologica Sinica,* 8:343–346 (in Chinese with English abstract).

Guo, S., Liu, S., Sun, S., Zhang, F., Zhou, S., Hao, X., Hu, R., Meng, W., Zhang, P., and Liu, J. (1991). Fission track dating of the 4th layer of the Peking Man site. *Acta Anthropologica Sinica,* 10:73–77 (in Chinese with English abstract).

Han, D. (1986). Fossils of Tragulidae from Lufeng, Yunnan. *Acta Anthropologica Sinica,* 5:68–78 (in Chinese with English abstract).

Han, D. and Xu, C. (1989). Quaternary mammalian faunas in south China. In *Early Humankind in China.* Wu, R., Wu, X., and S. Zhang, eds. Science Press, Beijing, pp.338–391 (in Chinese).

Han, D, and Zhang, S. (1978). A human fossil canine found in Jiande and new data of Quaternary mammals in Zhejiang. *Vertebrata PalAsiatica,* 16:255–263 (in Chinese with English abstract).

Hou, L. (1985). Fossil Birds from Zhoukoudian Loc. 1. In *Multi-disciplinary Study of the Peking Man Site at Zhoukoudian.* Institute of Vertebrate Paleontology and Paleoanthropology, ed. Science Press, Beijing, pp. 114–118 (in Chinese).

Howells, W. and Tsuchitani, P., eds. (1977). *Paleoanthropology in the People's Republic of China.* National Academy of Sciences, Washington D.C.

Hsu (Xu) C., Han, K., and Wang, L. (1974). Discovery of *Gigantopithecus* teeth and associated fauna in western Hupei. *Vertebrata PalAsiatica,* 12:293–309 (in Chinese).

Hsu, J. (Xu Ren) (1966). The climatic condition in North China during the time of *Sinanthropus. Scientia Sinica,* 15:410–414 (in Chinese).

Hu, Changkang. (1985). The history of mammalian fauna of locality 1 of Zhoukoudian and its recent advances. In *Multi-disciplinary Study of the Peking Man Site at Zhoukoudian.* Institute of Vertebrate Paleontology and Paleoanthropology, ed. Science Press, Beijing, pp. 107–113 (in Chinese).

Hu, Chengchih. (1973). Ape-man teeth from Yuanmou, Yunnan. *Acta Geologica Sinica,* (1):65–71 (in Chinese with English abstract).

Huang, B. and Cai, X. (1991). ESR dating of *Lamprotula* from Dingcun profile. In *The Late Pleistocene Environment of North China.* Chongqing Publishing House, Chongqing, pp. 36–46 (in Chinese with English abstract, pp. 179–180).

Huang, P., Jin, S., Liang, R., Lu, Z., Zheng, L., Yuan, Z., Fang, C., and Cai, B. (1991). Study of ESR dating for burying age of the first skull of Peking Man and chronological scale of the cave deposit in Zhoukoudian site Locality 1. *Acta Anthropologica Sinica,* 10:107–115 (in Chinese with English abstract).

Huang, W. and You, Y. (1987). On the problems of the karst cave and the deposits of the site of Jinniushan Man. *Carsologica Sinica,* 6:61–67 (in Chinese with English abstract).

Huang, W. and Zheng, S. (1982). An Upper Pleistocene human tooth and mammalian fossils from Zhangwu, Shaanxi. *Acta Anthropologica Sinica,* 1:14–17 (in Chinese with English abstract).

Huang, W., Fang, D., and Ye, Y. (1982). Preliminary study of the fossil hominid skull and fauna from Hexian. *Vertebrata PalAsiatica,* 20:248–256 (in Chinese with English abstract).

Huang, W., Ji, H., and Yang, D. (1991). Cave and stratigraphy. In *Wushan Hominid Site.* Haiyang Press, Beijing, pp. 10–15 (in Chinese with English abstract).

Huang, X. (1989). A calva of *Homo sapiens* from Chuandong, Puding, Guizhou. *Acta Anthropologica Sinica,* 8:379–380 (in Chinese).

Institute of Vertebrate Paleontology and Paleoanthropology, eds. (1980). *Atlas of Primitive Man in China.* Science Press, Beijing.

Jia, L. (1951). *Upper Cave Man.* Longmenlianhe Shudian, Shanghai (in Chinese).

Jia, L. (1975). *The Cave Home of Peking Man.* Foreign Languages Press, Peking.

Jia, L. (1978). On the climate of the Zhoukoudian region during the time of Peking Man. *Acta Stratigraphica Sinica,* 2:53–56 (in Chinese).

Jia, L. (1989). On problems of the Beijing Man site: A critique of new interpretations. *Current Anthropology,* 30:200–204.

Jia, L. and Huang, W. (1990). *The Story of Peking Man.* Oxford University Press, New York.

Jia, L., Gai, P., and You, Y. (1972). Excavation report of the Paleolithic site at Shiyu, Shanxi. *Kaogu Xuebao* (The Chinese Journal of Archaeology) (1):39–58 (in Chinese).

Jia, L., Wei, Q., and Li, C. (1979). Report on the excavation of the Hsuchiayao Man site in 1976. *Vertebrata PalAsiatica,* 17:277–293 (in Chinese with English abstract).

Jiang, C., Xiao, L., and Li, J. (1993). Hominoid teeth from Leilao in Yuanmou, Yunnan. *Acta Anthropologica Sinica,* 12:97–102 (in Chinese with English abstract).

Jiang, P. (1982). A fossil human tooth from Jilin Province. *Vertebrata PalAsiatica,* 20:64–71 (in Chinese with English abstract).

Kahlke, H. D. and Hu, C. (1957). On the distribution of *Megaceros* in China. *Vertebrata PalAsiatica,* 1:273–283 (in Chinese with English abstract).

Kong, Z. (1985). Vegetational and climatic changes since the Paleogene at Zhoukoudian and its adjacent regions. In *Multi-disciplinary Study of the Peking Man Site at Zhoukoudian.* Institute of Vertebrate Paleontology and Paleoanthropology, ed. Science Press, Beijing, pp. 119–154 (in Chinese).

Laboratory of the Institute of Archaeology of the Chinese Academy of Social Sciences. (1977). Report of radiocarbon dating (4). *Kaogu Xuebao,* 3:200–204 (in Chinese).

Leakey, L. (1966). *Homo hablis, Homo erectus,* and the australopithecines. *Nature,* 209:1279–1281.

Li, C. (1978). A Miocene gibbon-like primate from Shihung, Kiangsu Province. *Vertebrata PalAsiatica,* 16:187–192 (in Chinese with English summary).

Li, T. and Etler, D. (1992). New Middle Pleistocene hominid crania from Yunxian in China. *Nature,* 357:404–407.

Li, T., Wang, Z., Li, W., Fong, X., Hu, K., and Liu, W. (1991). Geological survey and excavation of the fossil site at Quyuan River Mouth, Yunxian, Hubei Province. *Jianghan Kaogu,* 1:1–14 (in Chinese).

Li, W., Zhang, Z., Gu, Y., Lin, Y., and Yan, F. (1982). A fauna from Lianhua Cave, Dantu, Jiangsu. *Acta Anthropologica Sinica,* 1:169–179 (in Chinese with English abstact).

Li, X. and Zhang, S. (1984). Paleoliths discovered in Ziyang Man Locality B. *Acta Anthropologica Sinica,* 3:215–224 (in Chinese with English abstract).

Li, X., Liu, G., Xu, G., Wang, F., Qiu, S., and Cai, L. (1985). Radiocarbon dating of fossil mammal bones from the Upper Cave and New Cave of Zhoukoudian. In *Multidisciplinary Study of the Peking Man Site at Zhoukoudian.* Institute of Vertebrate Paleontology and Paleoanthropology, Academia Sinica, ed. Science Press, Beijing, pp. 261–262 (in Chinese).

Li, Y. (1991) Paleolithic artifacts. In *Wushan Hominid Site.* Haiyang (Ocean) Press, Beijing, pp. 20–23 (in Chinese with English abstract).

Li, Y., and Ji, H. (1981). Environmental change in Peking Man's time. *Vertebrata PalAsiatica,* 19:337–347 (in Chinese with English summary).

Li, Y. H. (1960). The preservation of fossils in *Gigantopithecus* Cave of Liucheng, Guangxi. *Paleovertebrata et Paleoanthropologia,* 2:161–166 (in Chinese).

Li, Y. H. (1963). New material from Hetao. *Vertebrata PalAsiatica,* 7:376–377 (in Chinese).

Li, Y. H., Wu, M., Peng, S., and Zhou, S. (1984). Human tooth fossils and some mammalian remains from Tubo, Liujiang, Guangxi. *Acta Anthropologica Sinica,* 3:322–329 (in Chinese with English abstract).

Li, Y. H., Wu, M., Peng, S., and Zhou, S. (1985). Preliminary report on the investigation of Dingmo Cave in Tiandong County, Guangxi. *Acta Anthropologica Sinica,* 4:127–131 (in Chinese with English abstract).

Lian, C. (1981). On the occurence of fossil *Homo sapiens* in Taiwan. *Bulletin of the Department of Archaeology and Anthropology,* 42:53–70 (in Chinese with English abstract).

Liang, Z., Zhang, Z., and Zhang, H. (1991). Uranium series dates. In *The Late Pleistocene Environment of North China.* Chongqing Publishing House, Chongqing, pp. 46–52 (in Chinese with English abstract, pp. 181–182).

Licent, E., Teilhard de Chardin, P., and Black, D. (1926). On a presumably Pleistocene human tooth from the Sjara-osso-gol (Southeastern Ordos) deposits. *Bulletin of the Geological Society of China,* 5:285–290.

Lin, Q. and Zou, D. (1982). More *Gigantopithecus* teeth discovered from Jianshi, Hubei. *Acta Anthropologica Sinica,* 1:206 (in Chinese).

Lin, S., (1985). Large fossil mammals of Locality 1 of Zhoukoudian and the hunting behaviour of Peking Man. In *Multi-disciplinary Study of the Peking Man Site at Zhoukoudian.* Institute of Vertebrate Paleontology and Paleoanthropology, Academia Sinica, ed. Science Press, Beijing, pp. 95–101 (in Chinese).

Liu, C., Jin, Z., Zhu, R., Yang, H., and Wu, P. (1991). Dating. In *Wushan Hominid Site.* Haiyang Press, Beijing, pp. 156–163 (in Chinese with English abstract).

Liu, S., Zhang, F., Hu, R., Liu, J., Guo, S., Zhou, S., Meng, W., Zhang, P., Sun, S., and Hao, X. (1985). Dating Peking Man Site by the fission-track method. In *Multi-disciplinary Study of the Peking Man Site at Zhoukoudian.* Institute of Vertebrate Paleontology and Paleoanthropology, Academia Sinica, ed. Science Press, Beijing, pp. 241–245 (in Chinese).

Liu, T. and Ding, M. (1984). A tentative chronological correlation of early fossil horizons in China with the loess-deep sea records. *Acta Anthropologica Sinica,* 3:93–101 (in Chinese with English abstract).

Liu, Y. (1986). The paleoliths newly discovered at the Xigou site, Quwo County, Shanxi Province. *Acta Anthropologica Sinica,* 5:325–335 (in Chinese with English abstract).

Liu, Y. and Huang, W. (1984). Human fossil and Paleolithic remains from Jinchuan, Gansu. *Acta Anthropologica Sinica,* 3:11–18 (in Chinese with English abstract).

Lu, Q., Xu, Q., and Zheng, L. (1981). Preliminary research on the cranium of *Sivapithecus yunnanensis. Vertebrata PalAsiatica,* 19:101–106 (in Chinese with English abstract).

Lu, Z. (1989). Date of Jinniushan Man and his position in human evolution. *Liaohai Wenwu Xuekan,* (1):44–55 (in Chinese).

Lu, Z., Huang, Y., Li, P., and Meng, Z. (1989). Yiyuan fossil man. *Acta Anthropologica Sinica,* 8:301–313 (in Chinese with English abstract).

Management Committtee of Cultural Relics of Heilongjiang Province; Bureau of Cultural Affairs of Harbin Municipality, Northwestern Expedition of the Institute of Vertebrate Paleontology and Paleoanthropology, Academia Sinica. (1987). *Yanjiagang, A Late Paleolithic Campsite in Harbin.* Wenwu Press, Beijing (in Chinese with English abstract).

Museum of Liaoning Province and Museum of Benxi City. (1986). *Miaohoushan, A Site of the Early Paleolithic in Benxi County, Liaoning.* Wenwu Press, Beijing (in Chinese with English abstract).

Olsen, J. (1987). The practice of archaeology in China today. *Antiquity,* 61:282–290.

Pan, Y. and Wu, R. (1986). A new species of *Sinoadapis* from the hominoid site, Lufeng. *Acta Anthropologica Sinica,* 5:30–44 (in Chinese with English abstract).

Pan, Y., Li, Q., Lin, Y., and Jiang, C. (1991). Late Pleistocene fauna discovered in Yuanmou basin, Yunnan. *Acta Anthropologica Sinica,* 10:167–175 (in Chinese with English abstract).

Pei, J. (1985). Thermoluminescence dating of the Peking Man site and other caves. In *Multidisciplinary Study of the Peking Man Site at Zhoukoudian.* Institute of Vertebrate Paleontology and Paleoanthropology, Academia Sinica, ed. Science Press, Beijing, pp. 256–260 (in Chinese).

Pei, W. (Pei, Wenzhong). (1939). *The Upper Cave Industry of Choukoutien. Palaeontologia Sinica.* New Series C, No. 10. Geological Survey of China, Beijing.

Pei, W. (1940). *The Upper Cave Fauna of Choukoutien. Paleontologia Sincia.* New Series C, No. 10. Geological Survey of China, Beijing.

Pei, W. , ed. (1958). *Report on the Excavation of the Paleolithic Site at Tingtsun, Hsiangfenhsien, Shansi Province, China.* Science Press, Beijing (in Chinese with English summary).

Pei, W. and Li, Y. (1958). Discovery of a third mandible of *Gigantopithecus* in Liucheng, Kwangsi, South China. *Vertebrata PalAsiatica,* 2:193–200 (in Chinese with English abstract).

Pei, W. and Woo, J. (1956). New materials of *Gigantopithecus* teeth from South China. *Acta Palaeontologica Sinica,* 4:477–490 (in English with Chinese abstract).

Pei, W. and Woo, J. (1957). *Tzeyang Paleolithic Man.* Science Press, Beijing (in Chinese).

Pei, W. and Zhang, S. (1985). A Study of the Lithic Artifacts of *Sinanthropus. Palaeontologia Sinica.* New Series D, no. 12. Nanjing Institute of Geology and Paleontology, Institute of Vertebrate Paleontology and Paleoanthropology, Academia Sinica, eds. Science Press, Beijing (in Chinese with English summary).

Peng, S., Zhou, S., and Wang, W. (1987). Human fossils and associated fauna found in Nalai Cave, Longlin, Guangxi. *Shiqianyanjiu,* 4:43–46 (in Chinese).

Poirier, F. (1993). *Understanding Human Evolution,* 3rd ed. Prentice-Hall, Englewood Cliffs, N.J.

Poirier, F., Hu, H., and Chen, C. M. (1983). The evidence for Wildman in Hubei Province, The People's Republic of China. *Cryptozoology,* 2:25–39.

Pope, G. (1992). Craniofacial evidence for the origin of modern humans in China. *Yearbook of Physical Anthropology,* 35:243–298.

Pope, G. (1993). Ancient Asia's cutting edge. *Natural History,* May, 55–58.

Pope, G. and Keates, R. S. (1994). The evolution of human cognition and cultural capacity: A view from the Far East. In *Integrative Paths to the Past.* R. Corrucini and R. Ciochan, eds. Prentice Hall, Englewood Cliffs, N.J. pp. 531–568.

Qi, G. (1985). Stratigraphic summarization of *Ramapithecus* fossil locality, Lufeng, Yunnan. *Acta Anthropologica Sinica,* 4:55–69 (in Chinese with English abstract).

Qi, G. (1986). Fossils of Rhizomyidae from *Ramapithecus* fossil locality, Lufeng, Yunnan. *Acta Anthropologica Sinica,* 5:54–67 (in Chinese with English abstract).

Qi, G. (1989). Quaternary mammalian faunas and environment of fossil humans in north China. In *Early Humankind in China.* Wu, R., Wu, X., and S. Zhang, eds. Science Press, Beijing, pp. 276–337 (in Chinese).

Qian, F. (1991). Environment of Yuanmou Man. In *Quarternary Geology and Paleoanthropology of Yuanmou, Yunnan, China.* Science Press, Beijing, pp. 161–164 (in Chinese with English abstract, pp. 211–212).

Qian, F., Zhang, J., and Yin, W. (1985). Magnetic stratigraphy from the sediment of the west wall and the test pit of Locality 1 at Zhoukoudian. In *Multi-disciplinary Study of the Peking Man Site at Zhoukoudian.* Institute of Vertebrate Paleontology and Paleoanthropology, Academia Sinica, ed. Science Press, Beijing, pp. 251–255 (in Chinese).

Qian, F., Li, Q., Wu, P., Yuan, S., Xing, R., Chen, H., and Zhang, H. (1991). Lower Pleistocene, Yuanmou Formation. In *Quarternary Geology and Paleoanthropology of Yuanmou, Yunnan, China.* Science Press, Beijing, pp. 17–50 (in Chinese with English abstract, pp. 194–195).

Qiu, L., Song, F., and Wang, L. (1986). A fossil human tooth from Fengkai, Guangdong. *Acta Anthropologica Sinica,* 5:311–313 (in Chinese with English abstract).

Qiu, Z. D. and Han, D. (1986). Fossil lagomorpha from the hominoid locality of Lufeng, Yunnan. *Acta Anthropologica Sinica,* 5:41–53 (in Chinese with English abstract).

Qiu, Z. L., Gu, Y., Zhang, Y., and Zhang, S. (1973). New discovery of the fossils and cultural remains of *Sinanthropus pekinensis* from Zhoukoudian. *Vertebrata PalAsiatica,* 11:109–131 (in Chinese).

Qiu, Z. L., Xu, C., Zhang, W., Wang, R., Wang, J., and Zhao, C. (1982). A human fossil tooth and fossil mammals from Nanzhao, Henan. *Acta Anthropologica Sinica,* 1:109–117 (in Chinese with English abstract).

Qiu, Z. L., Zhang, Y., and Hu, S. (1985). Human tooth and paleoliths found at locality 2 of Longtanshan, Chengong, Kunming. *Acta Anthropologica Sinica,* 4:233–241 (in Chinese with English abstract).

Santa Luca, A. (1978). A re-examination of presumed Neanderthal fossils. *Journal of Human Evolution,* 7:619–636.

Schick, K. and Dong, Z. (1993). Early Paleolithic of China and Eastern Asia. *Evolutionary Anthropology,* 32:22–35.

Shapiro, H. (1971). The strange, unfinished saga of Peking Man. *Natural History,* 80:8,74.

Shapiro, H. (1974). *Peking Man.* Simon & Schuster, New York.

Shen, G. and Jin, L. (1991). Restudy of the upper age limit of Beijing Man Cave. *Acta Anthropologica Sinica,* 10:273–277 (in Chinese with English abstract).

Shikama, T., Lin, C. C., Shimoda, N., and Baba, H. (1976). Discovery of fossil *Homo sapiens* from Cho-chen in Taiwan. *Journal of the Anthropology Society of Nippon,* 84, pp. 131–138 (in English with Japanese abstract).

Simons, E. L. and Chopra, S. R. K. (1969). *Gigantopithecus* (Pongidae, Hominoidea) a new species from north India. *Postilla Peabody Museum Yale University,* 138:1–18.

Sohn, S. and Wolpoff, M. (1993). Zuttiyeh face: A view from the east. *American Journal of Physical Anthropology,* 91:325–347.

Song, F., Zhang, Z., Deng, Z., Zhen, X., and Cheng, Q. (1991). The study of the fauna and the human teeth from Tongzhongyan in Fengkai, Guangdong. In *Treatises in Commemoration of the 30th Anniversary of the Discovery of Huangyan Cave Site.* The Fengkai County Museum et al., eds. Guangzhou Travel and Tourism Press, Guandong, pp. 28–40 (in Chinese).

Stringer, C. (1984). The definition of *Homo erectus* and the existence of the species in Africa and Europe. In *The Early Evolution of Man.* P. Andrews and J. Franzen, eds. Senckenberg Museum, Frankfurt.

Stringer, C., Hublin, J., and Vandermeersch, B. (1984). The origin of anatomically modern humans in Western Europe. In *The Origin of Modern Humans.* F. Smith and F. Spencer, eds., Alan R. Liss, New York.

Szalay, F. and Li, C. (1986). Middle Paleocene euprimate from southern China and the distribution of primates in the Paleogene. *Journal of Human Evolution,* 18:387–398.

Tai, E. (Dai, Erjian) and Hsu, C. (Xu). (1973). New finds of paleoliths from Lantian. *Kaogu Xuebao* (Chinese Journal of Archaeology), (2)1–12 (in Chinese with English abstract).

Thorne, A. G. (1980). The arrival of man in Australia. In *The Cambridge Encyclopaedia of Archaeology.* A. Sherratt, ed. Cambridge University Press, Cambridge, pp. 96–100.

Turner, C. (1984). Advances in the dental search for Native American origins. *Acta Anthrogenetica,* 8:23–78.

Turner, C. (1986). The Native Americans: The dental evidence. *National Geographic Research,* 2:37–46.

von Koenigswald, G. H. R. (1935). Eine fossile Saugetierfauna mit Simia aus Sudchina. *Proc. K. Adad Weternsch, Amsterdam,* 38:872–879.

von Koenigswald, G. H. R. (1952). *Gigantopithecus blacki* von Koenigswald, a giant fossil hominid from the Pleistocene of South China. *Anthropological Papers of the American Museum of Natural History,* 43:295–325.

Wang, L. and Braüer, G. (1984). A multivariate comparison of the human calva from Huanglong County, Shaanxi Province. *Acta Anthropologica Sinica,* 3:313–321 (in Chinese with English summary).

Wang, L. and Li, Y. (1983). On a fossil human calva unearthed from Huanglong County, Shaanxi Province. *Acta Anthropologica Sinica,* 2:315–319 (in Chinese with English abstract).

Wang, L., Peng, S., and Chen, Y. (1982). On the human fossils and stone artifacts found in Baojiyan Cave, Guilin. *Acta Anthropologica Sinica,* 1:30–35 (in Chinese with English abstract).

Wang, Y. (1961). New materials of paleolithic cultural remains in Yimeng, Inner Mongolia. *Kaogu Xuebao,* (10):552–554 (in Chinese).

Weidenreich, F. (1935). The *Sinanthropus* population of Choukoutien (Locality 1) with a preliminary report on new discoveries. *Bulletin Geological Society of China,* 14:427–461.

Weidenreich, F. (1936a). The mandible of *Sinanthropus pekinensis:* A comparative study. *Palaeontologia Sinica,* New Series D, Vol. 7, Fascicle no. 3, 1–162.

Weidenreich, F. (1936b). Observation on the form and proportions of the endocranial casts of *Sinanthropus pekinensis,* other hominids, and the great apes: A comparative study of brain size. *Palaeontologia Sinica,* New Series D, Vol. 7, Fascicle no. 4, 1–50.

Weidenreich, F. (1937). The dentition of *Sinanthropus pekinensis:* A comparative odontography of the hominid. *Palaeontologia Sinica,* New Series D, No. 1:1–180.

Weidenreich, F. (1938). The ramification of the middle meningeal artery in fossil hominids and bearing upon phylogenetic problems. *Palaeontologia Sinica.* New Series D, No. 3:1–16.

Weidenreich, F. (1939). On the earliest representatives of modern mankind recovered on the soil of East Asia. *Bulletin Natural History Society Peking,* 13 Part 3:161–174.

Weidenreich, F. (1941). The extremity bones of *Sinanthropus pekinensis. Palaeontologia Sinica,* New Series D, No. 5:1–150.

Weidenreich, F. (1943). The skull of *Sinanthropus pekinensis:* A comparative study on a primitive hominid skull. *Palaeontologia Sinica,* New Series D, No. 10:1–184.

Weidenreich, F. (1945). Giant early man from Java and South China. *Anthropological Papers American Museum of Natural History,* 40:1–134.

Weidenreich, F. (1946). *Apes, Giants, and Early Man.* University of Chicago Press, Chicago.

Wen, B. (1978), Palaeolithic artefacts found in Yuanmou basin, Yunnan. In *Collected Papers on Paleoanthropology*. Institute of Vertebrate Paleontology and Paleoanthropology, ed. Science Press, Beijing, pp. 126–135 (in Chinese).

Woo, J. (Wu Rukang) (1957). *Dryopithecus* teeth from Keiyuan, Yunnan province. *Vertebrata PalAsiatica,* 1:25–32 (in English with Chinese abstract).

Woo, J. (1958a). New materials of *Dryopithecus* from Keiyuan, Yunnan. *Vertebrata PalAsiatica,* 2:38–43 (in English with Chinese abstract).

Woo, J. (1958b). The first discovery of *Dryopithecus* teeth in China and its significance. *Acta Palaeontologia Sinica,* 6:118–121 (in Chinese with English abstract).

Woo, J. (1958c). Fossil human parietal bone and femur from Ordos, Inner Mongolia. *Vertebrata PalAsiatica,* 2:208–212 (in Chinese with English abstract).

Woo, J. (1958d). Investigation of human teeth. In *Report on the Excavation of the Paleolithic Site at Tingtsun, Hsiangfenhsien, Shansi Province, China.* W. Pei, ed. Science Press, Beijing, pp. 15–20 (in Chinese with English abstract).

Woo, J. (1958e). Tzeyang Paleolithic Man—Earliest representative of modern man in China. *American Journal of Physical Anthropology,* 16:459–465.

Woo, J. (1959a). Human fossils found in Liukiang, Kwangsi, China. *Vertebrata PalAsiatica,* 3:109–118.

Woo, J. (1959b). Liukiang Man—Earliest representative of modern man in East Asia. *Science Record,* New Series, 3:165–167.

Woo, J. (1961). Fossil human humerus from Chienping, Liaoning Province. *Vertebrata PalAsiatica,* 5:287–289 (in Chinese with English abstract).

Woo, J. (1962a). *Gigantopithecus* and its phylogenetic significance. *Scientia Sinica,* 11:15–18.

Woo, J. (1962b). The mandibles and dentition of *Gigantopithecus. Palaeontologia Sinica* New Series D, 11:1–94 (in Chinese with English summary).

Woo, J. (1964a). Mandible of the *Sinanthropus*-type discovered at Lantian, Shensi. *Vertebrata PalAsiatica,* 8:1–12 (in Chinese with English abstract).

Woo, J. (1964b). Mandible of *Sinanthropus lantianensis. Current Anthropology,* 5:98–102.

Woo, J. (1965). The hominid skull of Lantian, Shensi. *Vertebrata PalAsiatica,* 10:1–22 (in Chinese with English abstract).

Woo, J. and Chao, T. (Zhao, Zikui). (1959). New discovery of *Sinanthropus* mandible from Choukoutien. *Vertebrata PalAsiatica,* 3:169–172 (in Chinese with English abstract).

Woo, J. and Chia, L. (Jia Lanpo). (1954). New discoveries of *Sinanthropus pekinensis* in Choukoutien. *Acta Paleontologia Sinica,* 2:267–288 (in Chinese with English abstract).

Woo, J., Wu, X., and Wang, C. (1959). New reconstruction of the physiognomy of *Sinanthropus* woman. *Vertebrata PalAsiatica,* 3:165–166 (in Chinese with English abstract).

Wood, B. (1984). The origin of *Homo erectus. Courier Forschungsinstitut Senckenberg,* 67:99–111.

Wu, M. (1980). Human fossils discovered at Xujiayao site in 1977. *Vertebrata PalAsiatica,* 18:227–238 (in Chinese with English abstract).

Wu, M. (1983). *Homo erectus* from Hexian, Anhui found in 1981. *Acta Anthropologica Sinica,* 2:109–115 (in Chinese with English abstract).

Wu, M. (1984). New discoveries of human fossil in Tongzi, Guihou. *Acta Anthropologica Sinica,* 3:195–201 (in Chinese with English abstract).

Wu, M. (1989). Late *Homo sapiens* of China. In *Early Humankind in China.* R. Wu, X. Wu, and S. Zhang, eds. Science Press, Beijing, pp. 42–61 (in Chinese).

Wu, M., Wang, L., Zhang, Y., and Zhang, S. (1975). Human fossils found at Tongzi, Guizhou and the cultural remains. *Vertebrata PalAsiatica,* 13:14–23 (in Chinese).

Wu, R. (1982). Paleoanthropology in China, 1949–1979. *Current Anthropology,* 23:473–477.

Wu, R. (1984). The crania of *Ramapithecus* and *Sivapithecus* from Lufeng, China. In *The Early Evolution of Man.* P. Andrews and J. Franzen, eds. Senckenberg Museum, Frankfurt.

Wu, R. (1987). A revision of the classification of the Lufeng great apes. *Acta Anthropologica Sinica,* 6:265–271 (in Chinese with English abstract).

Wu, R. (1988). The reconstruction of the fossil human skull from Jinniushan, Yinkou, Liaoning Province and its main features. *Acta Anthropologica Sinica,* 7:97–101 (in Chinese with English abstract).

Wu, R. and Dong, X. (1980). The fossil human teeth from Yunxian, Hubei. *Acta Anthropologica Sinica,* 18:142–149 (in Chinese with English abstract).

Wu, R. and Dong, Z. (1982). Preliminary study of *Homo erectus* remains from Hexian, Anhui. *Acta Anthropologica Sinica,* 1:2–13 (in Chinese with English abstract).

Wu, R. and Olsen, J. (eds.), (1985). *Palaeoanthropology and Palaeolithic Archaeology in the People's Republic of China.* Academic Press, Orlando.

Wu, R. and Wu, X. (1982). Human fossil teeth from Xichuan, Henan. *Vertebrata PalAsiatica,* 20:1–9 (in Chinese with English summary).

Wu, R., Xu, Q., and Lu, Q. (1983). Morphological features of *Ramapithecus* and *Sivapithecus* and their phylogenetic relationships—Morphology and comparison of the crania. *Acta Anthropologica Sinica,* 2:1–10 (in Chinese with English abstract).

Wu, R., Lu, Q., and Xu, Q. (1984). Morphological features of *Ramapithecus* and *Sivapithecus* and their phylogenetic relationships—Morphology and comparison of the mandibles. *Acta Anthropologica Sinica,* 3:1–10 (in Chinese with English abstract).

Wu, R., Xu, Q., and Lu, Q. (1985). Morphological features of *Ramapithecus* and *Sivapithecus* and their phylogenetic relationships—Morphology and comparison of the teeth. *Acta Anthropologica Sinica,* 4:197–204 (in Chinese with English abstract).

Wu, R., Xu, Q., and Lu, Q. (1986). Relationship between Lufeng *Sivapithecus* and *Ramapithecus* and their phylogenetic position. *Acta Anthropologica Sinica,* 5:1–30 (in Chinese with English abstract).

Wu, R., Wu, X., and Zhang, S., eds. (1989). *Early Humankind in China.* Science Press, Beijing (in Chinese)

Wu, X. (1961a). Study on the Upper Cave Man of Choukoutien. *Vertebrata PalAsiatica,* 3:182–204 (in Chinese with English abstract).

Wu, X. (1961b). On the racial types of the Upper Cave Man of Choukoutien. *Scientia Sinica,* 10:998–1005.

Wu, X. (1981). The well preserved cranium of an early *Homo sapiens* from Dali, Shaanxi. *Scientia Sinica,* 24:200–206.

Wu, X. (1987a). Relation between Upper Paleolithic man in China and their southern neighbors in Niah and Tabon. *Acta Anthropologica Sinica,* 6:180–183 (in Chinese with English abstract).

Wu, X. (1987b). The influence on human evolution of Australia from China. In *Collected Papers of China-Australia Symposium on Quarternary.* China-Australia Cooperative Project Team on Quaternary of Academia Sinica, eds. Science Press, Beijing, pp. 246–250 (in Chinese with English abstract).

Wu, X. (1988a). The relationship between Upper Paleolithic human fossils of China and Japan. *Acta Anthropologica Sinica,* 7:235–238 (in Chinese with English abstract).

Wu, X. (1988b). Comparative study of early *Homo sapiens* from China and Europe. *Acta Anthropologica Sinica,* 7:287–293 (in Chinese with English abstract).

Wu, X. (1989). Early *Homo sapiens* in China. In *Early Humankind in China.* X. Wu, and S. Shang, eds. Science Press, Beijing, pp. 24–41.

Wu, X. (1990). The evolution of humankind in China. *Acta Anthropologica Sinica,* 9:312–321 (in Chinese with English abstract).

Wu, X. (1992a). Origins and affinities of the Stone Age inhabitants of Japan. In *Japanese as a Member of the Asian and Pacific Population.* K. Hanihara, ed. International Research Center for Japanese Studies, Kyoto, pp. 1–8 (in English with Japanese abstract).

Wu, X. (1992b). The origin and dispersal of anatomically modern humans in East Asia and Southeast Asia. In *The Evolution and Dispersal of Modern Humans in Asia.* T. Akazawa et al., eds. Hokusen-Sha, Tokyo, pp. 373–378.

Wu, X. and Bräuer, G. (1993). Morphological comparison of archaic *Homo sapiens* crania from China and Africa. *Zeitschrift für Morphologie und Anthropologie* 79:241–259.

Wu, X. and You, Y. (1979). A preliminary observation of the Dali Man site. *Vertebrata PalAsiatica,* 17:294–303 (in Chinese with English abstract).

Wu, X. and Zong, G. (1973). Late Pleistocene human tooth and mammalian fossils from Wuzhutai, Xintai, Shandong. *Vertebrata PalAsiatica,* 11:105–106 (in Chinese).

Wu, X., Zhao, Z., Yuan, Z., and Shen, J. (1962). Report on the Paleoanthropological expedition of the northeastern part of Kwangsi. *Vertebrata PalAsiatica,* 6:408–414 (in Chinese with English abstract).

Wu, X., Yuan, Z., Han, D., Qi, T., and Lu, Q. (1966). Report of the excavation at Lantian Man locality of Gongwangling in 1965. *Vertebrata PalAsiatica,* 10:23–30 (in Chinese with English abstract).

Wu, X., Zhang, Z., and Zhang, J. (1984). Estimation of the stature of Liujiang Paleolithic Man. *Acta Anthropologica Sinica,* 3:210–211 (in Chinese with English abstract).

Xia, M. (1982). Uranium-series dating of fossil bones from Peking Man cave—Mixing model. *Acta Anthropologica Sinica,* 1:191–196 (in Chinese with English abstract).

Xie, J., Zhang, Z., and Yang, F. (1987). Human fossil found at Wushan, Gansu. *Shiqian-yanjiu,* 4:47–51 (in Chinese).

Xu, C. (1978). Excavation of *Homo erectus* fossil site in Yunxian, Hubei. In *Collected Papers of Paleoanthropology.* Institute of Vertebrate Paleontology and Paleoanthropology, Academia Sinica, ed. Science Press, Beijing, pp. 175–179 (in Chinese).

Xu, C. and Han, D. (1992). New materials found from Longgudong, Gaopin, Jianshi, Hubei. Paper presented at Annual Meeting of Chinese Society of Vertebrate Paleontology Dalian (in Chinese).

Xu, C. and Zhang, Y. (1986). Human fossil newly discovered at Chaoxian, Anhui. *Acta Anthropologica Sinica,* 5:305–310 (in Chinese with English abstract).

Xu, C., Zhang, Y., Chen, C., and Fang, D. (1984). Human occipital bone and mammalian fossils from Chaoxian, Anhui. *Acta Anthropologica Sinica,* 3:202–209 (in Chinese with English abstract).

Xu, Q. and Lu, Q. (1979). The mandibles of *Ramapithecus* and *Sivapithecus* from Lufeng, Yunnan. *Vertebrata PalAsiatica,* 17:1–13 (in Chinese with English abstract).

Xue, X. (1987). Human fossil tooth from Luonan, Shaanxi and its geological age. *Acta Anthropologica Sinica,* 6:284–288 (in Chinese with English abstract).

Yang, Z., Mo, Y., Qian, F., Wang, K., Niu, P., Chen, H., Yin, W., and Wei, X. (1985). Study of late Cenozoic strata at Zhoukoudian. In *Multi-disciplinary Study of the Peking Man Site at Zhoukoudian.* Institute of Vertebrate Paleontology and Paleoanthropology, ed. Science Press, Beijing, pp 1–85 (in Chinese).

You, Y., Dong, X., Chen, C., and Fang, X. (1989). A fossil human tooth from Qingliu, Fujian. *Acta Anthropologica Sinica,* 8:197–202 (in Chinese with English abstract).

Yu, H. (1988). A brief study of late Palaeolithic localities at Xuetian village of Wuchang County, Heilongjiang Province. *Acta Anthropologica Sinica,* 7:255–262 (in Chinese with English abstract).

Yu, J. (1984). Fossil man and cultural artifacts from Chuandong, Puding county, Guizhou province. *Journal of Nanjing University (Natural Science),* 1:145–168 (in Chinese with English abstract).

Yuan, S., Chen, T., and Gao, S. (1983). Uranium series dating of "Ordos Man" and "Sjara-osso-gol Culture." *Acta Anthropologica Sinica,* 2:90–94 (in Chinese with English abstract).

Yuan, S., Chen, T., and Gao, S. (1986). Uranium series chronological sequence of some Paleolithic sites in south China. *Acta Anthropologica Sinica,* 5:179–190 (in Chinese with English abstract).

Yunnan Provincial Museum. (1977). A preliminary study on a human skull from Lijiang, Yunnan *Vertebrata PalAsiatica,* 15:157–161 (in Chinese with English abstract).

Yunnan Provincial Museum. (1991). *On Materials of Human Origins and Prehistoric Culture in Yunnan.* Yunnan Renmin Press, Kunming (in Chinese).

Zhang, S. (1980). Introduction to Paleolithic culture in Guizhou. *Guiyang Shiyuan Xuebao (Shehui Kexue),* 2:1–11 (in Chinese).

Zhang, S. (1983). On some problems of the Upper Paleolithic culture in southern China. *Acta Anthropologica Sinica,* 2:218–230 (in Chinese with English abstract).

Zhang, S. (1987). *Paleolithic Culture of China.* Tianjin Science and Technology Press, Tianjin (in Chinese).

Zhang, S. (1990). Regional industrial gradual advance and cultural exchange of Paleolithic in north China. *Acta Anthropologica Sinica,* 9:322–333 (in Chinese with English abstract).

Zhang, S. and Zhou, C. (1984). A preliminary study of the second excavation of the Dali Man locality. *Acta Anthropologica Sinica,* 3:19–29 (in Chinese with English abstract).

Zhang, X. (1987). New materials of *Ramapithecus* from Keiyuan, Yunnan. *Acta Anthropologica Sinica,* 6:81–86 (in Chinese with English abstract).

Zhang, X. and Li, C. (1982). Human fossil discovered in Longlin, Guangxi. *Acta Anthropologica Sinica,* 1:199 (in Chinese).

Zhang, X., Hu, S., and Zheng, L. (1978). Late Pleistocene human teeth from Kunming, Yunnan. *Vertebrata PalAsiatica,* 16:288–289 (in Chinese).

Zhang, X., Lin, Y., Jiang, C., and Xiao, L. (1987a). New species of *Ramapithecus* from Yuanmou, Yunnan. *Sixiangzhanxian,* (3):54–56 (in Chinese).

Zhang, X., Lin, Y., Jiang, C., and Xiao, L. (1987b). A new species of hominid discovered from Yuanmou, Yunnan. *Sixiangzhanxian,* (3):57–60 (in Chinese).

Zhang, X., Zheng, L., Gao, F., Jiang, C., and Zhang, J. (1988). Preliminary study on a skull of *Ramapithecus hudiensis* from Yuanmou. *Sixiangzhanxian,* (5)55–61 (in Chinese).

Zhang, X., Zheng, L., Yang, L., and Bao, Z. (1991). Human fossils and the paleoculture from Mengzi. In *On Materials of Human Origin and Prehistoric Culture of Yunnan.* Yunnan Provincial Museum, eds. Yunnan Renming Press, Kunming, pp. 234–246 (in Chinese).

Zhang, Y. (1982). Variability and evolutionary trends in tooth size of *Gigantopithecus blacki. American Journal Physical Anthropology,* 59:21–52.

Zhang, Y. (1983). Variability in tooth size of *Gigantopithecus blacki* and the dietary hypothesis for australopithecines. *Acta Anthropologica Sinica,* 2:205–217 (in Chinese with English abstract).

Zhang, Y. (1984). The "*Australopithecus*" of West Hupei and some early Pleistocene hominids of Indonesia. *Acta Anthropologica Sinica,* 3:85–92 (in Chinese with English abstract).

Zhang, Y. (1987). Enamel hypoplasia of *Gigantopithecus blacki. Acta Anthropologica Sinica,* 6:175–179 (in Chinese with English abstract).

Zhang, Y. (1989). Tooth wear in early *Homo sapiens* from Chaohu and the hypothesis of use of anterior teeth as tools. *Acta Anthropologica Sinica,* 8:314–319 (in Chinese with English abstract).

Zhang, Y. (1991). An examination of temporal variation in the hominid dental sample from Zhoukoudian Locality 1. *Acta Anthropologica Sinica,* 10:15–25 (in Chinese with English summary).

Zhang, Y., Wu, M., and Liu, J. (1973). *Gigantopithecus* tooth fossils newly discovered from Wuming, Guangxi. *Kexue Tongbao,* 18:130–133 (in Chinese).

Zhang, Y., Wang, L., Dong, X., and Chen, W. (1975). Discovery of a *Gigantopithecus* tooth from Bama district in Kwangsi. *Vertebrata PalAsiatica,* 13:148–153 (in Chinese with English abstract).

Zhang, Y., Huang, W., Tang, Y., Ji, H., You, Y., Tong, Y., Ding, S., Huang, X., and Zheng, J. (1978). Cenozoic in Lantian Region, Shaanxi. *Institute of Vertebrate Paleontology and Paleoanthropology, Academia Sinica.* Memoir No. 14. Science Press, Beijing, pp. 1–64 (in Chinese).

Zhao, S., Sha, M., Zhang, C., Liu, M., Wang, S., Wu, Q., and Ma, Z. (1985). Uranium series dating of Peking Man Site. In *Multi-disciplinary Study of the Peking Man Site at Zhoukoudian.* Institute of Vertebrate Paleontology and Paleoanthropology, ed. Science Press, Beijing, pp. 246–250 (in Chinese).

Zhao, Z., Liu, X., and Wang, L. (1981). Human fossils and associated fauna of Jiulengshan Hill, Guangzi. *Vertebrata PalAsiatica,* 19:45–54 (in Chinese with English abstract).

Zheng, L. (1985). A fossil human tooth from Zhaotong, Yunnan. *Acta Anthropologica Sinica,* 4:105–108 (in Chinese with English abstract).

Zhou, G. and Hu, C. (1979). Supplementary notes on the teeth of Yuanmou Man with a discussion on morphological evolution of the mesial upper incisors in hominoids *Vertebrata PalAsiatica,* 17:149–162 (in Chinese with English abstract).

Zhou, G., Wang, B., and Zhai, H. (1991). New material of Yuanmou man. *Kaogu Yu Wenwu,* (1):56–61 (in Chinese).

Zong, G., Pan, Y., Jiang, C., and Xiao, L. (1991). Stratigraphic subdivision of hominoid fossil localities of Yuanmou, Yunnan. *Acta Anthropologica Sinica,* 10:155–166 (in Chinese).

Zou, G. and Zhang, X. (1993). New chapter of palaeontological research—the discovery of an ancient human skull near Nanjing. *Chinese Science News,* April 9th., p. 2 (in Chinese).

Index